T0256216

Artificial Intelligence Technologies for Computational Biology

This text emphasizes the importance of artificial intelligence techniques in the field of biological computation. It also discusses fundamental principles that can be applied beyond bio-inspired computing.

It comprehensively covers important topics including data integration, data mining, machine learning, genetic algorithms, evolutionary computation, evolved neural networks, nature-inspired algorithms, and protein structure alignment. This text covers the application of evolutionary computations for fractal visualization of sequence data, artificial intelligence, and automatic image interpretation in modern biological systems.

This text is primarily written for graduate students and academic researchers in areas of electrical engineering, electronics engineering, computer engineering, and computational biology.

This book:

- Covers algorithms in the fields of artificial intelligence, and machine learning useful in biological data analysis.
- Discusses comprehensively artificial intelligence and automatic image interpretation in modern biological systems.
- Presents the application of evolutionary computations for fractal visualization of sequence data.
- Explores the use of genetic algorithms for pair-wise and multiple sequence alignments.
- Examines the roles of efficient computational techniques in biology.

Artificial Intelligence Technologies for Computational Biology

Edited by
Ranjeet Kumar Rout
Saiyed Umer
Sabha Sheikh
Amrit Lal Sangal

CRC Press
Taylor & Francis Group
Boca Raton London New York

CRC Press is an imprint of the
Taylor & Francis Group, an **informa** business

First edition published 2023
by CRC Press
6000 Broken Sound Parkway NW, Suite 300, Boca Raton, FL 33487-2742

and by CRC Press
4 Park Square, Milton Park, Abingdon, Oxon, OX14 4RN

CRC Press is an imprint of Taylor & Francis Group, LLC

Library of Congress Cataloging-in-Publication Data

Names: Rout, Ranjeet Kumar, editor.
Title: Artificial intelligence technologies for computational biology /
edited by Ranjeet Kumar Rout, Saiyed Umer, Sabha Sheikh, Amrit Lal Sangal.
Description: First edition. | Boca Raton : CRC Press, 2023. | Includes
bibliographical references and index.
Identifiers: LCCN 2022022589 (print) | LCCN 2022022590 (ebook) | ISBN
9781032160009 (hardback) | ISBN 9781032160023 (paperback) | ISBN
9781003246688 (ebook)
Subjects: LCSH: Artificial intelligence--Biological applications. |
Computational biology--Data processing.
Classification: LCC QH324.25 .A79 2023 (print) | LCC QH324.25 (ebook) |
DDC 570.285--dc23/eng/20220808
LC record available at https://lccn.loc.gov/2022022589
LC ebook record available at https://lccn.loc.gov/2022022590

ISBN: 978-1-032-16000-9 (hbk)
ISBN: 978-1-032-16002-3 (pbk)
ISBN: 978-1-003-24668-8 (ebk)

DOI: 10.1201/9781003246688

Publisher's note: This book has been prepared from camera-ready copy provided by the authors.

Contents

Preface

Computational biology, which began in the early 1960s in the domains of protein biophysics and biochemistry, with a focus on modeling enzyme processes and other kinetic characteristics, has become one of the most important milestones in the analysis of biological and medical data in recent decades. The GenBank sequence database had 601,438 residues when it was originally released in 1982. By 2005, the number surpassed a million, and it continues to rise at an exponential rate. With the emergence of improved and more easily accessible computing systems came the ability to explore biological systems to a much greater depth, particularly related to platforms for large-scale analysis of biological samples, such as whole-genome sequencing tools, mass spectrometry, and many more. Such a sizeable volume of data, which was being captured in an accelerated and technology-dependent manner, required new ways of handling, managing, and analyzing information through improved data analysis flows achieved through the adoption and application of knowledge from other fields such as mathematics, statistics, and computer science to biology, medicine and disease analysis. This has resulted in a great expansion of the data sources and available computer tools that feed the reference databases and constantly improve our understanding of complex biological mechanisms. The ability to handle immense amounts of advanced knowledge permits us to integrate the varied data streams into a contextualized system through systems approaches, network analysis and modeling efficient methodologies.

The objective of this book is to help induce the convergence between computer scientists interested in language design, concurrency theory, software engineering, or program verification and physicists, mathematicians, and biologists interested in the systems-level understanding of cellular processes. This book captures new developments and summarizes what is known across the spectrum of mathematical and computational biology and medicine. It aims to promote the integration of mathematical, statistical, and computer-aided methods into biology. The content of this book is intended for students enrolled in courses in computational biology or bioinformatics as well as for molecular biologists, mathematicians and computer scientists, researchers, and professionals in the mathematical, statistical, and computational sciences, basic biology, and bioengineering, as well as interdisciplinary researchers working in this field. This book is aimed at both beginners and specialists in the field of computational biology and brings together a selection of state-of-the-art approaches and shows both how data and analytical procedures help us to expand our understanding of biology and how aberrant or modulated processes can lead to disease. The primary focus of this book is to provide the reader with a broad coverage of the concepts, themes, and instrumentalities of this important and evolving area of biological computation. In doing so, we hope to encourage an even wider adoption of biological computation methods for assisting in problem-solving and to stimulate research that may lead to additional innovations in this area of research. In writing this book, we have benefited from the expertise and support of a number of colleagues.

Contributors

Dev Arora
Computer Science and Information
   Systems
BITS Pilani, Pilani (Rajasthan), India

Umair Ayub
Computer Science & Engineering
National Institute of Technology,
   Srinagar, J&K, India

Shahid Azim
School of Computer and Systems
   Sciences
Jawaharlal Nehru University, Delhi,
   India

Sanghamitra Bandyopadhyay
Machine Intelligence Unit
Indian Statistical Institute, Kolkata,
   India

Hina Bansal
Amity Institute of Biotechnology
Amity University Uttar Pradesh,
   Sector 125, Noida-201303
Uttar Pradesh, India

Sudipto Bhattacharjee
Computer Science and Engineering
University of Calcutta, Kolkata, India

Lavanya Madhuri Bollipo
Computer Science and Engineering
National Institute of Technology
   Warangal, Andra Pradesh, India

Vijaya Gajanan Buddhavarapu
College of Information and Computer
   Sciences
University of Massachusetts Amherst,
   Amherst, MA 01003, United States

Ankur Chaurasia
Amity Institute of Biotechnology
Amity University Uttar Pradesh,
   Sector 125, Noida-201303
Uttar Pradesh, India

Prakash Choudhary
Computer Science and Engineering
National Institute of Technology
   Hamirpur, Himachal Pradesh, India

Monidipa Das
Machine Intelligence Unit
Indian Statistical Institute, Kolkata,
   India

Supantha Das
Information Technology
Academy of Technology, Hooghly,
   West Bengal, India

Sucheta Dawn
Machine Intelligence Unit
Indian Statistical Institute, Kolkata,
   India

Samridhi Dev
School of Computer and Systems
   Sciences
Jawaharlal Nehru University, Delhi,
   India

R.Vasundhara Devi
Department of Computer Science
Perunthalaivar Kamarajar Arts College,
   Puducherry, India

Amit Dua
Science and Information Systems
BITS Pilani, Pilani (Rajasthan), India

Soumadip Ghosh

Computer Science and Engineering
Institute of Engineering & Management,
    Kolkata, West Bengal, India

J. Angel Arul Jothi

Department of Computer Science
Birla Institute of Technology & Science
Pilani - Dubai Campus, Dubai, United
Arab Emirate

Kadambari K V

Computer Science and Engineering
National Institute of Technology
    Warangal, Andra Pradesh, India

Arshpreet Kaur

Computer Science and Engineering
DIT University, Dehradun, Uttarkhand,
    India

Monika Khandelwal

Computer Science & Engineering
National Institute of Technology,
    Srinagar, J&K, India

Vartika Kulshrestha

Computer Science and Engineering
DIT University, Dehradun, Uttarkhand,
    India

Sushil Kumar

School of Computer and Systems
    Sciences
Jawaharlal Nehru University, Delhi,
    India

Hiya Luthra

Amity Institute of Biotechnology
Amity University Uttar Pradesh, Sector
    125, Noida-201303
Uttar Pradesh, India

Venkata Maha Lakshmi N

Computer Science & Engineering
National Institute of Technology,
    Srinagar, J&K, India

Saurav Mallik

Center for Precision Health
School of Biomedical Informatics
The University of Texas Health Science
Center, Houston, Houston, TX, 77030,
USA

Sameer Mansuri

Computer Science and Engineering
National Institute of Technology
    Hamirpur, Himachal Pradesh, India

Nilarun Mukherjee

Department of Computer Science and
    Information Technology
Bengal Institute of Technology, Kolkata,
    India

Ranjeet Kumar Rout

Computer Science & Engineering
National Institute of Technology,
    Srinagar, J&K, India

Banani Saha

Computer Science and Engineering
University of Calcutta, Kolkata, India

Sudipto Saha

Division of Bioinformatics
Bose Institute, Kolkata, India

S. Siva Sathya

Department of Computer Science
Perunthalaivar Kamarajar Arts College,
    Puducherry, India

Souvik Sengupta

Computer Science & Engineering
Aliah University, West Bangal, India

Nazir Shabbir

Computer Science & Engineering
National Institute of Technology,
    Srinagar, J&K, India

Aditi Sharan

School of Computer and Systems
    Sciences
Jawaharlal Nehru University, Delhi,
    India

Kumar Shashvat

Computer Science and Engineering
DIT University, Dehradun, Uttarkhand,
    India

Hemant Kr. Soni

Computer Science and Engineering
National Institute of Technology, Delhi,
    India

Guimin Qin

Computer Science and Engineering
Xidian University, Xi'an,
    Shaanxi710071, China

Saiyed Umer

Computer Science & Engineering

Aliah University, West Bangal, India

List of Figures

List of Tables

1 Graph Representation Learning for Protein Classification

Sucheta Dawn and Monidipa Das
Machine Intelligence Unit, Indian Statistical Institute, Kolkata, India

CONTENTS

DOI: 10.1201/9781003246688-1

1.1 INTRODUCTION

Protein is a highly complex substance, involved in essential chemical processes that are required for all living organisms [3]. Proteins have great nutritional value and play many roles in our body, such as repairing and building our body's tissues, controlling metabolic reactions happening in our body [27]. In ecology, there is a transfer of matter and energy from organism to organism in the form of food [16]. This sequence of transfers existed in nature is known as food chain and protein is a part of this. Although protein is necessary for the animal body, animals cannot synthesize the proteins on their own. Therefore, they have to intake protein in terms of food. As shown in Figure 1.1, plants are at the base level of this food chain triangle. Plants can synthesize amino acids, the basic units of proteins, during photosynthesis using carbon dioxide present in the air. Ruminants such as cows, goats etc., at the second layer of the food chain triangle, eat plants to meet their amino-acid requirements. The top layer consists of non-ruminant animals, including human beings, who obtain the required protein primarily from other animals and their products like meat, milk, eggs, etc.

Proteins perform various duties in the cells, such as being the structural components of cells, enzymes, hormones, pigments, storage proteins and even some toxins in the cells [14]. Proteins are organ and species-specific, meaning that protein structures for muscle in the brain and heart are very much different from each other. Furthermore, proteins in different species are different. Therefore, the classification of proteins on their structure, composition and functions has become a widely used approach to simplify the problem of capturing relationships between groups of proteins and the complex nature of ecology and evolution.

Proteins are long chains of amino acids arranged as breads in a string, larger than a sugar or salt molecule [1]. There are about 20 types of amino acids commonly found in a protein chain [8]. The type of amino acids and their arrangement determines the protein's structure and function to a large extent [13]. However, it is not possible to explain all of the structures and functions of the proteins from

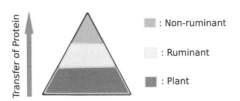

Figure 1.1   Food chain.

their amino acid sequences. Therefore, the classification of proteins to capture the relationship between amino acid sequence, structure and function of protein using different tools is a long-standing problem with significant scientific importance in molecular biology [12].

Graph Neural Network (GNN) is an emerging branch of deep learning in non-Euclidean space, especially performing well in different tasks where graph-structured data are involved [19]. With the increasing growth of biological network data, the GNN approach has become an important tool in bioinformatics [4]. GNN approaches can extract structural and feature-related information from the graph and find euclidean representations for non-euclidean data, which help to perform several tasks like classification, regression, link prediction etc., on graphs [18]. GNNs have achieved excellent results in many biological tasks using different hierarchical views of a graph, which can include the collective behaviour of the nodes in the graph. Therefore, we believe that GNNs are potential enough to bridge the gap between the structure along with the function of a protein and the properties of the amino acids present in it.

1.2 PROTEINS AS GRAPHS

Analysis of proteins as graphs can provide useful insights into the various functions, folding and global structural properties of the proteins. This section discusses the structural organizations of proteins along with a theoretical background for representing the same as graphs.

1.2.1 STRUCTURE OF PROTEINS

Proteins have four levels of structural organization [15], namely,

- Primary Structure

- Secondary Structure

- Tertiary Structure

- Quaternary Structure

1.2.1.1 Primary Structure

The primary structure of proteins speaks about the sequence of amino acids, present in the protein as the basic building blocks. As depicted in Figure 1.2, amino acids are small organic molecules that consist of an α(central) carbon atom linked to an amino group ($-NH_2$), a carboxyl group($-COOH$), a hydrogen atom and a variable component called a side chain. Although roughly 500 amino acids have been discovered in nature, only 20 amino acids among these can be found in protein. Peptide bonds link these amino acids together and form a long chain-like structure for protein molecule (refer to Figure 1.3). Peptide bonds are formed by a biochemical reaction

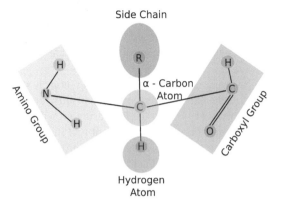

Figure 1.2 Structure of amino acid.

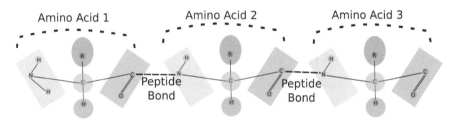

Figure 1.3 Chain of amino acids, linked together by peptide bond to form a long chain of protein.

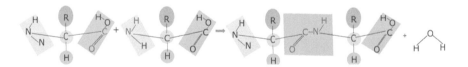

Figure 1.4 Release of water molecule during formation of peptide bond.

that extracts a water molecule while joining two neighbouring amino acids (refer to Figure 1.4). The number of amino acids in a protein chain differs based on organ and species. If a protein chain contains 100 amino acids of 20 different types, then 20^{100} many possible arrangements are possible.

1.2.1.2 Secondary Structure

In the peptide bond, the nitrogen and carbon atoms of two neighbouring amino acids cannot lie on a straight line as the bond angle between two atoms of the chain is about $110°$. Each nitrogen and carbon atom can rotate to a certain extent, resulting in a three-dimensional form of local segments of proteins. The two most commonly found secondary structural elements are α-helices and β-sheets.

1.2.1.3 Tertiary Structure

The tertiary structure of protein molecules occurs due to interaction between the side chains present in the amino acids. The common features of protein tertiary structure reveal much about the biological functions of the proteins and their evolutionary origins. If the tertiary structure of a protein is disrupted, it loses its activity.

1.2.1.4 Quaternary Structure

The quaternary structure of a protein is formed by the organization of several protein chains or subunits. Hydrogen bonds and Van der Waals forces between nonpolar side chains hold these subunits together. Proper arrangement of protein chains in the quaternary structure allows the protein to function properly.

1.2.2 WHAT IS GRAPH?

A graph is a mathematical way to represent a network that also includes the relationships between the entities. It consists of nodes, usually denoted as points and lines joining two nodes, called edges. The mathematical description of a graph is as follows.

Definition 1 *Graph.* *A graph $G = (V, E)$ consists of a set of vertices V and the set of edges $E \subseteq V \times V$. We call a node $v_i \in V$ is related to another node $v_j \in V$ if $e_{ij} = (v_i, v_j) \in E$. A graph can be directed or undirected based on the relationship defined among the nodes of the edges. The directed edges are usually denoted by an arrow showing the direction.*

Definition 2 *Neighbourhood of a node.* *v_i in a graph is defined as the set of nodes, which are connected by edges to the node v_i, i.e.*

$$\mathcal{N}(v_i) = \{v_j \in V \mid e_{ij} = (v_i, v_j) \in E\} \tag{1.1}$$

Definition 3 *Degree of a node.* *Degree of a node v_i, $d(v_i)$ is the number of nodes in its neighbourhood, i.e.*

$$d(v_i) = |\mathcal{N}(v_i)| \tag{1.2}$$

$|.|$ denotes cardinality of a set.

Definition 4 *Adjacency Matrix.* *Let the graph $G = (V, E)$ has N number of nodes, i.e., $|V| = N$. Then, adjacency matrix A is a $N \times N$ matrix, where*

$$A_{ij} = w_{ij}, \text{ if } e_{ij} \in E,$$
$$= 0, \text{ Otherwise} \tag{1.3}$$

For unweighted graph, all edges are given the same weightage, i.e., $w_{ij} = 1, \forall e_{ij} \in E$.

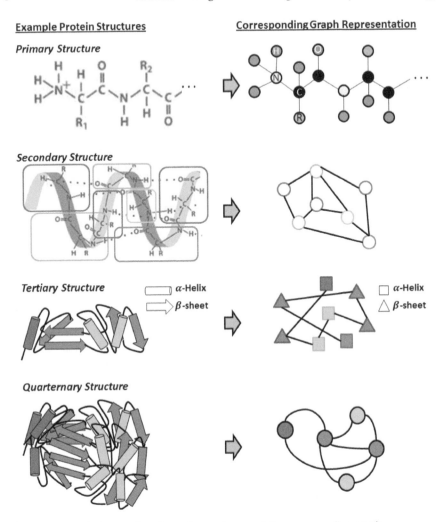

Figure 1.5 Typical examples of protein structures and the corresponding graph representations.

1.2.3 GRAPH REPRESENTATION FOR PROTEINS

One of the major areas of biological research today is to explore how proteins perform an incredible array of diverse tasks. As we have already discussed in the previous subsection, protein molecules have different hierarchical structures constructed from only 20 amino acids. A key concept in understanding how proteins work is derived from the three-dimensional structure formed by the amino acid sequences. Here, GNN can play an important role, capturing the relationship between the amino acid sequence and the three-dimensional structure and eventually with the function or the task the protein does. In this regard, one first needs to generate graphs from the protein data (refer to Figure 1.5). This processing of the data is done based on the specified task to be performed. For example, PROTEINS [2] is a dataset, classified into enzymes and non-enzymes, where graph nodes are secondary structure elements (SSEs) and two nodes share an edge if they are neighbours in the amino acid sequence or in 3D space. It has three discrete labels representing helix, sheet or turn. Here, the task is to predict the Quaternary structure of the protein molecule, which talks about the function of the protein based on the secondary structure of the protein. The D&D [5] is a dataset of 1178 protein structures, classified into enzymes and non-enzymes. Another dataset ENZYMES [2] consists of protein tertiary structures consisting of 600 enzymes from the BRENDA enzyme database. The task is to correctly assign each enzyme to one of the six groups categorized using the Enzyme Commission Number (EC Number) classification scheme.

1.3 PROTEIN CLASSIFICATION: A GRAPH REPRESENTATION LEARNING PERSPECTIVE

As the name indicates, Graph Representation Learning is an approach to represent the nodes of a graph in such a way that it can be easily handled by the existing machine learning algorithms. Graph structured data are full of non-uniformity; different graphs have a different number of nodes and every node has a variable number of neighbours. Therefore, the traditional neural network architecture, such as Convolutional Neural Network (CNN) or Recurrent Neural Network (RNN), which operates using a fixed-sized filter or accepts input of fixed size, cannot be directly applied on graphs. Graph Representational Learning takes care about the fact that how we can use machine learning on a graph. The graph has the power of focusing on the relationship between the entities together with the individual property of the entities. Note that, machine learning is not the only possible way to analyze graph data; the field of network analysis is an independent area of research to handle graph structure. However, due to the exponential growth and complexity of graph data, machine learning is expected to be far more potential for modelling, analyzing and understanding this type of data.

The task of protein classification can be viewed as a potential application of machine learning on graph-structured data. In this particular context, a graph structure $G = (V, E)$ is first formulated from the dataset, where each node in V denotes the Secondary Structure of the protein molecule, each edge denotes the relationship between a pair of structures, which are defined according to the specified task and the graph label are the functions of the protein. For instance, in the case of the PROTEINS and D&D dataset, the classification of a protein molecule as an enzyme or non-enzyme has been formulated as a graph classification problem. Another application of the graph classification task on the ENZYMES dataset is to predict the type of enzymes.

1.4 GRAPH REPRESENTATION LEARNING (GRL) MODELS FOR PROTEIN CLASSIFICATION

In this section, we present the categorization of existing approaches in the literature, which are used to perform the graph classification task. Incidentally, some existing works in the literature such as [17] focus on modelling interaction on biological graphs. However, in this work, our focus is on learning the interaction between the nodes in a biological graph in terms of low-dimensional vectors, which can help to perform the classification task. We categorize the existing graph representation learning approaches into the following four groups: i) Traditional approaches of incorporating kernel method for the graph classification task, ii) Graph Convolutional Network (GCN), iii) Graph Neural Network (GNN) with different pooling techniques and iv) approaches beyond Graph Convolutional Network (GCN), which includes several graph attention neural networks [23]. We provide a summary of some of the approaches under each category here.

1.4.1 TRADITIONAL APPROACH

Traditional approaches to graph classification follow the standard machine learning techniques, which was mostly used before the invention of deep learning. In this approach, several graph kernel methods have been designed using statistical and graphical properties to extract explicit feature representations of the graph.

1.4.1.1 The Weisfeiler-Lehman Kernel

The main idea of the graph kernel method is to decompose the graph into substructures and the Weisfeiler-Lehman kernel method is one of such approaches [20]. WL-kernel extracts node-level features using the iterative neighbourhood aggregation approach, which enables the model to gather neighbourhood information and accumulate this rich information into a graph level representation.

Weisfeiler-Lehman graph kernel follows a similar notion of the well-known 1-d Weisfeiler-Lehman test of isomorphism while designing the kernel [6]. Given two graphs and to test whether they are isomorphic or not, the key idea of the algorithm is to iteratively assign a new colour to a node from the sorted alphabet of colours, based on the node's current colour and the multi-set of colours of neighbouring nodes (refer to Figure 1.6). We summarize the steps of the method for an iteration i in the following algorithm.

Algorithm:

1. The nodes are initiated with node label. If node label is not given, node labeling function $f^{(0)}$ is used to assign label to each node. Possible node labeling functions are assigning degree, centralities or clustering coefficient of the node. Here, we have used node degree as initial label of the node.

2. The multiset of the node label its sorted with an order defined and the feature vector representation for the graph is the frequency distribution of the node label.

3. For iteration $i > 0$, the node labelling function $f^{(i)}$ will take the target node v's node label $l^{(i-1)}(v)$ and the multiset $\{l^{(i-1)}(u) : u \in \mathcal{N}(v)\}$ and $f^{(i)}$ outputs an

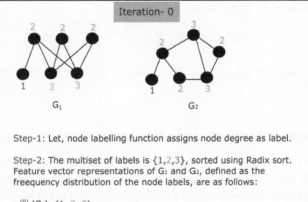

Step-1: Let, node labelling function assigns node degree as label.

Step-2: The multiset of labels is {1,2,3}, sorted using Radix sort. Feature vector representations of G_1 and G_2, defined as the freequency distribution of the node labels, are as follows:

$\varphi^{(0)} (G_1) = (1, 3, 2)$

$\varphi^{(0)} (G_2) = (1, 3, 2)$

Therefore, after 0-th iteration, both graph ended up having same vector representation.

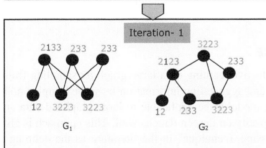

Step-1: Node labelling function assigns node label as concatenation of existing own label and the labels of the neighbouring nodes.

Step-2: The multiset of labels is {12, 233, 2123, 2133, 3223}, sorted using Radix sort and we give the colours to the labels as follows:

12	⇨	4		
233	⇨	5	2133 ⇨ 7	
2123	⇨	6	3223 ⇨ 8	

Feature vector representations of G_1 and G_2 after first iteration is the contanetation of existing feature vector and freequency distribution of the current node labels.

$\varphi^{(1)} (G_1) = (1, 3, 2, 1, 2, 0, 1, 2)$

$\varphi^{(0)} (G_2) = (1, 3, 2, 1, 2, 1, 0, 2)$

Therefore, after first iteration, both graphs ended up having different vector representation, which concludes that these two graphs are non-isomorphic.

Figure 1.6 Weisfeiler-Lehman test of graph isomorphism.

unique label $l^{(i)}(v)$ for the node v, i.e.,

$$l^{(i)}(v) = f^{(i)}(l^{(i-1)}(v), \{l^{(i-1)}(u) : u \in \mathcal{N}(v)\}) \tag{1.4}$$

4. Algorithm will repeat Steps 2 to 3 until i reaches the pre-defined iteration bound.

1.4.2   GRAPH CONVOLUTIONAL NETWORK

Motivated by the success of CNN in image data, this category of models aim to imitate CNN like operation for graph-structured data. GCN approaches can be categorized into two classes, the spectral-based approach and the spatial-based approach [23]. Spectral-based GCN approaches define the convolution operation graph from a signal processing perspective. But, due to high computational cost and domain dependency, a relatively newer approach, the spatial-based technique has gained the interest of the researchers. The spatial-based approach follows different aggregation techniques on the node features from the neighbourhood to formulate graph convolution.

1.4.2.1   GraphSAGE

GraphSAGE uses inductive learning on a large graph by sampling from the neighbourhood of a graph and aggregating information from the sampled nodes [9]. This approach enables the model to be applicable to large graphical data and improves the generalization capacity on unseen data as well. This approach is also termed as "Neural Message Passing Technique" in the literature as the defining property of these approaches is that the nodes exchange vector-shaped messages within their neighbourhood and their features get updated considering the existing feature and the messages that has been received from the neighbourhood using a neural network architecture.

The i-th iteration of the neural message passing technique can be expressed in two steps:

$$\mathcal{M}^{(i)}_{\mathcal{N}(v)} = AGGREGATE^{(i)}(\{h_u^{(i-1)} : u \in \mathcal{N}(v)\}) \tag{1.5}$$

$$h_v^{(i)} = COMBINE^{(i)}(h_v^{(i-1)}, \mathcal{M}^{(i)}_{\mathcal{N}(v)}) \tag{1.6}$$

$AGGREGATE^{(i)}$ and $COMBINE$ are two trainable neural networks.

$AGGREGATE^{(i)}$ accumulates the feature information about the neighbouring nodes ($\mathcal{N}(v)$) that has been gathered so far (i.e. upto $i-1$-th iteration) and cooks the message $\mathcal{M}^{(i)}_{\mathcal{N}(v)}$ for v in i-th iteration. The job of $COMBINE^{(i)}$ function is to accept the message $\mathcal{M}^{(i)}_{\mathcal{N}(v)}$ and based on the received message update the existing feature vector of v (see Figure 1.7). Therefore, building a powerful GNN ends up finding powerful $AGGREGATE$ and $COMBINE$ functions for the model.

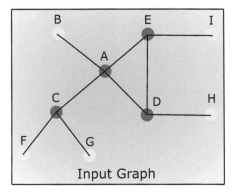

GraphSAGE has examined three candidate functions as aggregator, which are permutation invariant, trainable and high representational capacity, namely, *MEAN*, *LSTM* and *MAX* pooling aggregator. In the pooling variant of GraphSAGE, the message generating step, i.e. the aggregator function has been formulated as follows:

$$\mathcal{M}^{(i)}_{\mathcal{N}(v)} = MAX(\{ReLU(W.h_u^{(i-1)}), \forall u \in \mathcal{N}(v)\}) \qquad (1.7)$$

where W is a learnable matrix, i.e. a weight matrix of a fully connected neural network, *MAX* represents element-wise max operator.

The final update step, i.e., the *COMBINE* function is a linear mapping which takes the concatenation of $h_v^{(i-1)}$, the feature vector of v from the previous iteration and $\mathcal{M}^{(i)}_{\mathcal{N}(v)}$, the message vector and generate the feature vector $h_v^{(i)}$ after i-th iteration.

GraphSAGE is a model that can be learned in both supervised and unsupervised ways. In supervised learning, which is more applicable for protein graph

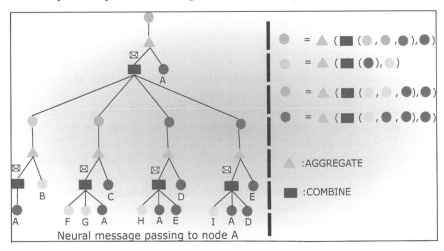

Figure 1.7 Representation learning using neural message passing technique.

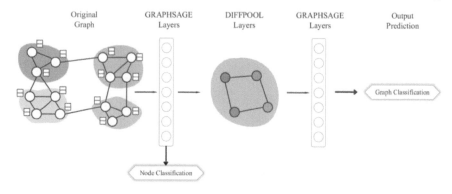

Figure 1.8   Hierarchical structure learning using DIFFPOOL.

classification, the model uses another permutation invariant function *READOUT* such as *SUM*, which will generate the embedding for the whole graph, based on the feature vectors of the nodes from the final iteration. In the protein classification task, the cross-entropy loss has been calculated and back-propagating the loss, the weight matrices in the model have been tuned via stochastic gradient descent technique. However, for the unsupervised setting, the loss function has been formulated in such a way that it will enforce the nearby nodes in the graph to have similar representation while the distinct nodes have very different representations.

1.4.3   GRAPH POOLING MODULES

The model, which falls into this category, is specially designed for the graph classification task. As we have discussed in earlier approaches, after finding low-dimensional vectors for nodes, to find the representation for the whole graph, a graph level pooling function is needed.

1.4.3.1   DIFFPOOL

Most commonly used graph pooling functions, such as *MEAN*, *MAX* or *SUM*, follow a flat information abstraction from node level to graph level. These methods fail to learn the hierarchical representation of graphs. DIFFPOOL [26] is a differentiable graph pooling module, which is capable of abstracting information from different hierarchical levels. As shown in Figure 1.8 [24], the model learns a differentiable soft cluster assignment for nodes at each layer of a deep GNN and maps the nodes to a set of clusters, which is treated as the input for the next layer.

At layer l, DIFFPOOL uses two seperate GNNs $GNN_{l,embed}$ and $GNN_{l,pool}$ to generate the cluster assignment matrix $Z^{(l)}$ and to find the embedding matrix $S^{(l)}$, respectively. l-th layer of DIFFPOOL architecture can be formalized as follows:

$$Z^{(l)} = GNN_{l,embed}(A^{(l)}, X^{(l)}) \qquad (1.8)$$

$$S^{(l)} = softmax(GNN_{l,pool}(A^{(l)}, X^{(l)})) \qquad (1.9)$$

At $l+1$-th layer, aggregation operation and adjacency matrix have been calculated based on these two matrices, $Z^{(L)}$ and $S^{(l)}$ as formulized below:

$$X^{(l+1)} = S^{(l)^T} Z^{(l)} \qquad (1.10)$$

$$A^{(l+1)} = S^{(l)^T} A^{(l)} S^{(l)} \qquad (1.11)$$

DIFFPOOL can be combined with any GCN model and this process will enable the model to extract hierarchical structures within the graph.

1.4.4 GRAPH ATTENTION MECHANISM

In deep learning, the Attention Mechanism was introduced to mimic the ability of the human brain to selectively concentrate on relevant things. Inspired by the success story in machine translation and natural language processing, several GNN-based models have incorporated attention mechanisms during aggregating from neighbouring nodes, generating random walks etc. We discuss one of the pioneering works of GNN architectures, who have used this attention mechanism in their model.

1.4.4.1 Graph Attention Network

This is a spatial approach, which computes the hidden representation for each node in the graph using a self-attention strategy during aggregating neighbours [22].

In order to find the importance or weightage of the neighbouring element v_j of a node v_i in a graph at t- iteration, Graph Attention Network (GANet) uses attention mechanism as follows:

$$\alpha(h_i^{(t)}, h_j^{(t)}) = \frac{exp\left(LeakyReLU\left(\vec{a}^T \left[W\vec{h_i}^{(t)} \| W\vec{h_j}^{(t)}\right]\right)\right)}{\sum_{l\in\mathcal{N}(v_i)} exp\left(LeakyReLU\left(\vec{a}^T \left[W\vec{h_i}^{(t)} \| W\vec{h_l}^{(t)}\right]\right)\right)} \qquad (1.12)$$

where W is weight matrix for a linear transformation, applied to every node. h_i^t and h_j^t denote the vector representations of the nodes v_i and v_j respectively, after t-th iteration. $\vec{a}$ is the single-layer feed-forward network. T and $\|$ denote the operations transposition and concatenation, respectively.

Then, the embedding for the node v_i for $t+1$-th iteration has been formulated using multi-head attention mechanism as follows:

$$h_i^{(t+1)} = \|_{k=1}^K \sigma\left(\sum_{j\in\mathcal{N}(v_i)} \alpha_k(h_i^{(t)}, h_j^{(t)}) W_k^{(t)} h_j^{(t)}\right) \qquad (1.13)$$

Based on this Graph Attention Mechanism, [7] designs graph attention networks, denoted as GANet, which uses a GCN layer for network embedding learning. However, using an attention mechanism is very much computationally expensive and not applicable to a large graph.

Table 1.1

Statistics of the benchmark protein datasets used in graph classification

Dataset	Size	Classes	Avg. nodes
D&D	1178	2	284.32
PROTEINS	1113	2	39.1
ENZYMES	600	6	32.6

1.4.5 PROTEIN DATASETS AND GRL MODEL PERFORMANCE

This section summarizes three datasets that are popularly used as benchmarks for graph classification task on proteins [25]. The details of the datasets can be found in Table 1.1.

1.4.5.1 D&D

This dataset was constructed from the Protein Data Bank (PDB), which consists of X-ray crystal structures with resolution ≤ 2.5 and R-factor ≥ 0.25 by [5]. Then, the dataset is split into Enzymes and non-Enzymes, based on Enzyme Commission Number, annotations in the PDB and Medline abstracts. D&D contains 1178 graphs of protein, among which 691 are enzymes and 487 are non-enzymes.

1.4.5.2 PROTEINS

In the experimental section in [5], while classifying enzymes and non-enzymes from protein structures, it has been noticed that there might be a correlation between Protein function and its helix context. Helix content distribution suggests reviewing this classification task performance incorporating the helix context of nodes. The dataset PROTEINS consists of graphs, where nodes represent secondary structure elements and are annotated by their type, i.e., helix, sheet or turn as well as several physical and chemical information. An edge connects two nodes if they are neighbours along the amino acid sequence or one of three nearest neighbours in space. Graph levels are given as enzymes and non-enzymes.

1.4.5.3 ENZYMES

Enzymes have unique three-dimensional structure, which acts as a biological catalysts. Enzyme Commission Number (EC Number) is a classification scheme for enzymes based on how they catalyze the chemical reactions. Enzymes are categorized into six groups, namely,

- EC 1 → Oxidoreductases
- EC 2 → Transferases
- EC 3 → Hydrolases

Table 1.2
Performance of the discussed models on different datasets

Dataset	WL-kernel	GraphSAGE	DIFFPOOL	GANet
D&D	78.29%	72.9%	80.64%	-
PROTEINS	74.99%	73.5%	76.25%	78.65%
ENZYMES	53.33%	58.2%	62.53%	64.32%

- EC 4 → Lyases
- EC 5 → Isomerases
- EC 6 → Ligases

The dataset ENZYMES was derived from the BRENDA database [2]. It contains 600 protein tertiary structures. Graph levels are given as the class of enzymes among six EC classes.

1.4.5.4   Comparative performance study of the GRL models

Table 1.2 shows the performances of the discussed model on datasets D&D, PRO-TEINS and ENZYMES, respectively. Although GraphSAGE is inspired by the WL-algorithm, it is unable to perform as good as the WL kernel method. It proves the fact that the AGGREGATE and COMBINE functions used in the GraphSAGE technique do not have equal representational power as that of the colour assigning technique of the WL kernel. Overall, DiffPOOL and GANet seem to be effective than the other two approaches. However, due to high computational complexity, GANet cannot be performed on large datasets, such as D&D.

1.5   CASE STUDY

This section illustrates the above-discussed four major approaches of graph representation learning for classifying the benchmark PROTEINS dataset. To discuss the working procedure of the baselines, we start with real-life graph data from the dataset Proteins as an example. Figure 1.9 shows the graph on which we want to apply the baselines to generate the embedding. As depicted in the figure, we have a graph with 10 nodes, each of which denotes Secondary Structural Elements (SSEs) of two types, Helix (Green coloured) and Sheet (Yellow coloured). It is an Enzyme molecule and the task is to predict this as a graph label.

1.5.1   WL KERNEL METHOD

As we have discussed in an earlier section, there are several options for node labelling function. However, in the PROTEINS dataset, nodes are classified into three classes, namely, Helix, Sheet and Turn, we use the class of the nodes, denoted as 1, 2 and 3,

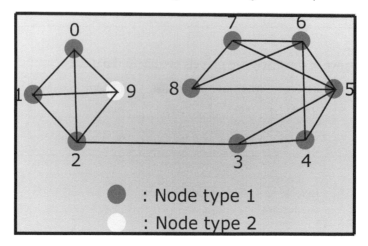

Figure 1.9 A graph example from the dataset PROTEINS.

as the node label function at the initialization step. In this particular graph example, nodes with labels 1 and 2 are present.

Therefore, at the beginning of the algorithm, we have the labels for the nodes as the node classes present in the dataset, i.e.,

$$
\begin{array}{ll}
l^{(0)}(0) = 1 & l^{(0)}(5) = 1 \\
l^{(0)}(1) = 1 & l^{(0)}(6) = 1 \\
l^{(0)}(2) = 1 & l^{(0)}(7) = 1 \\
l^{(0)}(3) = 1 & l^{(0)}(8) = 1 \\
l^{(0)}(4) = 1 & l^{(0)}(9) = 2
\end{array}
\tag{1.14}
$$

Therefore, after sorting the multiset of node labels using radix sort, the feature vector representation for the graph is the frequency distribution of the node labels, i.e., $G^{(0)} = (9, 1, 0)$.

For the next iteration, i.e., $i = 1$, every node's label will be formed by sorting the multi-set of the neighbours' labels and concatenating with its own label, i.e.,

$$
l^{(1)}(v) = l^{(0)}(v) \,\|\, sorted(\{l^{(0)}(u) : u \in \mathcal{N}(v)\})
\tag{1.15}
$$

Following this approach, we have got the nodes' label for first iteration as follows:

$$
\begin{array}{ll}
l^{(1)}(0) = 1112 & l^{(1)}(5) = 111111 \\
l^{(1)}(1) = 1112 & l^{(1)}(6) = 11111 \\
l^{(1)}(2) = 11112 & l^{(1)}(7) = 1111 \\
l^{(1)}(3) = 1111 & l^{(1)}(8) = 1111 \\
l^{(1)}(4) = 1111 & l^{(1)}(9) = 2111
\end{array}
\tag{1.16}
$$

Applying Radix sort on the multi-set of the node labels, we have the sorted list of node labels as follows:

$$\{1111, 1112, 2111, 11111, 11112, 111111\} \tag{1.17}$$

Therefore, the feature vector representation for the graph after first iteration is the concatenation of $G^{(0)}$ and the frequency distribution of the node labels after first iteration, i.e., $G^{(1)} = G^{(0)} \parallel (9, 1, 0, 4, 2, 1, 1, 1, 1)$.

This algorithm will continue to find the feature representation of the graph iteratively until i reaches the pre-defined bound for iteration.

1.5.2   GRAPHSAGE

Now, we will discuss about one iteration of GraphSAGE on the same graph. As discussed previously, in order to start the process, we have to initialize the node embeddings. We may initialize randomly but here to include the node-related information, we have initialized with one-hot encoding of the node class. Therefore, we have the initial embeddings of the nodes as follows:

$$
\begin{aligned}
h^{(1)}(0) &= (1,0,0) & h^{(1)}(5) &= (1,0,0) \\
h^{(1)}(1) &= (1,0,0) & h^{(1)}(6) &= (1,0,0) \\
h^{(1)}(2) &= (1,0,0) & h^{(1)}(7) &= (1,0,0) \\
h^{(1)}(3) &= (1,0,0) & h^{(1)}(8) &= (1,0,0) \\
h^{(1)}(4) &= (1,0,0) & h^{(1)}(9) &= (0,1,0)
\end{aligned}
\tag{1.18}
$$

GraphSAGE includes two MLPs in the process, one for the AGGREGATE function, while finding the message vector from the neighbourhood of a node and another one is for COMBINE function, which essentially takes two vectors, existing embedding of the node and the message vector and expresses as the new embedding. Weight matrices of the MLPs are randomly initialized and their sizes depend on the dimension of the input and output vectors.

In this case, we want to find the embedding of the nodes in the space $\mathscr{R}^3$. The matrices are initialized randomly as follows:

$$
W_{agg} = \begin{pmatrix} 2.14 & 1.33 & -2.78 \\ 0.39 & 3.88 & -1.54 \\ -2.78 & 0.26 & 1.86 \end{pmatrix}, \qquad b_{agg} = \begin{pmatrix} 5.58 \\ 0.24 \\ -4.76 \end{pmatrix}
$$

$$
W_{com} = \begin{pmatrix} 4.11 & 8.24 & 3.32 & 8.5 & 1.42 & 9.16 \\ 7.61 & 4.20 & 8.83 & 0.26 & 6.15 & 2.64 \\ 9.17 & 7.12 & 2.01 & 1.10 & 1.17 & 7.60 \end{pmatrix}, \qquad b_{com} = \begin{pmatrix} 1.96 \\ 2.77 \\ 6.09 \end{pmatrix}
$$

W_{agg} and b_{agg} are the weight and bias matrices, respectively, involved with AGGREGATE function. Similarly, W_{com} and b_{com} are the weight and bias matrices associated with COMBINE function. The GraphSAGE algorithm aggregates information from a sampled neighbourhood of a node, which enables the model to handle a large graph.

The algorithm uses a uniform sampler function to choose a set of fixed-sized neighbours during aggregation. However, as we are dealing with a small example, we take the full neighbourhood of every node while aggregating. With the matrices we have initialized, the AGGREGATE and COMBINE functions are defined as follows:

$$\mathscr{M}_{\mathscr{N}(v)}^{(i)^T} = MAX(\{ReLU(W_{agg}h_u^{(i-1)^T} + b_{agg}), \forall u \in \mathscr{N}(v)\}) \qquad (1.19)$$

$$h_v^{(i)^T} = W_{com}\left(\left[h_v^{(i-1)^T} \;\|\; \mathscr{M}_{\mathscr{N}(v)}^{(i)^T}\right] + b_{com}\right) \qquad (1.20)$$

Therefore, the vector representation of the 0_{th} node after 1_{st} iteration can be calculated as follows:

$$\mathscr{M}_{\mathscr{N}(0)}^{(1)^T} = MAX\{ReLU(W_{agg}h_1^{(0)^T} + b_{agg}), ReLU(W_{agg}h_2^{(0)^T} + b_{agg}), ReLU(W_{agg}h_9^{(0)^T}$$
$$+ b_{agg})\}$$

$$= MAX\left\{ ReLU\left(\begin{pmatrix} 2.14 & 1.33 & -2.78 \\ 0.39 & 3.88 & -1.54 \\ -2.78 & 0.26 & 1.86 \end{pmatrix}\begin{pmatrix} 1 \\ 0 \\ 0 \end{pmatrix} + \begin{pmatrix} 5.58 \\ 0.24 \\ -4.76 \end{pmatrix}\right), \right.$$

$$ReLU\left(\begin{pmatrix} 2.14 & 1.33 & -2.78 \\ 0.39 & 3.88 & -1.54 \\ -2.78 & 0.26 & 1.86 \end{pmatrix}\begin{pmatrix} 1 \\ 0 \\ 0 \end{pmatrix} + \begin{pmatrix} 5.58 \\ 0.24 \\ -4.76 \end{pmatrix}\right),$$

$$\left. ReLU\left(\begin{pmatrix} 2.14 & 1.33 & -2.78 \\ 0.39 & 3.88 & -1.54 \\ -2.78 & 0.26 & 1.86 \end{pmatrix}\begin{pmatrix} 0 \\ 1 \\ 0 \end{pmatrix} + \begin{pmatrix} 5.58 \\ 0.24 \\ -4.76 \end{pmatrix}\right)\right\}$$

$$= MAX\left\{ ReLU\left(\begin{pmatrix} 2.14 \\ 0.39 \\ -2.78 \end{pmatrix} + \begin{pmatrix} 5.58 \\ 0.24 \\ -4.76 \end{pmatrix}\right), \right.$$

$$\left. ReLU\left(\begin{pmatrix} 2.14 \\ 0.39 \\ -2.78 \end{pmatrix} + \begin{pmatrix} 5.58 \\ 0.24 \\ -4.76 \end{pmatrix}\right), \; ReLU\left(\begin{pmatrix} 1.33 \\ 3.88 \\ 0.26 \end{pmatrix} + \begin{pmatrix} 5.58 \\ 0.24 \\ -4.76 \end{pmatrix}\right)\right\}$$

$$(1.21)$$

$$= MAX\left\{ \begin{pmatrix} 7.72 \\ 0.63 \\ 0 \end{pmatrix}, \begin{pmatrix} 7.72 \\ 0.63 \\ 0 \end{pmatrix}, \begin{pmatrix} 6.91 \\ 4.12 \\ 0 \end{pmatrix} \right\}$$

$$= \begin{pmatrix} 7.72 \\ 4.12 \\ 0 \end{pmatrix} \qquad (1.22)$$

$$h_0^{(1)^T} = W_{com} \left(\left[h_0^{(0)^T} \parallel \mathcal{M}_{\mathcal{N}(v)}^{(1)^T} \right] + b_{com} \right)$$

$$= \begin{pmatrix} 4.11 & 8.24 & 3.32 & 8.5 & 1.42 & 9.16 \\ 7.61 & 4.20 & 8.83 & 0.26 & 6.15 & 2.64 \\ 9.17 & 7.12 & 2.01 & 1.10 & 1.17 & 7.60 \end{pmatrix} \left(\left[\begin{pmatrix} 1 \\ 0 \\ 0 \end{pmatrix} \parallel \begin{pmatrix} 7.72 \\ 4.12 \\ 0 \end{pmatrix} \right] \right)$$

$$+ \begin{pmatrix} 1.96 \\ 2.77 \\ 6.09 \end{pmatrix}$$

$$= \begin{pmatrix} 4.11 & 8.24 & 3.32 & 8.5 & 1.42 & 9.16 \\ 7.61 & 4.20 & 8.83 & 0.26 & 6.15 & 2.64 \\ 9.17 & 7.12 & 2.01 & 1.10 & 1.17 & 7.60 \end{pmatrix} \begin{pmatrix} 1 \\ 0 \\ 0 \\ 0 \\ 7.72 \\ 4.12 \\ 0 \end{pmatrix} + \begin{pmatrix} 1.96 \\ 2.77 \\ 6.09 \end{pmatrix}$$

$$= \begin{pmatrix} 75.59 \\ 19.89 \\ 22.48 \end{pmatrix} + \begin{pmatrix} 1.96 \\ 2.77 \\ 6.09 \end{pmatrix}$$

$$= \begin{pmatrix} 77.55 \\ 22.66 \\ 28.57 \end{pmatrix} \tag{1.23}$$

After normalizing the vector, we get

$$h_0^{(1)} = \frac{h_0^{(1)}}{\parallel h_0^{(1)} \parallel_2} = \begin{pmatrix} 0.9050 & 0.2644 & 0.3334 \end{pmatrix} \tag{1.24}$$

Calculating in similar way, we get vector representations for other nodes as follows:

$$
\begin{aligned}
h_1^{(1)} &= \begin{pmatrix} 0.9050 & 0.2644 & 0.3334 \end{pmatrix} \\
h_2^{(1)} &= \begin{pmatrix} 0.8536 & 0.4153 & 0.3145 \end{pmatrix} \\
h_3^{(1)} &= \begin{pmatrix} 0.9269 & 0.2077 & 0.3127 \end{pmatrix} \\
h_4^{(1)} &= \begin{pmatrix} 0.9269 & 0.2077 & 0.3127 \end{pmatrix} \\
h_5^{(1)} &= \begin{pmatrix} 0.9269 & 0.2077 & 0.3127 \end{pmatrix} \\
h_6^{(1)} &= \begin{pmatrix} 0.9269 & 0.2077 & 0.3127 \end{pmatrix}
\end{aligned}
\tag{1.25}
$$

$$
\begin{aligned}
h_7^{(1)} &= \begin{pmatrix} 0.9269 & 0.2077 & 0.3127 \end{pmatrix} \\
h_8^{(1)} &= \begin{pmatrix} 0.9269 & 0.2077 & 0.3127 \end{pmatrix} \\
h_9^{(1)} &= \begin{pmatrix} 0.9476 & 0.1587 & 0.2772 \end{pmatrix}
\end{aligned}
\tag{1.26}
$$

1.5.3 DIFFPOOL

DIFFPOOL follows more sophisticated graph pooling technique using two GNN models seperately to find an embedding of the entire graph, which captures the hierarchical structures of the graph. We explains DIFFPOOL architecture applying it on the graph example (Figure 1.9) from the PROTEINS dataset. The algorithm begins with the initial feature and adjacency matrices $X^{(0)}$ and $A^{(0)}$, respectively. In this example, $X^{(0)}$ and $A^{(0)}$ are as follows:

$$X^{(0)} = \begin{pmatrix} 1 & 0 & 0 \\ 1 & 0 & 0 \\ 1 & 0 & 0 \\ 1 & 0 & 0 \\ 1 & 0 & 0 \\ 1 & 0 & 0 \\ 1 & 0 & 0 \\ 1 & 0 & 0 \\ 1 & 0 & 0 \\ 0 & 1 & 0 \end{pmatrix}_{10 \times 3} \qquad A^{(0)} = \begin{pmatrix} 0 & 1 & 1 & 0 & 0 & 0 & 0 & 0 & 0 & 1 \\ 1 & 0 & 1 & 0 & 0 & 0 & 0 & 0 & 0 & 1 \\ 1 & 1 & 0 & 1 & 0 & 0 & 0 & 0 & 0 & 1 \\ 0 & 0 & 1 & 0 & 1 & 1 & 0 & 0 & 0 & 0 \\ 0 & 0 & 0 & 1 & 0 & 1 & 1 & 0 & 0 & 0 \\ 0 & 0 & 0 & 1 & 1 & 0 & 1 & 1 & 1 & 0 \\ 0 & 0 & 0 & 0 & 1 & 1 & 0 & 1 & 1 & 0 \\ 0 & 0 & 0 & 0 & 0 & 1 & 1 & 0 & 1 & 0 \\ 0 & 0 & 0 & 0 & 0 & 1 & 1 & 1 & 0 & 0 \\ 1 & 1 & 1 & 0 & 0 & 0 & 0 & 0 & 0 & 0 \end{pmatrix}_{10 \times 10}$$

We denote the GNN that finds the embedding for the next level by GNN_{emb} and the GNN for cluster assignment by GNN_{pool}. For sake of simplicity, we use a single layer GNN rather than using a fancy one. The number of clusters and the embedding size in every layer are hyper parameters. At the zero-th level, every single node is a cluster. Therefore, the number of clusters at zero-th level, $n^{(0)} = 10$. We want to divide the graph into three clusters in the first level and embedding space to be $\mathcal{R}^5$. Therefore, $n^{(1)} = 3$ and $d^{(1)} = 5$. The size of the weight matrices between i-th and $i+1$-th level will be $n^{(i)} \times n^{(i+1)}$ for the weight matrix W_{emb} associated with the embedding generation and $n^{(i+1)} \times d^{(i+1)}$ for the cluster assignment matrix W_{pool}. The steps of DIFFPOOL for finding the first layer are as follows:

$$\begin{aligned} Z^{(0)} &= ReLU(A^{(0)} \times X^{(0)} \times W_{emb}) \\ S^{(0)} &= softmax(ReLu(A^{(0)} \times X^{(0)} \times W_{pool})) \\ X^{(1)} &= S^{(0)^T} \times Z^{(0)} \\ A^{(1)} &= S^{(0)^T} \times A^{(0)} \times S^{(0)} \end{aligned} \qquad (1.27)$$

We randomly initialize the matrices W_{emb} and W_{pool} as follows:

$$W_{emb} = \begin{pmatrix} 0.0975 & 0.9575 & 0.9706 & 0.8003 & 0.9157 \\ 0.2785 & 0.9649 & 0.9572 & 0.1419 & 0.7922 \\ 0.5469 & 0.1576 & 0.4854 & 0.4218 & 0.9595 \end{pmatrix} \quad \text{and}$$

$$W_{pool} = \begin{pmatrix} 0.6557 & 0.9340 & 0.7431 \\ 0.0357 & 0.6787 & 0.3922 \\ 0.8491 & 0.7577 & 0.6555 \end{pmatrix}$$

Then, using equation 1.27, we get $Z^{(0)}, S^{(0)}, X^{(1)}, A^{(1)}$ as follows:

$$Z^{(0)} = \begin{pmatrix} 0.4736 & 2.8799 & 2.8984 & 1.7424 & 2.6237 \\ 0.4736 & 2.8799 & 2.8984 & 1.7424 & 2.6237 \\ 0.5711 & 3.8374 & 3.8689 & 2.5427 & 3.5394 \\ 0.2926 & 2.8725 & 2.9118 & 2.4008 & 2.7472 \\ 0.2926 & 2.8725 & 2.9118 & 2.4008 & 2.7472 \\ 0.4877 & 4.7875 & 4.8530 & 4.0014 & 4.5787 \\ 0.3902 & 3.8300 & 3.8824 & 3.2011 & 3.6629 \\ 0.2926 & 2.8725 & 2.9118 & 2.4008 & 2.7472 \\ 0.2926 & 2.8725 & 2.9118 & 2.4008 & 2.7472 \\ 0.2926 & 2.8725 & 2.9118 & 2.4008 & 2.7472 \end{pmatrix}$$

$$S^{(0)} = \begin{pmatrix} 0.1661 & 0.5512 & 0.2826 \\ 0.1661 & 0.5512 & 0.2826 \\ 0.1381 & 0.6054 & 0.2564 \\ 0.2172 & 0.5005 & 0.2823 \\ 0.2172 & 0.5005 & 0.2823 \\ 0.1523 & 0.6121 & 0.2357 \\ 0.1831 & 0.5572 & 0.2597 \\ 0.2172 & 0.5005 & 0.2823 \end{pmatrix}$$

$$X^{(1)} = \begin{pmatrix} 0.7875 & 6.8984 & 6.9828 & 5.4531 & 6.5360 \\ 2.1160 & 17.7510 & 17.9579 & 13.7013 & 16.7536 \\ 1.0434 & 8.7893 & 8.8922 & 6.8000 & 8.2986 \end{pmatrix}$$

$$A^{(1)} = \begin{pmatrix} 1.4940 & 3.9546 & 1.9297 \\ 3.9546 & 10.0662 & 5.0093 \\ 1.9297 & 5.0093 & 2.4716 \end{pmatrix}$$

1.5.4 GANET

To explain the multi-head attention mechanism, we use two attention mechanisms on our example data. For sake of simplicity, we give equal initial weightage to every node. For the attention mechanisms, the weight vectors of the feed-forward networks are initialized randomly as follows:

$$\vec{a}_1 = \begin{pmatrix} 0.95 \\ -5.37 \\ 8.52 \\ 1.93 \\ -4.73 \\ 7.34 \end{pmatrix} \quad \text{and} \quad \vec{a}_2 = \begin{pmatrix} 6.69 \\ -1.5 \\ 1.05 \\ -8.31 \\ -9.53 \\ 3.48 \end{pmatrix}$$

We calculate the attention or weightage of one of the neighbouring nodes of the node 0, 1 extensively using these two attention mechanisms and denoted as

$\alpha_1(h_0^{(0)}, h_1^{(0)})$ and $\alpha_2(h_0^{(0)}, h_1^{(0)})$, respectively. We denote the function *LeakyReLU* by $\mathcal{L}$

$$\alpha_1(h_0^{(0)}, h_1^{(0)}) = \frac{exp\left(LeakyReLU\left(\vec{a_1}^T\left[Wh_0^{(0)} \| Wh_1^{(0)}\right]\right)\right)}{\sum_{l \in \mathcal{N}(0)} exp\left(LeakyReLU\left(\vec{a_1}^T\left[Wh_0^{(0)} \| Wh_l^{(0)}\right]\right)\right)}$$

$$= \frac{exp\left(\mathcal{L}\left(\vec{a_1}^T \cdot \begin{pmatrix}1\\0\\0\\1\\0\\0\end{pmatrix}\right)\right)}{exp\left(\mathcal{L}\left(\vec{a_1}\begin{pmatrix}1\\0\\0\\1\\0\\0\end{pmatrix}\right)\right) + exp\left(\mathcal{L}\left(\vec{a_1}\begin{pmatrix}1\\0\\0\\1\\0\\0\end{pmatrix}\right)\right) + exp\left(\mathcal{L}\left(\vec{a_1}\begin{pmatrix}1\\0\\0\\1\\0\\0\end{pmatrix}\right)\right)}$$

$$= \frac{exp\left(\mathcal{L}(2.88)\right)}{exp\left(\mathcal{L}(2.88)\right) + exp\left(\mathcal{L}(2.88)\right) + exp\left(\mathcal{L}(-3.78)\right)}$$

$$= \frac{exp(2.88)}{exp(2.88) + exp(2.88) + exp(-0.0378)}$$

$$= \frac{17.814}{17.814 + 17.814 + 0.963}$$

$$= \frac{17.814}{36.591}$$

$$= 0.487$$

$$(1.28)$$

Similarly, we get

$$\alpha_2(h_0^{(0)}, h_1^{(0)}) = \frac{exp\left(LeakyReLU\left(\vec{a_2}^T\left[Wh_0^{(0)} \| Wh_1^{(0)}\right]\right)\right)}{\sum_{l \in \mathcal{N}(0)} exp\left(LeakyReLU\left(\vec{a_2}^T\left[Wh_0^{(0)} \| Wh_l^{(0)}\right]\right)\right)} \qquad (1.29)$$

$$= \frac{exp\left(\mathscr{L}\left(\vec{a}_2^T \cdot \begin{pmatrix} 1 \\ 0 \\ 0 \\ 1 \\ 0 \\ 0 \end{pmatrix}\right)\right)}{exp\left(\mathscr{L}\left(\vec{a}_2 \begin{pmatrix} 1 \\ 0 \\ 0 \\ 1 \\ 0 \\ 0 \end{pmatrix}\right)\right) + exp\left(\mathscr{L}\left(\vec{a}_2 \begin{pmatrix} 1 \\ 0 \\ 0 \\ 1 \\ 0 \\ 0 \end{pmatrix}\right)\right) + exp\left(\mathscr{L}\left(\vec{a}_2 \begin{pmatrix} 1 \\ 0 \\ 0 \\ 1 \\ 0 \\ 0 \end{pmatrix}\right)\right)}$$

$$= \frac{exp\left(\mathscr{L}\left(-1.62\right)\right)}{exp\left(\mathscr{L}\left(-1.628\right)\right) + exp\left(\mathscr{L}\left(-1.62\right)\right) + exp\left(\mathscr{L}\left(-2.81\right)\right)}$$

$$= \frac{exp\left(-0.0162\right)}{exp\left(-0.0162\right) + exp\left(-0.0162\right) + exp\left(-0.0281\right)}$$

$$= \frac{0.984}{0.984 + 0.984 + 0.972}$$

$$= \frac{0.984}{2.94}$$

$$= 0.335 \tag{1.30}$$

These $\alpha_1(h_0^{(0)}, h_1^{(0)})$ and $\alpha_2(h_0^{(0)}, h_1^{(0)})$ will be used as weightage during aggregating information for node 0 from node 1.

1.5.5 PERFORMANCES

Figures 1.10, 1.11 and 1.12 show the performances of the discussed model on datasets D&D, PROTEINS and ENZYMES, respectively, results of our experiments. Overall, DiffPOOL and GANet seem to be effective than other two approaches. However, due to high computational complexity, GANet cannot be performed on large datasets, such as D&D.

1.6 SUMMARY AND FUTURE RESEARCH DIRECTIONS

Proteins belong to one of the three major groups of macro-nutrients (the other two being carbohydrate and fat), which participate in almost all the cellular activities of a living being and are considered to be the building blocks of life. Each protein is comprised of a chain of amino acids, the arrangement of which essentially determines the structure and the functions of the protein. However, the relationships between amino acid sequences and the various structural/functional aspects of proteins are still not satisfactorily revealed, and thus, the relevant area continues to remain interesting to researchers in the present era. Further, the recent exponential growth of protein structure databases has offered enormous opportunities to explore deep learning methods

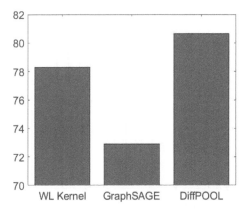

Figure 1.10 Graph classification test set performance in percentage on the dataset D&D. The experiment of GANet on D&D cannot be performed due to memory bound.

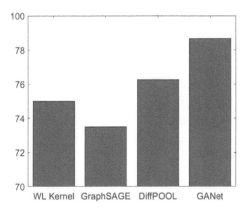

Figure 1.11 Graph classification test set performance in percentage on the dataset PRO-TEINS.

for protein quality assessment, structure prediction and relationship modelling. This chapter exclusively focuses on the various approaches of graph representation learning for protein classification. We have broadly categorized these approaches into four major groups: i) traditional graph kernel-based approaches, ii) GCN models, iii) GNN with various pooling strategies, and iv) GCN with an incorporated attention mechanism. Each of these approaches has been thoroughly discussed along with the theoretical details and the relevant illustrations.

Huge scopes remain in future to employ encoding learning on graphs, which allows storing graphical data as a set of low dimensional vectors. This unsupervised encoding learning for protein would enable us to perform different tasks using the

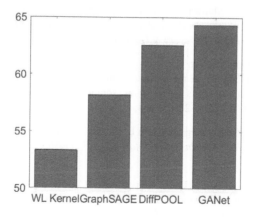

Figure 1.12 Graph classification test set performance in percentage on the dataset ENZYMES.

same set of encoding [21]. As a future direction of work, contrastive learning on graphs may also improve the standard of the protein classification task. Further, multiple views of a protein molecule using a composition of data augmentations may also help in extracting important hierarchical information for protein classification [10]. Hence, another interesting branch of study would be to define different relationships among the nodes and learn multi-modal embedding [11] to perform protein classification.

REFERENCES

1. Jayanth R Banavar, Trinh X Hoang, Amos Maritan, Flavio Seno, and Antonio Trovato. Unified perspective on proteins: A physics approach. *Physical Review E*, 70(4):041905, 2004.

2. Karsten M Borgwardt, Cheng Soon Ong, Stefan Schönauer, SVN Vishwanathan, Alex J Smola, and Hans-Peter Kriegel. Protein function prediction via graph kernels. *Bioinformatics*, 21(suppl_1):i47–i56, 2005.

3. Carl Ivar Branden and John Tooze. *Introduction to Protein Structure*. Garland Science, 2012.

4. Michaël Defferrard, Xavier Bresson, and Pierre Vandergheynst. Convolutional neural networks on graphs with fast localized spectral filtering. In *Advances in Neural Information Processing Systems*, pages 3844–3852, 2016.

5. Paul D Dobson and Andrew J Doig. Distinguishing enzyme structures from non-enzymes without alignments. *Journal of Molecular Biology*, 330(4):771–783, 2003.

6. Brendan L Douglas. The weisfeiler-lehman method and graph isomorphism testing. *arXiv preprint arXiv:1101.5211*, 2011.

7. Hongyang Gao and Shuiwang Ji. Graph representation learning via hard and channel-wise attention networks. In *Proceedings of the 25th ACM SIGKDD International Conference on Knowledge Discovery & Data Mining*, pages 741–749, 2019.

8. Qiang Gao, Wujun Xu, Yao Xu, Dong Wu, Yuhan Sun, Feng Deng, and Wanling Shen. Amino acid adsorption on mesoporous materials: influence of types of amino acids, modification of mesoporous materials, and solution conditions. *The Journal of Physical Chemistry B*, 112(7):2261–2267, 2008.

9. Will Hamilton, Zhitao Ying, and Jure Leskovec. Inductive representation learning on large graphs. In *Advances in Neural Information Processing Systems*, pages 1024–1034, 2017.

10. Kaveh Hassani and Amir Hosein Khasahmadi. Contrastive multi-view representation learning on graphs. In *International Conference on Machine Learning*, pages 4116–4126. PMLR, 2020.

11. Wengong Jin, Kevin Yang, Regina Barzilay, and Tommi Jaakkola. Learning multimodal graph-to-graph translation for molecular optimization. *arXiv preprint arXiv:1812.01070*, 2018.

12. Jumper, J., Evans, R., Pritzel, A., Green, T., Figurnov, M., Ronneberger, O., Tunyasuvunakool, K., Bates, R., Žídek, A., Potapenko, A. and Bridgland, A., 2021. Highly accurate protein structure prediction with AlphaFold. *Nature*, 596(7873), pp. 583–589.

13. Arthur Lesk. *Introduction to Protein Science: Architecture, Function, and Genomics*. Oxford University Press, 2010.

14. Hanne K Mhre, Lars Dalheim, Guro K Edvinsen, Edel O Elvevoll, and Ida-Johanne Jensen. Protein determination—method matters. *Foods*, 7(1):5, 2018.

15. Cristian Micheletti. Comparing proteins by their internal dynamics: Exploring structure–function relationships beyond static structural alignments. *Physics of Life Reviews*, 10(1):1–26, 2013.

16. David M Post, Michael L Pace, and Nelson G Hairston. Ecosystem size determines food-chain length in lakes. *Nature*, 405(6790):1047–1049, 2000.

17. Ranjeet Kumar Rout, Santi P Maity, Pabitra Pal Choudhury, Jayanta Kumar Das, Sk Hassan, Hari Mohan Pandey, et al. Analysis of boolean functions based on interaction graphs and their influence in system biology. *Neural Computing and Applications*, 32(12):7803–7821, 2020.

18. Franco Scarselli, Marco Gori, Ah Chung Tsoi, Markus Hagenbuchner, and Gabriele Monfardini. Computational capabilities of graph neural networks. *IEEE Transactions on Neural Networks*, 20(1):81–102, 2008.

19. Franco Scarselli, Marco Gori, Ah Chung Tsoi, Markus Hagenbuchner, and Gabriele Monfardini. The graph neural network model. *IEEE Transactions on Neural Networks*, 20(1):61–80, 2008.

20. Shervashidze N, Schweitzer P, Van Leeuwen EJ, Mehlhorn K, Borgwardt KM. Weisfeiler-lehman graph kernels. *Journal of Machine Learning Research*. 2011 Sep 1;12(9).

21. Fan-Yun Sun, Jordan Hoffmann, Vikas Verma, and Jian Tang. Infograph: Unsupervised and semi-supervised graph-level representation learning via mutual information maximization. *arXiv preprint arXiv:1908.01000*, 2019.

22. Petar Veličković, Guillem Cucurull, Arantxa Casanova, Adriana Romero, Pietro Lio, and Yoshua Bengio. Graph attention networks. *arXiv preprint arXiv:1710.10903*, 2017.

23. Zonghan Wu, Shirui Pan, Fengwen Chen, Guodong Long, Chengqi Zhang, and S Yu Philip. A comprehensive survey on graph neural networks. *IEEE Transactions on Neural Networks and Learning Systems*, 2020.

24. Yu Xie, Maoguo Gong, Yuan Gao, AK Qin, and Xiaolong Fan. A multi-task representation learning architecture for enhanced graph classification. *Frontiers in Neuroscience*, 13:1395, 2020.

25. Pinar Yanardag and SVN Vishwanathan. Deep graph kernels. In *Proceedings of the 21th ACM SIGKDD International Conference on Knowledge Discovery and Data Mining*, pages 1365–1374, 2015.

26. Rex Ying, Jiaxuan You, Christopher Morris, Xiang Ren, William L Hamilton, and Jure Leskovec. Hierarchical graph representation learning with differentiable pooling. *arXiv preprint arXiv:1806.08804*, 2018.

27. Marketa J Zvelebil and Jeremy O Baum. *Understanding Bioinformatics*. Garland Science, 2007.

2 Extraction of Sequence-Based Features for Prediction of Methylation Sites in Protein Sequences

Monika Khandelwal and Nazir Shabbir
Department of Computer Science & Engineering,
National Institute of Technology, Srinagar, J&K, India

Saiyed Umer
Department of Computer Science & Engineering,
Aliah University, West Bangal, India

CONTENTS

DOI: 10.1201/9781003246688-2

2.1   INTRODUCTION

Protein methylation shows an important role in several biological and cellular processes, comprising signal transduction, gene regulation, metabolism, RNA processing and gene activation [32, 11]. Protein methylation is a reversible method of post translational modifications (PTM) of proteins. This is a process where methyl groups are added into the proteins by changing the protein sequence to encode more information. There are also other forms of PTM, for instance, ubiquitination [24], sumoylation [51] and phosphorylation [49]. The protein methylation usually occurs at N-terminal side chains of arginine (R) and lysine (K), two main protein methylation sites. This can also occur at histidine (H), asparagine (N), proline (P) and alainine (A). A protein family, methyltransferases, carried out these additions by using S-adenosylmethionine to transfer a methyl group. We are mainly focusing on R and K amino residues, for which publicly data are available and methylation mechanism is best understood.

In arginine methylation, methyl group is added to nitrogen within an arginine in a polypeptide. R methylation can happen once (monomethylated) or twice (dimethylated). Dimethylated arginine configuration can be symmetric or asymmetric: symmetric dimethylarginine and asymmetric dimethylarginine. Arginine methylation commonly takes place at R and glycine (G) abundant regions and it effects the protein–protein interactions and protein structure [37]. Arginine methylation involved in numerous functional methods, including signal transduction, transcriptional regulation, genome stability, cell-type differentiation, DNA repair, cancer and RNA processing [1, 2, 36]. S-adenosylmethionine is used by a protein class named histone methyltransferases (HMT) to add a methyl group to the lysine amino residues. Lysine methylation can be done one time (monomethyllysine), two times (dimethyllysine), or three times (trimethyllysine) by lysine methyltransferases. Actually, lysine methylation is primarily studied in H4 and H3 histone proteins, playing vital roles in various biological processes, including transcriptional silencing, X-chromosome inactivation and heterochromatin compaction [25, 29].

Protein methylation is related to several human disorders and diseases, such as multiple sclerosis [30], coronary heart disease [8], cancers [46], neurodegenerative disorders [28] and SARS coronavirus [19, 20, 17, 18]. Identification of methylation sites is crucial for understanding of molecular structure of methylation sites in protein. A precondition to disclosure of molecular methods involved in protein methylation is to correctly recognize methylation sites in proteins. However, experimental procedures, including methylation-specific antibodies [3], mutagenesis of possible methylated residues, mass spectroscopy [48], have been useful for recognition of protein methylation sites. These experimental methods to recognize methylation sites are costly and laborious. Hence, computational methods designed for identification of methylation sites in protein are required for its fast speed and its convenience.

Recently, machine learning (ML)-based techniques have been developed for protein methylation sites and these methods not only offer accurate, robust and fast calculation, and also predict methylation sites from protein sequences only without having any prior knowledge. These techniques are vary in representation of features, which is exactly how protein primary sequences are represented as fixed dimensional feature vectors. Actually, numerous computational techniques have been established to guess methylation sites from protein primary sequences only. Daily et al. [10] intended a method which specified features to capture disorder information from primary sequences and applied supervised learning method – Support Vector Machine (SVM) for estimation of protein methylation sites. Similarly, Chen et al. [7] designed a method to predict protein methylation sites called MeMo, which used orthogonal binary feature descriptor to represent the information from protein sequences and used SVM algorithm to predict lysine and arginine methylation sites. Shao et al. [41] developed a prediction model for lysine and arginine methylation called BPB-PPMS by combining bi-profile Bayes-based feature extraction method with SVM algorithm to recognize methylation sites from protein primary sequences. Shi et al. [43] developed an improved method named PMeS to predict lysine and arginine methylation in protein sequences by combining SVM with improved feature encoding technique comprised of position weight amino acid composition, surface area, normalized van der Waals volume and sparse property coding.

Further for improving the performance of the prediction model, different informative features are integrated in recent methods. Shien et al. [44] designed a method named MASA, by combining SVM with fundamental characteristics of proteins such as accessible surface area (ASA), sequence information and secondary structure to predict methylation locations on arginine (R), lysine (K), asparagine (N) and glutamate (E). Hu et al. [22] designed a method to predict methyllysine and methylarginine positions based on multiple sequence features such as physicochemical properties [35, 38], sequence conservation, and structural disorder and applied nearest neighbour algorithm for prediction of protein methylation positions. Additionally, Qiu et al. [33] constructed a predictor named iMethyl-PseAAC by combining amino acid composition, physicochemical properties, structural disorder information and sequential evolution with SVM to predict methylation sites. Deng et al. [12] built a predictor named GPS-MSP to estimate arginine methylation locations and lysine methylation types from the protein primary sequences. Wei et al. [47] proposed a model named MePred-RF by combining various sequence-based features with Random Forest (RF) algorithm to predict lysine methylation and arginine methylation.

The prediction performance of already existing predictors for methylation sites is not satisfactory in case of sensitivity, accuracy and specificity. Most of existing methods requires disorder, structural and evolutionary information for extracting features and sometimes this information may not be available. So, we developed a novel decision tree based model to predict protein methylation sites. Our proposed model exploits protein sequence information to extract features from multiple perspectives for encoding protein sequences. To improve feature representation, MRMD (Maximal Relevance Maximal Distance) feature selection procedure is used for maximizing relevance between target class and features, and concurrently minimizes dependency between features. Comparative results specify that our proposed model overtakes various state-of-the-art models for recognition of methylation sites from protein primary sequences.

Rest of the paper is arranged as follows. In Section 2, we will illustrate about the dataset used in the study and feature representation. Section 3 will describe about the feature selection technique, and Section 4 will represent the model used for methylation sites prediction. The results are discussed in Section 5 and also compared the performance of our model with other classifiers. Finally, Section 6 will summarize the work done in this paper.

2.2   MATERIALS AND METHODS

In this section, we will illustrate about the dataset and its specifications. We used the dataset described by Qiu et al. [33] in this study. The dataset is downloaded from https://www.hindawi.com/journals/bmri/2014/947416/ #supplementary-materials.

2.2.1   DATASETS AND ITS SPECIFICATIONS

The dataset used in this study was formulated by using the Chou's peptide scheme [9] to represent the peptide sequence into non-methylated or methylated peptides. Based on Chou's peptide structure, a peptide sequence is represented by:

$$P_\omega(\Theta) = P_{-\omega}...P_{-2}P_{-1}\Theta P_1 P_2......P_\omega \qquad (2.1)$$

where Θ can be one of the two amino acid residues: R and K. P_ω indicates the ωth upstream residue and $P_{-\omega}$ indicates the ωth downstream residue from centre. The peptide sequence length will be $2\omega + 1$. If the residues in a peptide are less than ω, then lacking residues will be filled with closest neighbor residue. The peptide $P_\omega(\Theta)$ further divided into two types: methylated (positive) dataset if centre is methylation site and non-methylated dataset (negative) if centre is not methylation site.

In this dataset, $\omega = 5$ so peptide sequence length will be 11. We combine the positive and negative dataset to form a standard dataset for R-methylation and K-methylation. The standard dataset for arginine and lysine methylation is represented as:

$$\begin{cases} D_5(R) = D_5^+(R) \cup D_5^-(R) \\ D_5(K) = D_5^+(K) \cup D_5^-(K) \end{cases} \qquad (2.2)$$

where $D_5^+(R)$ depicts the true methylated dataset for arginine and $D_5^-(R)$ depicts the negative dataset for arginine methylation, whereas $D_5^+(K)$ depicts the true methylated dataset for lysine methylation and $D_5^-(K)$ depicts the negative dataset for

Table 2.1

The datasets used to study Arginine and Lysine Mmethylation

Dataset	Arginine Methylation		Lysine Methylation	
	Positive Samples	Negative Samples	Positive Samples	Negative Samples
D1	185	185	226	217
D2	185	185	226	217
D3	185	185	226	217
D4	185	185	226	217
D5	185	185	226	217
D6	185	185	226	217
D7	185	185	226	217

lysine methylation. $D_5(R)$ defines the complete dataset for R methylation, consisting of 1481 samples, out of these, 185 samples are positive and 1296 samples are negative to study R methylation. Similarly, $D_5(K)$ defines the complete dataset to study K methylation, consisting of 1744 samples, out of these, 226 samples are positive and 1518 samples are negative to study K methylation.

It is important to note that the quantity of positive and negative data in the complete datasets is extremely imbalanced, with proportion of positive to negative samples at around 1: 7 which is impractical. To solve this issue, the approach proposed by Qiu et al. [33] is utilized, which divides the negative dataset into seven parts. Each one of the seven parts is combined to positive samples for creating a small subset. Thus, seven small subsets are formed for arginine and lysine methylation, and designated by D1 through D7. The summary of all small datasets is given in Table 2.1.

2.3 FEATURE REPRESENTATION

Feature representation is the process through which each protein sequences is mapped into some feature vectors There are various feature extractor to encode the protein sequences, among those which best describe the protein sequences are- Information Theory Features (ITF), Twenty-Bit Features (TBF), Overlapping Property Features (OPF), TwentyOne-Bit Features (TOBF), Conjoint Triad Features(CTF) and Skip Dipeptide Composition Features (SDCF). By fusing all these six feature descriptor sequentially, we will get a best prediction model. The method which is used by all feature descriptor is mentioned below.

2.3.1 INFORMATION THEORY FEATURES

In information theory features, information gain is calculated by measuring the difference between shannon entropy and relative shannon entropy.

2.3.1.1 Shannon Entropy

Degree of uncertainty in the variable outcomes is called shannon entropy(SE) and here SE calculate the preservation of each amino acids present in the peptide sequences, and has been used in the prediction of protein post translational modification [5]. Shannon entropy can be calculated by the below equation:

$$SE = -\sum_{j=1}^{20} P_j log_2(P_j) \tag{2.3}$$

where P_j denotes the frequency of amino acid j in given peptide sequence.

2.3.1.2 Relative Shannon Entropy

The Relative entropy(RSE) is the modified version of shannon entropy when the variable outcomes values are continuous. RSE calculate the amino acids quantity in each, with respect to the background distribution. Relative Shannon entropy can be formulated by the following equation:

$$RSE = \sum_{j=1}^{20} P_j log_2 \frac{P_j}{P_0} \tag{2.4}$$

where P_0 is the number of amino acid uniformly distribution in the sequence

2.3.1.3 Information Gain Score

How much information is transformed from any position to other position which the predictor variable is related to, is called Information gain score (IGS). It is formulated in the protein sequence by the following equation:

$$IGS = SE - RSE \tag{2.5}$$

2.3.2 OVERLAPPING PROPERTY FEATURES

To extract the OPF features from the sequences, we have to arrange the amino acids into groups based on the ten physicochemical properties [13] given below in Table 2.2.

Here, different amino acid is masked by using 10 bits, where each bit represents one in ten group. The bits place is fixed to one, if the amino acid related to that particular group, and zero if that amino acid not belong to that group.

2.3.3 TWENTY BIT FEATURES

In the Twenty bit features(TBF) feature extraction, each amino acids are represented with 1-Dimensional feature vector with values 0 or 1 such as:

$$Bit_{20}(A_j) = [f_1(j), f_2(j), \ldots, f_{20}(j)] \tag{2.6}$$

Table 2.2

Classification of amino acids according to ten physicochemical properties

Physicochemical Properties	Amino Acids
Polar	{C,E,D,N,Q,K,S,R,H,T,W,Y}
Positive	{R,K,H}
Negative	{E,D}
Charged	{E,K,D,H,R}
Hydrophobic	{C,A,G,I,V,L,H,K,F,T,Y,M,W}
Aliphatic	{L,I,V}
Aromatic	{H,F,W,Y}
Small	{C,A,D,P,G,V,S,T,N}
Tiny	{C,A,G,S}
Proline	{P}

where A_j shows the jth amino acid in the sequence, $f_k(j)$ is formulated as

$$f_k(j) = \begin{cases} 1 & if \ k = j \\ 0 & if \ k \neq j \end{cases} \tag{2.7}$$

Feature vector for every amino acid can be described as:

$$First \ amino \ acid(Bit_{20}(A_1)) = \begin{bmatrix} 1 \\ 0 \\ \vdots \\ 0 \end{bmatrix}, \ Second \ amino \ acid(Bit_{20}(A_2)) \begin{bmatrix} 0 \\ 1 \\ \vdots \\ 0 \end{bmatrix} \cdots \tag{2.8}$$

Finally for a peptide sequence P, a TBF feature extraction is encoded as a concatenation of all amino acid's feature vector sequentially such as:

$$TBF(P) = [Bit_{20}(A_1), Bit_{20}(A_2), \ldots\ldots, Bit_{20}(A_N)] \tag{2.9}$$

where N is the length of the given peptide sequence, and the size of TBF(P) matrix is 20×N.

For example:

$$\begin{bmatrix} 1 & 0 & 0 & . & . & 0 \\ 0 & 1 & 0 & . & . & 0 \\ 0 & 0 & 1 & . & . & 0 \\ . & . & . & . & . & . \\ . & . & . & . & . & . \\ 0 & 0 & 0 & 0 & 0 & 1 \end{bmatrix} \tag{2.10}$$

2.3.4   TWENTY ONE BIT FEATURES

In the Twenty One Bit Features (TOBF) feature extraction, amino acids classification are based on seven physicochemical properties [15]: secondary structures, solvent accessibility, polarity, polarizability, charge, hydrophobicity, and normalized Van der Waals volume. Each properties are denoted as a θ_k(k=1, 2, ..., 7). For every property θ_k, it forms a reduced alphabet and clustered into the three group, like for the charge physicochemical property, the standard amino acids are divided into three groups: C_1 ={A, C, G, F, I, L, I, H, P, N, M, Q, T, S, Y, V, W}, C_2 ={E, D}, and C_3 ={R, K}. By doing this, the standard alphabet is changed into a reduced set with three components {C_1 , C_2 , C_3}.

Amino acids can have multiple properties simultaneously. To show the correlation of different properties, there is a need to construct new alphabet that reflect this correlation.

Therefore twenty amino acids alphabet converted into the new 21 alphabet such as

$$New_{alphabet} = G_1(\Theta_1), G_2(\Theta_1), G_3(\Theta_1)......G_1(\Theta_7), G_2(\Theta_7), G_3(\Theta_7) \qquad (2.11)$$

where $G_t(\theta_k)$ shows a tth group related to the kth property TOBF features can be represented as a:

$$\begin{bmatrix} Bit_{21}(P_1) \\ Bit_{21}(P_2) \\ . \\ . \\ Bit_{21}(P_N) \end{bmatrix} \qquad (2.12)$$

where $Bit_{21}(P_i)$; is represented as,

$$Bit_{21}(P_i) = [f_1(i), f_2(i), ..., f_{21}(i)] \qquad (2.13)$$

Thus, TOBF has in total 21 × N dimensional features

2.3.5   SKIP DIPEPTIDE COMPOSITION FEATURES

Frequency of two residues which are adjacent in a sequence is called as a Dipeptide Composition. Through the dipeptide composition we can find the correlation between two adjacent residues, but correlation can't be find out in the intervening residues (non-adjacent). Therefore to find correlation in intervening residues, there is a method called adaptive skip dipeptide composition (ASDCF), which is the proportion of every two residues with less than or equal to L intervening residues within a sequence.

Hence ASDCF feature vector for a sequence can be denoted as following:

$$FV = \begin{bmatrix} fv_1 \\ fv_2 \\ \vdots \\ fv_{400} \end{bmatrix} \qquad (2.14)$$

Table 2.3

Amino acids classification according to dipolar strength and side chain volume

Groups according to dipolar strength	Amino Acids
G1	{A, V, G}
G2	{I, F, L, P}
G3	{Y, M, S, T}
G4	{N, H, Q, W}
G5	{R, K}
G6	{E, D}
G7	{C}

where,

$$fv_i = \frac{O^i}{\sum_{k=1}^{L} n(k)} \tag{2.15}$$

and where O^i denotes the observed total number of ith two residues with less than or equal to L intervening residues, and n(k) shows the total number of all those two residues with less than or equal to k intervening residues

2.3.6 CONJOINT TRIAD FEATURES

In the Conjoint Triad Features(CTF) [42] feature extractor, twenty amino acids are clustered into the seven groups according to dipole strength and side chain volumes, below Table 2.3 shows the categorization of amino acid group

In CTF, the representation for the sequence are as follows- the peptide sequence S is changed to another new sequence of groups. After that, conjoint triads are searched across the new changed sequence S'. In total there are 343 conjoint triad types. Every inspected triad are added and counted in a frequency/occurence vector, here unit/element shows the frequency of specific triad perceived in S'. Finally, sequence wind up by showing a total 343 coloumn feature vector.

We extracted feature representation from different feature extractor, then all are concatenated with fixed dimension into the matrix form.

2.4 FEATURE SELECTION

There are number of feature vectors which are not contributed in the prediction model. To decrease the computational time, feature selection process optimally chooses the features which are best in prediction. So, we need to rank a feature vector according to its relevance among the features and their target classes and also the redundancy between the features themselves. To utilize this concept we utilize Maximal Relevance Maximal Distance (MRMD) method [52]. Features which are in higher rank amount to the maximum relevance and minimum dependency. MRMD method measures the relevance in features by using a Pearson's correlation coefficient(PCC) technique which is employed to detect the interrelationship between features and predictor classes. MRMD method measures the redundancy between features by measuring the similarity between any of two feature vectors with the help of different distance functions. Euclidean distance(ED), Tanimoto coefficient(TC), cosine similarity(CS) are the distance function that is used in measuring a redundancy between the features. After getting a ranked feature set through MRMD method, feature which are lower index in rank and higher index in rank are informative and non-informative feature vector respectively. Now feature selection algorithm is used to pick the finest collection of features from ranked features group which gives the best performance. In the Sequential backward search , the features which is in MRMD based ranked feature group, is omitted one at a time from the lower index rank features , and are employed to construct the model.

2.5 MODEL

After getting a ranked feature set, classifier such as random forest algorithm [4] and decision tree algorithm [34] are trained to fit and evaluate the model. Random forest algorithm [4] is the algorithm which uses multiple decision trees to generate a model and both algorithm has been widely applied in the major fields of bioinformatics and computational biology, such as protein structure class prediction, protein methylation sites [50].

2.5.1 FRAMEWORK OF PROPOSED MODEL

We have designed a decision tree based method to predict a protein methylation sites. The proposed model framework as shown in Figure 2.1. First the protein sequences are separated into the same length peptide sequences of eleven residues by sliding window. The peptide sequences, whose center is either K or R, are taken and others are rejected. Next, features are extracted from the peptide sequences to represent them as a fixed-dimensional feature vectors.

The extracted features are based on sequence only: overlapping property features (OPF), information theory features (ITF), twenty-one bit features (TOBF), twenty bit features (TBF), conjoint triad features (CTF), and skip dipeptide composition features (SDCF). Next, a feature selection method is used for ranking the features and optimal features are selected using Sequential Backward Search approach. Finally, the methylation sites are predicted by the trained decision tree model.

2.5.2 DECISION TREE ALGORITHM

Decision tree algorithm is proposed by John Ross Quinlan [34] in 1986 used for classification and regression problems. Decision tree is a supervised algorithm, and it is generated by using the Iterative Dichotomiser 3 algorithm (ID3) or CART algorithm (Classification and Regression Tree). Decision tree uses decision nodes to split the

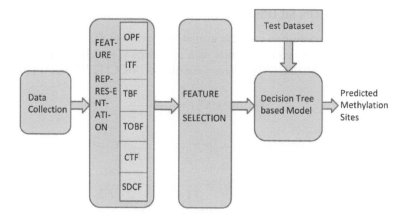

Figure 2.1 The proposed Model Framework. OPF means Overlapping Property Features; ITF means Information Theory Features; TBF means Twenty Bit Features; TOBF means Twenty One Bit Features; CTF means Conjoint Triad Features; SDCF means Skip Dipeptide Composition Features.

dataset into smaller subsets based on information gain or Gini index. Decision tree has been useful in various areas of computational and system biology [14], comprising protein secondary structure prediction [21], protein coding regions in DNA [39], protein–protein interaction (PPI) [45], protein fold recognition [23], to find genes in DNA [40]. In this study, decision tree is used as the predictor to identify methylation and non-methylation sites in a peptide sequence and is implemented in python programming language(Version-3.6) with ML library.

2.6   RESULTS AND DISCUSSION

In this section, we will see patterns of amino acid composition in the peptide sequence, then evaluation metrics to measure a model, after that model analysis using multiple feature subset.

2.6.1   AMINO ACID COMPOSITION ANALYSIS

In the given datasets, glycine (G) is significantly in generous amount in positive R-based methylated peptides that shows that glycine (G) details in the sequence are benefitting in separating a positive R-based methylated peptides from negative R-based methylated ones. But there is no difference between those positive K-based methylated and negative K-based methylated peptides. All 20 amino acids frequency are almost the same in number between positive K-based methylated and negative K-based methylated peptides.

2.6.2   EVALUATION METRICES

There are various evaluation metrices available to evaluate the model performance. We have used four evaluation metrics such as accuracy, sensitivity (also called as recall), specificity, Matthew's correlation coefficient (MCC) [27] [26] [16] [31].

$$Sensitivity(SE) = \frac{T1}{T1 + F2} \tag{2.16}$$

$$Specificity(SP) = \frac{T2}{T2 + F1} \tag{2.17}$$

$$Accuracy(ACC) = \frac{T1 + T2}{T1 + F1 + T2 + F2} \tag{2.18}$$

$$MCC = \frac{T1 \times T2 - F1 \times F2}{\sqrt{(T1 + F1) \times (T1 + F2) \times (F1 + T2) \times (T2 + F2)}} \tag{2.19}$$

T1 represents a true-positive which specifies the count of methylation examples which are appropriately recognized as methylation sites,

F1 represents a false-positive which specifies the count of non-methylation examples which are falsely recognized as methylation sites,

T2 represents a true-negative which specifies the count of non-methylation examples which are appropriately recognized as non-methylation sites,

F2 represents a false-negative which specifies the count of methylation examples which are falsely recognized as non-methylation sites.

2.6.3   MODEL ANALYSIS USING OPTIMAL FEATURE SUBSET

In this section, we will see the performance of our model for R-methylation and K-methylation site on different datasets, comparing with other state-of-the-art predictors performance.

For finding the effect of feature selection in the R-methylation and K-methylation protein sequences, Leave one out cross-validation (LOOCV) results are shown in Tables 2.4 and 2.5 for each metrics with respect to all seven datasets. There are 1305 coloumn feature vectors extracted with the help of six feature descriptor. For R-methylation site and K-methylation dataset, [370×1305] and [443×1305] matrix are generated, respectively. After applying MRMD method by tuning a parameter as 50, then top 50 ranked features are extracted from the 1305 coloumns feature vector. After that, sequential backward selection feature selector is used in this 50 ranked features by tuning the parameter as 14. Finally, the model optimally selects the 14 subsets of features and generated a value of sensivity, specificity, accuracy, mathew corelation coefficient for each dataset

Computing the performance of Decision tree based classification model after using MRMD method and selecting a optimal feature subset on the benchmark datasets are:

Table 2.4

Results of decision tree model to predict R-METHYLATION SITE after using MRMD method

Datasets	SE	SP	ACC	MCC
Datasets1	84.9%	79.5%	82.2%	64.4%
Datasets2	74.1%	87.6%	80.8%	62.2%
Datasets3	73.0%	84.3%	78.6%	57.7%
Datasets4	82.2%	90.8%	86.5%	73.2%
Datasets5	73.5%	90.3%	81.9%	64.7%
Datasets6	80.5%	87.6%	84.1%	68.3%
Datasets7	77.8%	87.6%	82.7%	65.7%
Average	**78%**	**86.8%**	**82.4%**	**65.1%**

Table 2.5

Results of decision tree model to predict K-METHYLATION SITE after using MRMD method

Datasets	SE	SP	ACC	MCC
Datasets1	87.2%	46.5%	67.3%	37.0%
Datasets2	68.1%	88.5%	78.1%	57.7%
Datasets3	70.4%	73.7%	72.0%	44.1%
Datasets4	76.1%	76.0%	76.1%	52.1%
Datasets5	82.3%	65.4%	74.0%	48.5%
Datasets6	76.5%	69.1%	72.9%	45.8%
Datasets7	43.4%	92.6%	67.5%	41.2%
Average	**72.0%**	**73.1%**	**72.5%**	**46.6%**

2.6.3.1 Performance of Decision Tree and Random Forest Classifier

As shown in Figure 2.2, there are three classifier (LibSVM [6], random forest and decision tree) which is trained on total number of seven datasets to predict the R-methylation site, and it is evident that decision tree outperforms the other two classifiers by around 2–3% in all evaluation metrics. However, for specificity evaluation metrics, the LibSVM performs better as compared to the Random forest.

On predicting the K-methylation site on all seven dataset, again it is shown in Figure 2.3 that decision tree outperforms the other two classifiers by around 3–4% in all evaluation metrics. Bur for sensivity metrics, LibSVM performs well as compared to Random forest, and for MCC metrics LibSVM [6] degrading its performance.

Computational time – Training time to train the decision tree model is approximately 677 seconds which is very less as compared to other two Random forest and LibSVM classifier.

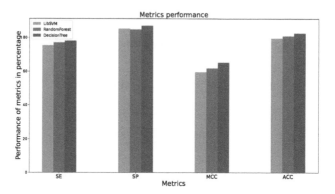

Figure 2.2 Performance on R-methylated sites datasets by using LibSVM, decision tree and random forest algorithm.

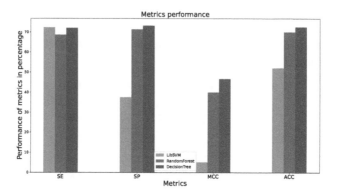

Figure 2.3 Performance on K-methylated sites datasets by using LibSVM, decision tree and random forest algorithm.

2.7 CONCLUSION

In this paper, we have analyzed a protein methylation sites, and results show that methylation sites can be predicted accurately from protein sequences. The novelty of our proposed method is that our decision tree-based predictor uses only a sequence information to encode proteins. In this paper, six informative features such as overlapping property features, ITF, TOBF, TBF, CTF, and SDCF are all integrated to enhance the performance of the predictor. To avoid redundant and irrelevant features, MRMD feature selection method is adopted for selection of best features. Comparative results shows that, our predictor gives better performance than existing predictors. Our predictor gives the accuracy of 82.4% for R-methylation and 72.5% for K-methylation, specificity of 86.8% for R-methylation and 73.1% for K-methylation, sensitivity of 78% for R-methylation and 72% for K-methylation.

The existing methods for predicting a protein methylation sites, used only single classifier for building the model. So, in future, we can further improve the performance by using ensemble classifier to make predictions of both R and K methylation sites.

REFERENCES

1. Mark T Bedford and Stéphane Richard. Arginine methylation: an emerging regulatorof protein function. *Molecular Cell*, 18(3):263–272, 2005.

2. Francois-Michel Boisvert, Carol Anne Chenard, and Stephane Richard. Protein interfaces in signaling regulated by arginine methylation. *Science Signaling*, 2005(271):re2–re2, 2005.

3. François-Michel Boisvert, Jocelyn Côté, Marie-Chloé Boulanger, and Stéphane Richard. A proteomic analysis of arginine-methylated protein complexes. *Molecular & Cellular Proteomics*, 2(12):1319–1330, 2003.

4. L Breiman. Random Forests Machine Learning, vol. 45, 2001.

5. John A Capra and Mona Singh. Predicting functionally important residues from sequence conservation. *Bioinformatics*, 23(15):1875–1882, 2007.

6. Chih-Chung Chang and Chih-Jen Lin. Libsvm: A library for support vector machines. *ACM Transactions on Intelligent Systems and Technology (TIST)*, 2(3):1–27, 2011.

7. Hu Chen, Yu Xue, Ni Huang, Xuebiao Yao, and Zhirong Sun. Memo: a web tool for prediction of protein methylation modifications. *Nucleic Acids Research*, 34(suppl_2):W249–W253, 2006.

8. Xiaobo Chen, F Niroomand, Z Liu, A Zankl, HA Katus, L Jahn, and CP Tiefenbacher. Expression of nitric oxide related enzymes in coronary heart disease. *Basic Research in Cardiology*, 101(4):346–353, 2006.

9. Kuo-Chen Chou. Using subsite coupling to predict signal peptides. *Protein Engineering*, 14(2):75–79, 2001.

10. Kenneth M Daily, Predrag Radivojac, and A Keith Dunker. Intrinsic disorder and prote in modifications: building an svm predictor for methylation. In *2005 IEEE Symposium on Computational Intelligence in Bioinformatics and Computational Biology*, pages 1–7. IEEE, 2005.

11. Jayanta Kumar Das, Suvankar Ghosh, Ranjeet Kumar Rout, and Pabitra Pal Choudhury. A study of p53 gene and its regulatory genes network. In *2018 8th International Conference on Cloud Computing, Data Science & Engineering (Confluence)*, pages 14–19. IEEE, 2018.

12. Wankun Deng, Yongbo Wang, Lili Ma, Ying Zhang, Shahid Ullah, and Yu Xue. Computational prediction of methylation types of covalently modified lysine and arginine residues in proteins. *Briefings in Bioinformatics*, 18(4):647–658, 2017.

13. Yongchao Dou, Bo Yao, and Chi Zhang. Phosphosvm: prediction of phosphorylation sites by integrating various protein sequence attributes with a support vector machine. *Amino Acids*, 46(6):1459–1469, 2014.

14. Pierre Geurts, Alexandre Irrthum, and Louis Wehenkel. Supervised learning with decision tree-based methods in computational and systems biology. *Molecular Biosystems*, 5(12):1593–1605, 2009.

15. Geetha Govindan and Achuthsankar S Nair. Composition, transition and distribution (ctd) – A dynamic feature for predictions based on hierarchical structure of cellular sorting. In *2011 Annual IEEE India Conference*, pages 1–6. Ieee, 2011.

16. Shou-Hui Guo, En-Ze Deng, Li-Qin Xu, Hui Ding, Hao Lin, Wei Chen, and Kuo-Chen Chou. inuc-pseknc: a sequence-based predictor for predicting nucleosome positioning in genomes with pseudo k-tuple nucleotide composition. *Bioinformatics*, 30(11):1522–1529, 2014.

17. SK Sarif Hassan, Atanu Moitra, Ranjeet Kumar Rout, Pabitra Pal Choudhury, Prasanta Pramanik, and Siddhartha Sankar Jana. On spatial molecular arrangements of sars-cov2 genomes of indian patients. *bioRxiv*, 2020.

18. SK Sarif Hassan and Ranjeet Kumar Rout. A quantitative comparisons of β-coronavirus genomes and their associated genes. 2020.

19. SK Sarif Hassan, Ranjeet Kumar Rout, and Vipul Sharma. A quantitative genomic view of the coronaviruses: Sars-cov2. 2020.

20. SS Hassan and RK Rout. Spatial distribution of amino acids of the sars-cov2 proteins. 2020.

21. Jieyue He, Hae-Jin Hu, Robert Harrison, Phang C Tai, and Yi Pan. Rule generation for protein secondary structure prediction with support vector machines and decision tree. *IEEE Transactions on Nanobioscience*, 5(1):46–53, 2006.

22. Le-Le Hu, Zhen Li, Kai Wang, Shen Niu, Xiao-He Shi, Yu-Dong Cai, and Hai-Peng Li. Prediction and analysis of protein methylarginine and methyllysine based on multisequence features. *Biopolymers*, 95(11):763–771, 2011.

23. Mahshid Khatibi Bardsiri and Mahdi Eftekhari. Comparing ensemble learning methods based on decision tree classifiers for protein fold recognition. *International Journal of Data Mining and Bioinformatics*, 9(1):89–105, 2014.

24. David Komander. The emerging complexity of protein ubiquitination. *Biochemical Society Transactions*, 37(5):937–953, 2009.

25. David Y Lee, Catherine Teyssier, Brian D Strahl, and Michael R Stallcup. Role of protein methylation in regulation of transcription. *Endocrine Reviews*, 26(2):147–170, 2005.

26. Wen-Chao Li, En-Ze Deng, Hui Ding, Wei Chen, and Hao Lin. iori-pseknc: a predictor for identifying origin of replication with pseudo k-tuple nucleotide composition. *Chemometrics and Intelligent Laboratory Systems*, 141:100–106, 2015.

27. Hao Lin, Wei-Xin Liu, Jiao He, Xin-Hui Liu, Hui Ding, and Wei Chen. Predicting cancerlectins by the optimal g-gap dipeptides. *Scientific Reports*, 5(1):1–9, 2015.

28. Valter D Longo and Brian K Kennedy. Sirtuins in aging and age-related disease. *Cell*, 126(2):257–268, 2006.

29. Cyrus Martin and Yi Zhang. The diverse functions of histone lysine methylation. *Nature Reviews Molecular Cell Biology*, 6(11):838–849, 2005.

30. Fabrizio G Mastronardi, D Denise Wood, Jiang Mei, Reinout Raijmakers, Vivian Tseveleki, Hans-Michael Dosch, Lesley Probert, Patrizia Casaccia-Bonnefil, and Mario A Moscarello. Increased citrullination of histone h3 in multiple sclerosis brain and animal models of demyelination: a role for tumor necrosis factor-induced peptidylarginine deiminase 4 translocation. *Journal of Neuroscience*, 26(44):11387–11396, 2006.

31. Brian W Matthews. Comparison of the predicted and observed secondary structure of t4 phage lysozyme. *Biochimica et Biophysica Acta (BBA)-Protein Structure*, 405(2):442–451, 1975.

32. Woon Ki Paik, David C Paik, and Sangduk Kim. Historical review: the field of protein methylation. *Trends in Biochemical Sciences*, 32(3):146–152, 2007.

33. Wang-Ren Qiu, Xuan Xiao, Wei-Zhong Lin, and Kuo-Chen Chou. imethyl-pseaac: identification of protein methylation sites via a pseudo amino acid composition approach. *BioMed Research International*, 2014, 2014.

34. J. Ross Quinlan. Induction of decision trees. *Machine Learning*, 1(1):81–106, 1986.

35. Ranjeet Rout, Hari Pandey, Sanchit Sindhwani, Saiyed Umer, and Smitarani Pati. Investigation of protein sequence similarity based on physio-chemical properties of amino acids. *Authorea Preprints*, 2020.

36. Ranjeet Kumar Rout, SK Sarif Hassan, Sanchit Sindhwani, Hari Mohan Pandey, and Saiyed Umer. Intelligent classification and analysis of essential genes using quantitative methods. *ACM Transactions on Multimedia Computing, Communications, and Applications (TOMM)*, 16(1s):1–21, 2020.

37. Ranjeet Kumar Rout, Pabitra Pal Choudhury, Santi Prasad Maity, BS Daya Sagar, and Sk Sarif Hassan. Fractal and mathematical morphology in intricate comparison between tertiary protein structures. *Computer Methods in Biomechanics and Biomedical Engineering: Imaging & Visualization*, 6(2):192–203, 2018.

38. Ranjeet Kumar Rout, Saiyed Umer, Sabha Sheikh, Sanchit Sindhwani, and Smitarani Pati. Eightydvec: a method for protein sequence similarity analysis using physicochemical properties of amino acids. *Computer Methods in Biomechanics and Biomedical Engineering: Imaging & Visualization*, pages 1–11, 2021.

39. Steven Salzberg. Locating protein coding regions in human dna using a decision tree algorithm. *Journal of Computational Biology*, 2(3):473–485, 1995.

40. Steven Salzberg, Arthur L Delcher, Kenneth H Fasman, and John Henderson. A decision tree system for finding genes in dna. *Journal of Computational Biology*, 5(4):667–680, 1998.

41. Jianlin Shao, Dong Xu, Sau-Na Tsai, Yifei Wang, and Sai-Ming Ngai. Computational identification of protein methylation sites through bi-profile bayes feature extraction. *PloS one*, 4(3):e4920, 2009.

42. Juwen Shen, Jian Zhang, Xiaomin Luo, Weiliang Zhu, Kunqian Yu, Kaixian Chen, Yixue Li, and Hualiang Jiang. Predicting protein–protein interactions based only on sequences information. *Proceedings of the National Academy of Sciences*, 104(11):4337–4341, 2007.

43. Shao-Ping Shi, Jian-Ding Qiu, Xing-Yu Sun, Sheng-Bao Suo, Shu-Yun Huang, and Ru-Ping Liang. Pmes: prediction of methylation sites based on enhanced feature encoding scheme. *PloS one*, 7(6):e38772, 2012.

44. Dray-Ming Shien, Tzong-Yi Lee, Wen-Chi Chang, Justin Bo-Kai Hsu, Jorng-Tzong Horng, Po-Chiang Hsu, Ting-Yuan Wang, and Hsien-Da Huang. Incorporating structural characteristics for identification of protein methylation sites. *Journal of Computational Chemistry*, 30(9):1532–1543, 2009.

45. Aisha Sikandar, Waqas Anwar, Usama Ijaz Bajwa, Xuan Wang, Misba Sikandar, Lin Yao, Zoe L Jiang, and Zhang Chunkai. Decision tree based approaches for detecting protein complex in protein protein interaction network (ppi) via link and sequence analysis. *IEEE Access*, 6:22108–22120, 2018.

46. Radhika A Varier and HT Marc Timmers. Histone lysine methylation and demethylation pathways in cancer. *Biochimica et Biophysica Acta (BBA)-Reviews on Cancer*, 1815(1):75–89, 2011.

47. Leyi Wei, Pengwei Xing, Gaotao Shi, Zhiliang Ji, and Quan Zou. Fast prediction of protein methylation sites using a sequence-based feature selection technique. *IEEE/ACM Transactions on Computational Biology and Bioinformatics*, 16(4):1264–1273, 2017.

48. Christine C Wu, Michael J MacCoss, Kathryn E Howell, and John R Yates. A method for the comprehensive proteomic analysis of membrane proteins. *Nature Biotechnology*, 21(5):532–538, 2003.

49. Yu Xue, Fengfeng Zhou, Minjie Zhu, Kashif Ahmed, Guoliang Chen, and Xuebiao Yao. Gps: a comprehensive www server for phosphorylation sites prediction. *Nucleic Acids Research*, 33(suppl_2):W184–W187, 2005.

50. Chang-Jian Zhang, Hua Tang, Wen-Chao Li, Hao Lin, Wei Chen, and Kuo-Chen Chou. iori-human: identify human origin of replication by incorporating dinucleotide physicochemical properties into pseudo nucleotide composition. *Oncotarget*, 7(43):69783, 2016.

51. Fengfeng Zhou, Yu Xue, Hualei Lu, Guoliang Chen, and Xuebiao Yao. A genome-wide analysis of sumoylation-related biological processes and functions in human nucleus. *FEBS Letters*, 579(16):3369–3375, 2005.

52. Quan Zou, Jinjin Li, Li Song, Xiangxiang Zeng, and Guohua Wang. Similarity computation strategies in the microrna-disease network: a survey. *Briefings in Functional Genomics*, 15(1):55–64, 2016.

3 A Taxonomy of e-Healthcare Techniques and Solutions: Challenges and Future Directions

Dev Arora and Amit Dua
Department of Computer Science and Information Systems,
BITS Pilani, Pilani (Rajasthan), India

Umair Ayub
Department of Computer Science and Engineering,
National Institute of Technology, Srinagar, J&K, India

CONTENTS

3.1 INTRODUCTION

Advancements in technology have benefited and improved human lives beyond measure. Agriculture, automobile, education and all the other industries have used technology in some way or the other to enhance the quality of lives of people. In this era

DOI: 10.1201/9781003246688-3

of healthcare, everyone requires economical and fast healthcare services facilitated by developing technologies. The e-Healthcare enables healthcare services like remote diagnosis/surgery to the country's inaccessible areas through improved technology. e-Healthcare is a socio-technical ecosystem, including health-service providers, health-professionals, technology providers, citizens and patients. The e-Healthcare is an umbrella term, which includes medical informatics, public health and business. Integrating technology with healthcare has made significant improvements in diagnosis, prognosis, treatment and medicine, contributing to the growth in average life expectancy for humans.

3.1.1   NEED FOR E-HEALTHCARE

E-healthcare grew as a need to effectively track and record the patient's health history. Paper records were traditionally used to document patient history and follow the patient's health condition for further treatment. Electronic health records (EHR), a key component of e-healthcare, came into vogue for keeping a record of patient's healthcare information with the rise of technology.

As technology advanced, developments like telemedicine, artificial intelligence-based surgery robots, gene therapy and smart-medicine came into existence, broadening the scope and application of e-healthcare. This has significantly improved the quality as well as ease of access to healthcare. Access to telemedicine has immense importance in natural disasters and global pandemics like COVID-19 by assisting in providing remote consultation and treatment to patients. The use of the enhanced infrastructure supported by technology can handle the healthcare systems' pressure in such situations.

According to the Population Reference Bureau, due to an increase in life expectancy, the world population is growing at a rate of 31%, and it is expected to reach 9.8 billion by 2050 [1]. The people above age 65 is estimated to reach 1.6 billion worldwide, which will be 16.7% of the total world population [36]. This will increase the people's vulnerability towards medical problems, which will increase the pressure on healthcare resources. WHO predicted that this increase in demand for healthcare resources will cause a shortage of roughly 12.9 million healthcare workers by 2035 [9]. Technological advancements can support the healthcare system to prepare for the rising needs. e-Healthcare emerges as a solution to meet this shortage by creating new healthcare applications [64] and eradicating the existing challenges which conventional healthcare system is facing [34].

3.1.2   OVERVIEW OF E-HEALTHCARE

The healthcare industry has conventionally been slow in adapting to newer technologies. However, with the healthcare 4.0 revolution, there has been a change in this trend. For example, the healthcare sector is estimated to have 161 million connected internet-of-things (IoT) devices worldwide until 2020, compared to the 46 million devices in 2015 [8]. Figure 3.1 shows the growth and the estimated increase of Internet of healthcare things (IoHT) devices from 2015 to 2020. In a study by

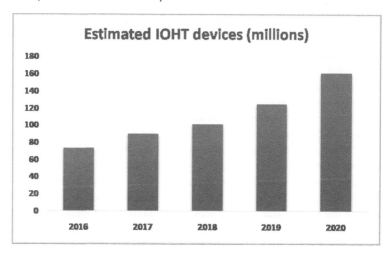

Figure 3.1   Estimated IoHT devices (in millions) for the period between 2016 and 2020 [8].

medium.com, a market-research firm, it was estimated that IoHT will save $300 billion in the medical industry by reducing cost through monitoring the patients remotely, thus saving on the hospital admission fees and travel costs [17]. The IoT-related healthcare revenue equaled $24 billion in 2016, and it is expected to grow to $135 billion worldwide till 2025 [14].

There is an enormous amount of healthcare data present. In 2013, 153 exabytes of healthcare data were produced. It is forecasted that 2314 exabytes of new healthcare data would be generated in 2020 [21]. Mckinsey, a leading management consulting firm, estimated that using big data analytics can save up to 12% to 17% of the healthcare costs amounting to approximately $400 billion in 2013 [6].

The healthcare data can be used for various purposes like creating machine learning (ML) models for disease diagnosis, disease prognosis, remotely monitoring patient's vital information, health-related discoveries and even for business analytics in healthcare. For example, an algorithm can be trained using images from various prior patients suffering from a medical problem, which can be used to detect this same medical problem in future patients. Similarly, a remote wearable device can collect data and draw inferences based upon a pre-trained model. Thus, it can be used for timely and better medical care, diagnostic accuracy [35], reduction in costs due to a decrease in patient care time and remote monitoring [38].

This vast amount and complex nature of healthcare data pose a challenge for healthcare providers in storing and processing it. It is not economically possible for healthcare organizations to keep this vast amount of data on their data servers. The ML models that are used to analyse and draw useful insights from the healthcare data require high computational power, which makes it economically viable for the healthcare provider to host their data centres on a cloud platform. A report by West Monroe Partner's depicted that 35% of the healthcare organizations had more than half of their healthcare data and infrastructure on the cloud platform. Although the healthcare industry has been slow in the adoption of newer technologies, it has fared quickly in cloud adoption compared to the financial services, energy and utility sector [15]. Cloud has also promoted the adoption of telemedicine, which is a crucial component of e-healthcare.

The global telemedicine market in 2019 was valued at an estimated \$41.4 billion [19], and it makes up for a quarter of the IT-health market [20]. Telemedicine solutions improve clinical management and the effectiveness of care delivery by providing access to healthcare remotely. Telemedicine solutions have a great scope in situations where sudden outbreaks of diseases like Ebola and Corona-virus make it a necessity for healthcare providers to avoid direct contact with the patients. Telemedicine was mentioned as an essential service by the WHO to tackle the COVID-19 outbreak. Telemedicine has relieved the burden on healthcare providers by reducing the hospitalization rate, which has led to the adoption of this technology significantly. The total number of people using telemedicine was projected to increase to 7 million in 2018 from 0.35 million in 2013 [13].

Artificial intelligence (AI) is a key technology and component of the upcoming healthcare 4.0 revolution [18]. AI has been used for genomics and gene editing, which can diagnose and treat genetic diseases. AI models have successfully diagnosed lung cancer with higher accuracy than human doctors. Although an AI-based healthcare physician is not trusted right now, AI will soon make it possible that AI-based health physicians cater to the rising number of patients, thus reducing healthcare facilities' burden [67]. Robotic surgery is also a very widely used application of AI's integration with healthcare. Over a million procedures were performed by more than 5000 surgical robots in 2018. The scenarios ranged from general surgery, orthopedics, neurology, gynecology to simpler hair and dental implants [11].

Electronic Health Record (EHR) is a vital component of the e-healthcare system. EHR is the term used for the structured collection of the patient's medical data regularly updated every time they visit a doctor from birth until death. Obtaining healthcare data can be very problematic since the patient can consult many doctors and have gone through a series of tests. The use of EHR to store patient's records solves this problem. EHR is a far better way to obtain a patient's summary than traditional paper records. This also leads to reduced cost, saving time and higher efficiency in the treatment process [5]. Healthcare system and providers are not connected with each other in terms of sharing data, which can be a significant barrier for future research in healthcare and patient's ease of access to healthcare. Keeping track of records on a centralized system makes it much easier for the patient. It increases the amount of data present with healthcare providers for research in the medical domain. EHR thus helps in integrating the healthcare system across the nation and the globe. EHR is also a key component required in the telemedicine industry's efficient functioning, allowing healthcare providers to retrieve the patient's records remotely. EHR uses technologies like cloud computing for storage, blockchain and cryptographic techniques to ensure the secure storage and retrieval of patient data. A person's EHR is created when he is born and is updated every time the person visits a doctor [55]. The main advantage of shifting to the EHR system is that all the medical records are created and stored in a central database and can be accessed when needed [60].

3.1.3 NEED OF TAXONOMY

Since healthcare and data related to the patient involve a wide range of personal information, this sector is not quick to adapt to technological changes. Many challenges need to be acknowledged to fully utilize the potential of the integration of technology with healthcare. One major challenge that needs to be addressed is trust. Despite the advancements in AI, a digital physician assistant is not a very popular concept. Telemedicine is also yet to utilize its full potential due to the same reason. Thus patient acceptance is the first step that needs to be achieved for e-healthcare to reach its full potential. Apart from this, there are numerous technical aspects as well that hinder the growth of e-healthcare.

The available healthcare data is very complex in nature, making it very difficult for algorithms to draw useful insights from them. Although the amount of healthcare data is enormous, the open-source data that can be used to draw inferences remains limited. Privacy and security concerns of healthcare data remain to be a significant obstacle in the growth of e-healthcare. Any healthcare data breach can be very damaging for the healthcare organizations as the patient's trust is compromised due to the data breach. Such violations can make people lose their faith in the healthcare system and can make them question the more profound ethics associated with the healthcare system [16]. Patient data is highly sensitive and should be protected at all costs. Sharing sensitive data among multiple organizations increases the possibility of data breaches due to a lack of control over data by the owner. Security concerns pertaining to third-party data sharing, compliance to security norms (i.e. In USA, Health Information Technology for Economic and Clinical Health (HITECH), Act,) and dealing with the system downtimes remain the significant challenges of adopting cloud platform for healthcare [2]. Data sharing among healthcare organizations to integrate the e-healthcare database and for better research is a practice that is lacking and needs to be done [3].

Several surveys based on e-healthcare exist in the literature. Sawand *et al.* [52] discussed the data collection methodologies at the patient side that are the fundamental basis to achieve an efficient, secure and robust health monitoring system. They also discussed the security threats, which e-health monitoring systems face, and the limitations of the existing implemented solutions. Abbas and Khan [22] discussed and reviewed state-of-the-art privacy-preserving approaches in the e-health cloud environment. The paper divided the privacy-preserving approaches into cryptographic and non-cryptographic approaches. It presented the strengths and weaknesses of the documented approaches. Saleem *et al.* [51] enlisted the communication security issues in e-health related environments. The authors explored the security prototypes, their security features and discussed the challenges and future scope in these prototypes. Shishvan *et al.* [59] had classified the ML work in healthcare into the following six categories – clinical diagnosis, prognosis, assistive technologies, personal health monitoring, health-related discoveries and business analytics. Shickel *et al.* [58] discussed different deep learning techniques and frameworks that were being applied in various clinical applications using the EHR data. Table 3.1 represents the existing surveys on various e-healthcare techniques.

Table 3.1

Classification of existing surveys on e-healthcare techniques

References	Machine learning	Privacy	Cloud computing	Data analytics
Sawand *et al.* [52]	no	no	no	yes
Abbas *et al.* [22]	no	yes	yes	no
Saleem *et al.* [51]	no	yes	no	no
Shishvan *et al.* [59]	yes	no	no	no
Shickel *et al.* [58]	yes	no	no	no

From the analysis of the existing work, it is evident that most of the literature has focused on a particular aspect of e-healthcare like data analytics, security or ML. On careful examination, a gap in the current literature in terms of providing an overall view of the e-healthcare paradigm was found, and the primary focus of this paper is to bridge that gap. Many highly cited, latest, and state-of-the-art publications from highly reputed journals were carefully studied, analyzed and categorized to present this information in a form that is beneficial to the readers. The research papers are organized based on the state-of-the-art technologies used. We carefully selected articles considering their relevance shortly and the amount of focus they have received in state-of-the-art publications.

3.1.4   CONTRIBUTION AND ORGANIZATION OF THIS PAPER

The significant findings and contributions of this survey paper are as follows:

- We discuss healthcare history and growth across the centuries and the technical aspects related to each healthcare revolution.

- We elaborate on the healthcare scenario across the globe and the disparities in resources and wealth dedicated to the healthcare sector.

- We classify the techniques used for e-healthcare in four broad categories, i.e., machine learning techniques, privacy techniques, cloud computing techniques and data analytics techniques.

- We propose the future direction of study and research in the field of e-healthcare based upon our findings.

This survey paper aims to represent the e-healthcare techniques that will be of prime importance in the healthcare 4.0 revolution. The purpose of e-healthcare is to improve the quality of patient care and ease of access to healthcare and prepare for the high pressure that the healthcare sector will face in the coming years. The paper analyses the work is done in the e-healthcare domains and suggests future outlooks for the e-healthcare sector.

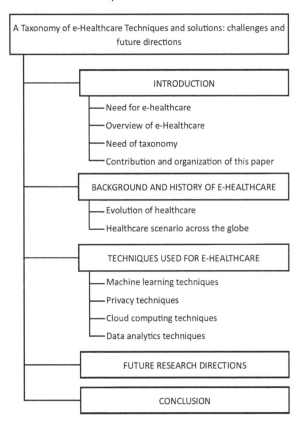

Figure 3.2   Organization of the paper.

The rest of the paper is organized as follows: Section 2 represents the history of healthcare evolution from healthcare 1.0 to healthcare 4.0 and illustrates the scenario of healthcare across the globe. Section 3 provides a detailed analysis of the techniques used in e-healthcare. These are further analysed to segregate the work done in this domain. Section 4 discusses the challenges in the e-healthcare sector for future research directions. Section 5 concludes the survey paper. A brief organization of the survey paper is shown in Figure 3.2.

3.2   BACKGROUND AND HISTORY OF E-HEALTHCARE

Like other industries, i.e., information technology, telecommunication, automobile, and the financial sector, healthcare has grown and transformed over the years. Healthcare has supported the advances in science and information technology to transform into the uprising system that we see today in 2020. The nineteenth-century marked the beginning of healthcare 1.0, which paved its way into healthcare-2.0 at the onset of the twentieth century. 1980 saw the start of healthcare 3.0 with the advent of

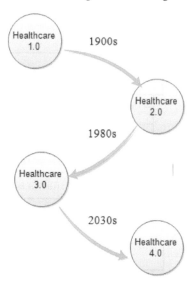

Figure 3.3   The timeline of transformation from healthcare 1.0 to healthcare 4.0.

computers and information technology. Now, currently, in the twenty-first century, we are in the era of healthcare 4.0. Figure 3.3 shows the evolution of healthcare from healthcare 1.0 to healthcare 4.0. The improvement in healthcare is directly correlated with the average life expectancy of humans. Humans' average life expectancy was 28.5 years at the beginning of the nineteenth century, the period that marked the start of the industrial revolution and healthcare-1.0 [12]. It increased to 32 years in the coming century, which marked the beginning of healthcare 2.0, and in 2018 it rose to 72.4 years [12]. This rise in life expectancy is attributed to the growth of research and development in the healthcare sector and its integration with technology.

3.2.1   EVOLUTION OF HEALTHCARE

Healthcare 1.0 began in the 1800s, with the prevention of expansion of infectious diseases. This prevention was brought about by the British government's effort to spread water pipelines to individual homes [42]. During this period only, vaccine immunology and germ theory was established scientifically [27]. During healthcare 1.0, vaccines were developed and made available. Thus the impact of global pandemics and epidemics was controlled, and the damage enrolled was reduced [68]. Thus, Healthcare 1.0 consisted of vaccination, germ control, sanitation efforts of the government and individuals along with epidemiology surveys to better understand the disease and thus aid in its prevention. This improved the healthcare facilities in the nineteenth century. Healthcare 1.0, therefore, was a way to tackle health problems using smart public health approaches.

Healthcare 2.0 was brought at the beginning of the twentieth century with the industrial revolution, the industries proliferated very quickly. Industries grew bigger and heavy machinery was installed. Mass production of things also marked that time in sectors like automobiles, which saw a considerable expansion in production. Similar was the case with healthcare. Hoffmann-La Roche, the second-largest pharmaceutical company of that time, was established during this period, in 1896 [47]. From the period between 1890 and 1910, the Mayo clinic emerged out as an international medical center. Using the capacities of industries for mass production, various antibiotics were developed and brought in the market [72]. Hospitals grew in size, the number of healthcare professionals increased, and doctors were trained and specialized in specific fields to treat complicated cases and attend more patients. Healthcare 2.0 was defined by the exponential growth in infrastructure.

The 1980s marked the beginning of healthcare 3.0. Initially, computers were massive in size, so they were owned only by large universities, corporations and government agencies. With the introduction of microcontrollers in the computer industry in the 1980s, the production of small and cheap controllers was made possible that offered large data storage capacity and fast, speedy computation power [61]. This hugely changed the healthcare industry also. The Internet widely transformed the learning process for healthcare. Medical literature was made available online through e-libraries and journals, while earlier, medical professionals and researchers had to go to the library to access medical literature articles. This advancement immensely helped in speeding up the process of evidence-based medicine. Diseases could be identified earlier, as reconstructed images were made available in the tomography instead of the single-images used earlier, which gave the doctors additional information for diagnosis and prognosis of the disease [33]. Information technology was thus the prominent feature of healthcare 3.0.

Healthcare 4.0, the era of smart medicine, has been brought along with industrial revolution 4.0. Industrial revolution 4.0 has developed new concepts like cloud computing, IoT, AI, automation, data exchange techniques and cyber-physical systems, which has transformed healthcare 3.0 to healthcare 4.0. Healthcare 4.0 has been conceptualized into brains and hands. The brain consists of components like treatment using molecular diagnosis [41], using AI and big data techniques to improve diagnostic and treatment procedures. The hands component of healthcare 4.0 consists of components like robots, wearable devices using IoT, mini-laboratory and 3D printing techniques. Diseases nowadays can be diagnosed within minutes, a joint implant for bones can be customized and 3D printing can be used to prepare bone scaffold [63]. IoT sensors can keep track of a patient's physiological information and inform the doctors of their patient's condition, making prognosis accurate and reliable. However, healthcare 4.0 hasn't utilized all the potential of AI yet. In the 2030s, we could witness dramatic shifts in how healthcare is delivered to people [67].

3.2.2 HEALTHCARE SCENARIO ACROSS THE GLOBE

Different countries have various policies for healthcare. Subsequently, the government expenditure on healthcare also varies according to the developmental stage of each country. Figure 3.4 shows the percentage of total Gross Domestic Product (GDP) spent on healthcare by the governments of various countries across the world.

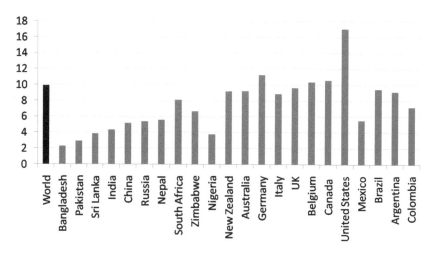

Figure 3.4 Percentage of GDP dedicated to healthcare by various countries [7].

The world average is 9.89%, which has been marked as a reference in Figure 3.4. The world average had grown from 8.62% in 2000 to 9.89% in 2017. This growth can be attributed to the increased research and development of newer technologies like AI, gene editing and techniques like robots being used for surgeries in the healthcare domain. It can be observed from Figure 3.4 that the countries in Asia have relatively less portion of GDP dedicated to healthcare. The average GDP contribution for South Asia is 3.46%, which is significantly lower than the world average. Countries in South Asia are still developing nations, with quite a significant number of people living below the poverty line, and good healthcare is not accessible to all. As compared to Asia, countries of the European Union (EU) have an average of 9.87%, which is almost equal to the world average. Nearly all countries of Europe have universal healthcare. Schiza *et al.* [54] articulated a methodology to implement a national healthcare system across Europe, with different member states having separate policies. This is based on the use of EHR interoperable, aimed at cooperation across EU by individual member states and funding research and development work in the e-healthcare domain and technologies. The paper provided a recommendation to achieve this framework, including legal, financial, managerial and technical concerns for developing this patient-centered healthcare system depending upon the country's situations.

In contrast, the USA has dedicated the highest portion of national GDP to healthcare, 17.06%, which is significantly higher than the world average. The USA has also made advances in the field of e-healthcare with enormous speed. USA has implemented laws and regulations that reward the healthcare provider for shifting to the EHR model, and a hefty penalty is imposed on those healthcare providers who haven't. This has dramatically increased the EHR adoption rate among hospitals with an increase of 5 times, from 16.1% in 2009 to 83.2% in 2014. Financial incentives

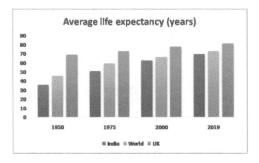

Figure 3.5 Average life expectancy comparison for the period between 1950 and 2019 [12].

are given to healthcare providers under the HITECH scheme. Hospitals and the medical staff are given millions of dollars as rewards and funds for adopting and using the EHR for meaningful purposes [39].

This variation of percentage of GDP contribution to healthcare has a direct impact on the average life expectancy of that nation's population. Figure 3.5 illustrates the comparison of the life expectancy growth of India, United Kingdom and the World. We can observe from Figure 3.5 that the average life expectancy of the United Kingdom has always been at an edge compared to India. This disparity can be attributed to more significant research in the healthcare sector, followed by more significant healthcare infrastructure and resources.

3.3 TECHNIQUES USED FOR e-HEALTHCARE

This section classifies the work done in the field of e-healthcare into the following four broad categories based upon the analysis of existing proposals and field of study: ML techniques, privacy techniques, cloud computing techniques and data analytics techniques. The taxonomy of e-healthcare techniques is shown in Figure 3.6. Research in these areas is of utmost importance for the e-healthcare platform. ML techniques are primarily being implemented for disease diagnosis and prognosis. The cloud platform is used to store a large amount of healthcare data, which facilitates data sharing between healthcare providers. Security measures are further implemented on the cloud platform to meet the privacy issue. Efficient query search, retrieval and storage measures are essential to improve the efficiency and performance of the e-healthcare platform. The detailed analysis of existing proposals in the e-healthcare techniques used in the healthcare sector is shown in Table 3.2.

3.3.1 MACHINE LEARNING TECHNIQUES

ML is a subset of AI that allows software applications to predict outcomes with much higher accuracy without being explicitly programmed. ML models learn from user data, find patterns and then predict new output values. ML has seen a wide range of applications in various fields like image processing, analysing and

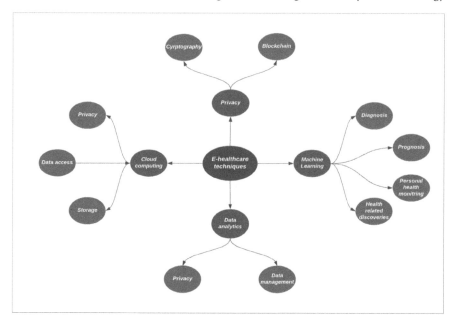

Figure 3.6 Taxonomy of e-healthcare techniques.

predicting financial data, weather forecasting, automatic chat-bots, etc. Similarly, ML has multiple healthcare applications like disease diagnosis, prognosis, treatment plan recommendation, etc. Yu *et al.* [71] divided the applications of ML in the healthcare domain into the four subclasses – deep learning for auxiliary diagnosis, deep-learning for prognosis, deep learning for early warning and deep learning for other tasks.

There is an enormous volume of healthcare data available, and thus it is essential that the data is collected and organized suitably for the ML models. The healthcare data has high dimensionality, which leads to less processing efficiency. Ramasamy *et al.* [48] proposed a neural network model to guess the class of healthcare data in which the high dimensionality of the healthcare data is removed using principal component analysis (PCA). The model produced 95.5% accuracy for diagnosing heart disease on the datasets from the University of California Irvine (UCI) ML repository.

The type of ML-based healthcare application also depends on the type of healthcare data present. Sharma *et al.* [56] discussed the techniques of ML used according to the type of healthcare data available and their clinical application. The authors explained that supervised learning can be used with examples with class labels, which can help make models for diagnosing diseases. Unsupervised learning can be used when the given criteria don't have class labels. In this case, methods like clustering can be used and can be applied to identify patterns in healthcare data. Statistical summarization is used when valuable insights about the patient's healthcare information can be drawn from summarizing.

Table 3.2

Comparison of proposals based on different e-healthcare techniques

Reference	Machine learning	Privacy	Cloud computing	Data analytics
Schiza et al. [54]	no	yes	yes	yes
Yu et al. [71]	yes	no	no	no
Ramasamy et al. [48]	yes	no	no	no
Sharma et al. [56]	yes	yes	yes	no
Ahamed and Farid [24]	yes	no	no	no
Wang et al. [66]	yes	yes	no	no
Mohan et al. [43]	yes	no	no	no
Sayeed et al. [53]	yes	no	no	no
Frunza et al. [31]	yes	no	no	no
Zhang et al. [74]	no	yes	yes	no
Zeb et al. [73]	no	yes	no	no
Garkoti et al. [32]	no	yes	yes	no
Kazantsev and Ponomareva [37]	no	no	yes	no
Shende et al. [57]	no	yes	yes	yes
Rahman et al. [25]	no	no	yes	no
Saha et al. [49]	no	yes	yes	no
Zheng and Lu [75]	no	yes	yes	yes
Viceconti et al. [62]	no	no	no	yes
Yang 2010 [70]	no	no	no	yes
Li et al. [40]	no	yes	no	yes
Wang et al. [65]	no	yes	no	yes
Wilson et al. [28]	no	no	no	yes
Nepal et al. [46]	no	yes	yes	yes
Droes et al. [29]	yes	no	no	no
Muthiah et al. [45]	no	no	no	no
Xu et al. [69]	no	yes	no	no
Saleem et al. [50]	no	yes	no	no
Bruce et al. [26]	no	yes	no	no
Ahad et al. [23]	no	no	no	no
Fan et al. [30]	no	yes	yes	no

Collecting, organizing and finding healthcare data are not easy as the amount of data is substantial and complex, and there is no open-source collection of medical data. Ahamed and Farid [24] point out the issues and challenges of obtaining medical data. According to the authors, the healthcare data obtained could be biased and not diverse enough to cover all the possible scenarios. The data can be biased in terms of the dataset being old or the doctor's opinion towards a particular scenario, which leads to wrong data points in the training dataset and lesser accuracy for the model.

Another application of ML in healthcare is disease diagnosis and prognosis. Using models that have learned from previous data can predict disease and a patient's treatment regime. Wang et al. [66] proposed to use a parallel healthcare system based on artificial systems integrated with computational experiments and the concepts of parallel execution for this purpose. In this method, the real doctor and the artificial doctor interact with the real patient, and then simulations are run on the artificial patient to select the best treatment regime. Discrepancies between the artificial system and the real doctor are used to further enhance the accuracy. While this is a general model for predicting any disease, many models have been developed specifically for a particular disease, which provides greater accuracy of prediction. Mohan *et al.* [43] proposed a hybrid random forest with a linear model (HRFLM) to accurately predict cardiovascular diseases. HRFLM uses an artificial neural network (ANN) coupled with backpropagation and takes 13 features as input. The model is trained and tested on the UCI Cleveland dataset, which contained 297 patient records, out of which 252 data points are used for training the model while the remaining data points are used for testing the model. 88.7% accuracy is achieved for the UCI dataset for detecting cardiovascular diseases. Similarly, Sayeed *et al.* [53] proposed an electroencephalogram (EEG)-based system used for detecting seizures using a ML classifier. In this method, the features are extracted from the sub-bands. These sub-bands are obtained by decomposing the EEG signals. The obtained features are then fed to a deep neural network (DNN) classifier consisting of two hidden layers with 10 neurons each. The model achieved 100% accuracy for classifying the testing data into ictal and normal EEG, while with ictal and interictal, the accuracy achieved was 98.6%. Sometimes it is also essential to learn the semantic relations between diseases and their treatment plans. For this purpose, Frunza *et al.* [31] proposed an NLP (natural language processing) and ML-based approach to draw semantic relations existing between diseases and their corresponding treatments. The authors further classified the semantic relations into the three categories: cure, prevent and side effects. Six ML models have been used to compare the results of their model. SVM (support vector machine) and Naive Bayes classifier outperformed all other algorithms and gave the best results for the proposed approach.

ML technology has found a significant use case in the healthcare sector. It has primarily been used to diagnose diseases and suggests a treatment regime, and makes useful insights into a patient's health through personal wearable devices with the aid of IoT technology. Table 3.3 classifies the existing proposals in e-healthcare using ML technology into five broad categories. As it is evident from Table 3.4, disease diagnosis and prognosis are the major research areas in this field. Research in natural language processing can also enhance and speed up the developments in this field. A digital doctor is not trusted right now, but future research in this area can soon pave the way for an artificial health physician.

3.3.2   PRIVACY TECHNIQUES

With the adoption of EHR technology in healthcare, attacks on healthcare data have become more prominent. EHR let the patients be the owner of their respective data and makes healthcare a patient-centric approach. Protecting patient's data is of utmost importance as it can be used for many maligned uses. Healthcare data frauds range from simple financial identity thefts to obtaining medical data for medical care

Table 3.3
Classification of existing proposals in e-healthcare using machine learning technology

Reference	NLP	Disease diagnosis	Prognosis	Personal health monitoring	Health related discoveries
Yu et al. [71]	yes	yes	yes	no	no
Ramasamy et al. [48]	no	yes	no	no	yes
Sharma et al. [56]	yes	yes	yes	no	yes
Ahamed and Farid et al. [24]	no	yes	yes	yes	no
Wang et al. [66]	yes	yes	yes	yes	yes
Mohan et al. [43]	no	yes	no	no	yes
Sayeed et al. [53]	no	yes	no	no	yes
Frunza et al. [31]	yes	no	yes	no	yes

Table 3.4
Classification of existing proposals on privacy preserving techniques used in e-healthcare

Reference	Data access	Storage	QoS	Cryptography	Blockchain	Edge computing
Sharma et al. [56]	no	no	no	no	no	no
Wang et al. [66]	yes	no	no	no	yes	no
Zhang et al. [74]	no	no	no	yes	no	no
Zeb et al. [73]	yes	no	no	yes	no	no
Garkoti et al. [32]	no	no	no	yes	no	no
Shende et al. [57]	yes	yes	no	yes	no	no
Saha et al. [49]	no	no	yes	yes	no	no
Zheng and Lu [75]	yes	yes	no	yes	no	no
Li et al. [40]	yes	yes	no	no	no	yes
Wang et al. [65]	no	no	yes	no	no	no

by various healthcare stakeholders. These incentives have made healthcare data very attractive in the black markets, where this sensitive information can be traded. The impact of such fraud on the patient can be life-threatening if some privacy attack tampers with their medical records.

In the USA, the Health Insurance Portability and Accountability Act (HIPAA) states a penalty of US $1.5 million per data breach incident. It also states that the impacted individuals have to be informed within 60 days of the breach event. If more than 500 people are affected by the breach, the healthcare provider must report it to the US Department of Health and Human Services (HHS) and the US media, who then publish it online [39]. Figure 3.7 reflects the increase in the number of healthcare data breaches in the US from 2015 onward. There is an 89% increase in

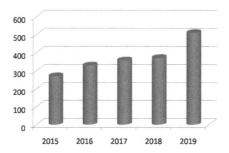

Figure 3.7   Healthcare data breaches of 500 or more records in US [10].

data breaches in the past 5 years. Thus, healthcare organizations must improve their privacy measures for the sensitive healthcare data they possess.

There are many approaches and techniques that can be used to preserve the privacy of healthcare data. The type of technique used depends upon the data present and the resources available with the healthcare provider. Sharma *et al.* [56] analysed the types of healthcare data at risk across various platforms and has categorized the privacy solutions into the three categories based on the users' preference and computational complexity: Expressive but expensive methods, efficient but less expressive methods and efficient methods with weaker privacy notion. Expressive methods allow constructing any big-data algorithms. While efficient methods provide high computational efficiency with limited algorithms support. Expensive methods impose a high implementation cost. This analysis is based on k-Health, a digital healthcare system aimed at disease monitoring. All the healthcare data that is available does not have the same importance level. Some data is more confidential than others. Thus classifying and securing all data at the same level increases the computational costs as well as delay in data access.

Cryptographic methods have gained importance as privacy-preserving techniques for securing healthcare data. Cryptography is the science of using mathematical concepts and rule-based algorithms to encrypt messages so that they are protected from outside attacks. These messages are then relayed to the other end-user and then decrypted using the cryptographic key. The cryptographic methods have been widely discussed and used to ensure the privacy of healthcare data. Zhang *et al.* [74] proposed PASH, a privacy-aware smart health access control system, to address patient's privacy and data security issues. Cipher text-policy attribute-based encryption (CP-ABE) can be used for securing data in smart-healthcare, but there are two flaws in doing so. First is that the access policies reveal the sensitive e-healthcare data since they are in unencrypted form. Second is that usually, a small attribute-based universe is supported in CP-ABE. This makes it unscalable for practical deployment as the universe's size grows linearly with the size of its public parameters. The authors address these issues in the proposed model by hiding the attribute values in the access policies and revealing only the attribute name and supporting an exponentially large universe. The authors have shown that their system is more secure, expressive and efficient than other related works. Zeb *et al.* [73] proposed a U-Prove technology-based security system for mobile device authorization and authentication in both offline and online modes in the e-healthcare environment. U-Prove is a token-based cryptographic concept in which the user can securely limit the amount of disclosed information used for authentication and verification. In the online mode, the server is contacted to get the patient's real-time history, and if the communication fails, the device receives in offline mode and fetches data from the e-health kit and the implanted devices.

It is also imperative to detect attacks and attempts of data breaches. Along with detecting these attacks, the responsible person needs to be identified. Garkoti *et al.* [32] proposed an approach to detect insider attacks on e-healthcare data. Cryptographic techniques fail to provide security against attacks by an insider at the endpoints. Watermarking can be used to detect such insider attacks and modifications, but it fails to deliver accountability. The authors proposed a methodology to combine cryptographic techniques with watermarking and then integrate with an accountability framework to detect insider attacks on a cloud-based e-healthcare system. Cryptographic methods are also vulnerable to certain weaknesses like hackers can get into the systems responsible for encryption and decryption or can hack through the weaker implementations like using default keys for encryption.

Blockchain is a decentralized way of recording transactions and securing data. Once a data record is entered in a blockchain, it can't be changed without altering all the continuing blocks. Blockchain is also being used as a security solution for preserving the privacy of healthcare data. Wang *et al.* [66] used blockchain technology to link hospitals, patients and healthcare communities for efficient sharing and protection of sensitive patient healthcare data in their proposed parallel healthcare system (PHS). Blockchain is an emerging decentralized technique that can provide privacy solutions and can be used in the future for securing healthcare information.

Due to patient data sensitivity, privacy measures are of utmost importance in the cloud-based e-healthcare platform. The privacy technique used depends upon the type of application and resources present with the healthcare provider. Table 3.4 reflects that cryptography is the most widely used technique to implement privacy solutions for e-healthcare data.

3.3.3   CLOUD COMPUTING TECHNIQUES

Cloud computing is a way for organizations to carry out their work and services without owning the infrastructure that requires an expensive network, hardware, computational power and technical professionals to maintain the infrastructure. The cloud offers diverse IT solutions on-demand as per requirements at a reduced cost for an organization and optimizes resource utilization [44]. Moving healthcare data to an e-health cloud has many advantages for the healthcare organizations, healthcare professionals and the patients, such as integration and exchange of medical data across various healthcare organizations, lower operating costs due to not owning the infrastructure, increase in efficiency of disease diagnosis, increased medical research activities. The adaptation of cloud-platform in healthcare can improve availability, flexibility and scalability of data and services [54]. As mentioned earlier, the benefits and the advantages of the cloud platform have led healthcare organizations to shift to the cloud platform.

The realization of an AI algorithm requires enormous computing power. Kazantsev and Ponomareva [37] proposed a cloud-oriented e-health management system for reducing cost and adding extra resources for the model using personal mobile web monitors and AI. The proposed solution halves the perinatal morbidity and mortality indices and decreases the cost associated with pregnancy care. For the ease of data access and improving the quality-of-service (QoS), efficient communication and data retrieval is crucial in the cloud platform. Shende *et al.* [57] suggested a methodology for accessing healthcare data stored on the e-healthcare cloud. To efficiently search for the encrypted data, the author proposed to use the server-side encryption (SSE) algorithm. The proposed approach solves the data storage problem and critical management for keywords search through a secure and efficient indexing method while preserving data privacy. Rahman *et al.* [25] proposed a five-layered cloud-based IoHT, a heterogeneous communication framework to efficiently handle and route both live and offline e-healthcare data. The framework consists of the following five layers: cloud, mist, perception, application and fog layers. The proposed framework achieves efficient resource utilization and optimal resource allocation using link-adaptation-based load balancing and software-defined networking (SDN). The simulation showed that the framework achieves high QoS through its design with a reduced packet drop rate and end-to-end delay.

Despite the advantages like scalability, flexibility, availability and low cost required, cloud computing faces many challenges like security, privacy and reliability. Improving the privacy and security aspects of the e-cloud platform is crucial for ensuring sensitive healthcare data safety. Saha *et al.* [49] proposed a framework to preserve the privacy issues of fog layer-based e-healthcare framework. Fog layers are introduced to enhance the capabilities of IoT-based healthcare systems. These lead to low latency and faster response time, but these are not privacy-preserving. The authors have categorized security threats into the following seven categories: identifiability, information disclosure, linkability, detectability, content unawareness, non-repudiation and policy and consent non-compliance. The proposed framework's performance is compared to recent works against response-time and latency and shown to be privacy-preserving for all the security threats, as mentioned earlier.

Zheng and Lu [75] proposed a privacy-preserving and efficient K nearest neighbor (K-NN) scheme for querying data in the e-healthcare cloud. The model used Kd-trees and then integrates with homomorphic encryption techniques to increase the efficiency of storage and retrieval of healthcare data in the e-healthcare cloud. The proposed approach works only when the cloud environment is considered honest but curious, i.e., the cloud environment provides reliable data storage and query services, but it may get curious about the data stored and the query content. The proposed model is shown to be efficient in computational complexity and privacy-preserving.

Table 3.5

Classification of existing proposals on cloud techniques in e-healthcare

Reference	Privacy	Data access	Storage	QoS	Assistive technology
Garkoti *et al.* [32]	yes	no	no	no	no
Kazantsev and Ponomareva [37]	no	no	no	no	yes
Shende *et al.* [57]	yes	yes	yes	no	no
Rahman *et al.* [25]	no	no	no	yes	no
Saha *et al.* [49]	yes	no	no	yes	no
Zheng and Lu [75]	yes	yes	yes	no	no

Cloud can be an ideal platform for storing a large amount of e-healthcare data. Table 3.5 reflects that the majority of the work in this field focuses on privacy, ease of data access, storage and quality-of-service. Thus improving privacy measures and performance of data access, quality-of-service and storage in the cloud has high importance and provides scope for future research.

3.3.4 DATA ANALYTICS TECHNIQUES

Although there is a tremendous amount of healthcare data present, data is not open source and very complex. The enormous amount of e-healthcare data poses a problem concerning improving computational efficiency and privacy. Healthcare has not fully used the potential of big data compared to other industries like insurance and e-commerce. There is a great need to develop our healthcare management systems to utilize big-data analytics benefits. Viceconti *et al.* [62] proposed the application and combination of big data analytics with virtual physiological human (VPH) technologies to create computer-medicine solutions. The authors have discussed five significant obstacles that need to be overcome to integrate VPH technologies and big data analytics in the healthcare domain. These are 1) analysis of complex and diverse data, 2) working with sensitive crucial data, 3) distributed data management having performance and security constraints, 4) precise analysis to integrate systems biology information with bioinformatics, and 5) specialized analysis to define the physiological envelope in everyday life.

This massive amount of healthcare data has to be efficiently stored and retrieved when required. Thus efficiency in storage and performance is a fundamental requisite for e-healthcare systems. Yang 2010 [70] proposed Replay EHR at Any-Point-In-Time (REAPIT), a data storage architecture that provides secure and efficient data retrieval for the e-healthcare environment. Internet small computer systems interface (iSCSI) protocol has been used in the architecture, and the parity due to every data change by the healthcare providers is recorded. Using real-time data or the backup data present, the EHR can be replayed as it was at any point in time in the past through computing reverse parities. The parties are significantly smaller than the data itself. Thus the proposed model provides faster encryption and higher performance.

Table 3.6

Classification of existing proposals on data analytics techniques in e-healthcare

Reference	Data access	Storage	Data management	Privacy preserving	Big data
Shende et al. [57]	yes	yes	no	yes	no
Zheng and Lu [75]	yes	yes	no	yes	no
Viceconti et al. [62]	no	yes	yes	no	yes
Yang 2010 [70]	no	yes	no	no	no
Li et al. [40]	yes	yes	yes	yes	no
Wang et al. [65]	yes	no	no	yes	no
Wilson et al. [28]	yes	no	no	yes	no

Edge computing is a decentralized data storage technique which finds application in healthcare systems as well. Li et al. [40] proposed a data management system for mobile healthcare systems, called EdgeCare. It is a decentralized data management system that uses edge computing. The proposed system performed better than centralized mechanisms that suffer from application challenges due to the ever-increasing amount of health data. The authors had demonstrated with numerical results and security analysis that EdgeCare had a good advantage in security aspects of healthcare data and supported efficient data trading. To increase data relay efficiency, Wang et al. [65] proposed a privacy-preserving priority classification scheme for the patient in remote e-healthcare. The privacy-preserving classification is done at the wireless body area network (WBAN) gateway, and the medical-data packets are relayed ahead according to their priorities. The efficiency of computational cost and communication overhead is tested using an android application and two java server programs.

Healthcare data is sensitive and can be used for many maligned uses ranging from simple financial identity thefts to selling medical data for monetary benefits. EHR gives the patients right over their data. Wilson et al. [28] proposed a protocol eCert to address the issues surrounding e-certification and securely manage e-healthcare data as a distributed e-health document termed as eHealth-eCert. A system is proposed which supports both eCert and the unique features of e-healthcare. The protocol is user-centric, giving the patients control and right over their data and deciding who can view it. The authors also propose a verification service for this distributed system.

Privacy-preserving solutions for healthcare data slow down the data transfer process, and thus the overall efficiency of the healthcare management system decreases. An efficient e-healthcare data management system plays a pivotal role in the e-healthcare sector. Table 3.6 reflects that the significant work done in the field of e-healthcare data has been aimed to increase efficiency in terms of data access, storage, and privacy. Healthcare is yet to utilize big data analytics techniques, and future work can be focused in this direction to improve the efficiency of the e-healthcare platform.

3.4   FUTURE RESEARCH DIRECTIONS

E-healthcare has been slow in the adoption of technology compared to other industries. Healthcare 4.0 is on the way, and along with it, it brings a lot of new advancements in the field of e-healthcare like gene therapy, smart medicine, robotic surgery and much more. There are a lot of areas that need to be researched to fully explore the potential of e-healthcare. This section provides the challenges, issues and directions for future research in this area.

- *Streamlining healthcare data collection for training ML models*: The healthcare data is typically very complex in nature, and there isn't a proper open-source repository for obtaining data to train ML models. Healthcare organizations coming together by integrating with EHR can solve this problem.
- *Exploring blockchain technology for securing the healthcare data*: Cryptographic techniques fail to provide full proof security in case of weaker implementations like using default keys for encryption or when hackers get into the systems responsible for encryption and decryption. It also fails to detect insider attacks on the data. Blockchain is an emerging technology that can help provide a solution in such situations, and future research can be done in this direction.
- *Improving privacy measures across cloud platforms*: High scalability, flexibility and availability of resources make cloud the ideal platform to store e-healthcare data. Third-party data sharing and privacy are some challenges in adopting cloud platforms and could provide research scope in this area.
- *Big-data analytics for e-healthcare data*: Big-data methods have less used in the healthcare domain. There are significant obstacles like the complexity, data's sensitivity and diversity of the data. Distributed data management techniques have performance and security constraints. These issues should be solved for better analytical capabilities of healthcare data.
- *Integrating 5G technology with IoHT devices and mobile healthcare applications*: Currently, 4G based network is used to facilitate smart healthcare. Smart healthcare is the analysis and use of health data collected through smart devices (i.e., smartwatch, wireless blood pressure monitor, smartphones, etc.). But as the number of applications connected to this network increase, the demands for bandwidth, ultra-low latency, ultra-high reliability, high density and high energy efficiency will also increase. Currently, there are more than 1,00,000 apps available for healthcare [4]. Efficiently integrating these with the healthcare system and using a combination of 5G and IoT devices can help solve the network-related problems.
- *Training healthcare professionals*: In the United States, 29% of the physicians fell in the age range 55–65. This is the most extensive distribution of doctors belonging to a particular age group [4]. Not every healthcare worker is well equipped with technology to benefit from the advantages of e-healthcare. Proper research should be focused on the training of healthcare workers to reap the full benefits of the e-healthcare platform.

- *Sharing medical knowledge and resources*: Countries across the globe need to share their resources and expertise in healthcare. This knowledge can be in terms of new medicines, new surgical techniques, new methods of treatment, or medical data to train algorithms and better understand the diseases.
- *Educating the older population and people's willingness to adapt*: Research has to be done to find out the user's opinions and views on the e-health platform. The people, particularly the older population, need to be educated about the technological advantages that the e-healthcare platform has and how to use it. Chadborn *et al.* [76] conducted a study to investigate the benefits and risks associated with smart health technologies. The authors found out that older people's views were usually not considered while creating healthcare applications and technologies. Organizations devising smart-health systems need to consider the ease of access, use and governance of these systems by the older population.

However, despite the growing ubiquity of smart health, innovations are often technology-driven, and the older user does not usually have input into the design. The purpose of the current study was to debate the positive and negative perceptions and attitudes towards digital health technologies. Methods: We conducted citizens' juries to enable a deliberative inquiry into the benefits and risks of smart-health technologies and systems. Transcriptions of group discussions were interpreted from a perspective of life-worlds versus systems-worlds. Results: Twenty-three participants of diverse demographics contributed to the debate. Views of older people were felt to be frequently ignored by organizations implementing systems and technologies. Participants demonstrated various levels of digital literacy and a range of concerns about the misuse of technology. Conclusion: Our interpretation contrasted the life-world of experiences, hopes and fears with the systems-world of surveillance, efficiencies and risks. This interpretation offers new perspectives on involving older people in the co-design and governance of smart-health and smart-homes.

3.5 CONCLUSION

This survey paper presents a comprehensive survey of e-healthcare techniques. The survey broadly classifies the e-healthcare techniques into four broad categories: ML techniques, cloud computing techniques, privacy techniques, and data analytics techniques. The research work done in the field of each of these techniques is thoroughly discussed. A critical analysis of the existing work has been done, and comparison has been made for each e-healthcare technique in the form of comparison table. A brief history of healthcare growth and the disparities in the healthcare scenario across the globe have also been discussed. Finally, future research directions and suggestions have been suggested.

REFERENCES

1. 2017 world population data sheet with focus on youth. `https://www.prb.org/2017-world-population-data-sheet/`. Accessed: 2020-04-28.

2. 5 ways cloud computing is impacting healthcare. `https://www.healthitoutcomes.com/doc/ways-cloud-computing-is-impacting-healthcare-0001`. Accessed: 2020-04-28.

3. 7 major challenges facing ehealth. `https://www.hansonzandi.com/7-major-challenges-facing-ehealth/`. Accessed: 2020-04-28.

4. 7 major challenges facing ehealth. `https://www.hansonzandi.com/7-major-challenges-facing-ehealth/`. Accessed: 2020-04-28.

5. Benefits of electronic health records. `http://www.usfhealthonline.com/resources/healthcare/benefits-ofehr/`. Accessed: 2020-04-28.

6. The big-data revolution in us health care: Accelerating value and innovation. `https://www.mckinsey.com/industries/healthcare-systems-and-services/our-insights/the-big-data-revolution-in-us-health-care`. Accessed: 2020-04-28.

7. Current health expenditure (% of gdp). `https://data.worldbank.org/indicator/SH.XPD.\:CHEX.GD.ZS`. Accessed: 2020-04-27.

8. Estimated healthcare iot device installations worldwide from 2015 to 2020. `https://www.statista.com/statistics/735810/healthcare-iot-installations-global-estimate/`. Accessed: 2020-04-28.

9. Global health workforce shortage to reach 12.9 million in coming decades. `https://www.who.int/mediacentre/news/releases/2013/health-workforce-shortage/en/`. Accessed: 2020-04-28.

10. Healthcare data breach statistics. `https://www.hipaajournal.com/healthcare-data-breach-statistics/`. Accessed: 2020-04-28.

11. How robots and ai are creating the 21st-century surgeon. `https://www.roboticsbusinessreview.com/health-medical/how-robots-and-ai-are-creating-the-21st-century-surgeon/`. Accessed: 2020-04-28.

12. Life expectancy. `https://ourworldindata.org/life-expectancy`. Accessed: 2020-04-28.

13. Master guide to telehealth statistics for 2019. `https://www.wheel.com/blog/master-guide-to-telehealth-statistics-for-2019/`. Accessed: 2020-04-28.

14. Projected size of the internet of things (iot) in healthcare market worldwide from 2016 to 2025. `https://www.statista.com/statistics/997959/worldwide-internet-of-things-in-healthcare-market-size/`. Accessed: 2020-04-28.

15. Report: Healthcare industry leads in cloud adoption. `https://www.beckershospital review.com/healthcare-information-technology/report-healthcare-industry-leads-in-cloud-adoption.html`. Accessed: 2020-04-28.

16. The special challenges of data analytics with health care. `https://www.stitchdata.com/blog/special-challenges-data-analytics-with-health-care/`. Accessed: 2020-04-28.

17. Staggering stats on healthcare iot innovation. `https://medium.com/datadriven investor/7-staggering-stats-on-healthcare-iot-innovation-fe6b92774a5c`. Accessed: 2020-04-28.

18. Surgical robots, new medicines and better care: 32 examples of ai in healthcare. `https://builtin.com/artificial-intelligence/artificial-intelligence-healthcare`. Accessed: 2020-04-28.

19. Telemedicine market size, share & trends analysis report by component, by delivery model, by technology, by application (teleradiology, telepsychiatry), by type, by end use, by region, and segment forecasts, 2020 - 202. `https://www.grandviewresearch.com/industry-analysis/telemedicine-industry`. Accessed: 2020-04-28.

20. Telemedicine statistics you should know. `https://blog.evisit.com/36-telemedicine-statistics-know`. Accessed: 2020-04-28.

21. Total amount of global healthcare data generated in 2013 and projections for 2020 (in exabytes)*. `https://www.statista.com/statistics/1037970/global-healthcare-data-volume/`. Accessed: 2020-04-28.

22. Assad Abbas and Samee U Khan. A review on the state-of-the-art privacy-preserving approaches in the e-health clouds. *IEEE Journal of Biomedical and Health Informatics*, 18(4):1431–1441, 2014.

23. Abdul Ahad, Mohammad Tahir, and Kok-Lim Alvin Yau. 5g-based smart healthcare network: Architecture, taxonomy, challenges and future research directions. *IEEE Access*, 7:100747–100762, 2019.

24. Farhad Ahamed and Farnaz Farid. Applying internet of things and machine-learning for personalized healthcare: issues and challenges. In *2018 International Conference on Machine Learning and Data Engineering (iCMLDE)*, pages 19–21. IEEE, 2018.

25. Md Asif-Ur-Rahman, Fariha Afsana, Mufti Mahmud, M Shamim Kaiser, Muhammad R Ahmed, Omprakash Kaiwartya, and Anne James-Taylor. Toward a heterogeneous mist, fog, and cloud-based framework for the internet of healthcare things. *IEEE Internet of Things Journal*, 6(3):4049–4062, 2018.

26. Ndibanje Bruce, Gi-Hyun Hwang, and Hoon Jae Lee. A hybrid and fast authentication protocol for handoff support in e-healthcare systems among wsns. In *2013 International Conference on ICT Convergence (ICTC)*, pages 72–77. IEEE, 2013.

27. Harry Burns. Germ theory: invisible killers revealed. *BMJ*, 334(suppl 1):s11–s11, 2007.

28. Lisha Chen-Wilson, Xin Wang, Gary B Wills, David Argles, and Charles Shoniregun. Healthcare data management issues and the ecert solution. In *International Conference on Information Society (i-Society 2011)*, pages 114–119. IEEE, 2011.

29. Rose-Marie Droes, Maurice Mulvenna, Chris Nugent, Marius Mikalsen, Stale Walderhaug, Tim Van Kasteren, Ben Krose, Stefano Puglia, Fabio Scanu, Marco Oreste Migliori, et al. Healthcare systems and other applications. *IEEE Pervasive Computing*, 6(1):59–63, 2007.

30. Kai Fan, Shanshan Zhu, Kuan Zhang, Hui Li, and Yintang Yang. A lightweight authentication scheme for cloud-based rfid healthcare systems. *IEEE Network*, 33(2):44–49, 2019.

31. Oana Frunza, Diana Inkpen, and Thomas Tran. A machine learning approach for identifying disease-treatment relations in short texts. *IEEE Transactions on Knowledge and Data Engineering*, 23(6):801–814, 2010.

32. Gaurav Garkoti, Sateesh K Peddoju, and R Balasubramanian. Detection of insider attacks in cloud based e-healthcare environment. In *2014 International Conference on Information Technology*, pages 195–200. IEEE, 2014.

33. Evidence-Based Medicine Working Group et al. Evidence-based medicine. a new approach to teaching the practice of medicine. *Jama*, 268(17):2420, 1992.

34. Omessaad Hamdi, Mohamed Aymen Chalouf, Dramane Ouattara, and Francine Krief. ehealth: Survey on research projects, comparative study of telemonitoring architectures and main issues. *Journal of Network and Computer Applications*, 46:100–112, 2014.

35. Moeen Hassanalieragh, Alex Page, Tolga Soyata, Gaurav Sharma, Mehmet Aktas, Gonzalo Mateos, Burak Kantarci, and Silvana Andreescu. Health monitoring and management using internet-of-things (iot) sensing with cloud-based processing: Opportunities and challenges. In *2015 IEEE International Conference on Services Computing*, pages 285–292. IEEE, 2015.

36. Wan He, Daniel Goodkind, Paul R Kowal, et al. An Aging World: 2015. 2016.

37. Alexander Kazantsev, Julia Ponomareva, and Pavel Kazantsev. Development and validation of an ai-enabled mhealth technology for in-home pregnancy management. In *2014 International Conference on Information Science, Electronics and Electrical Engineering*, volume 2, pages 927–931. IEEE, 2014.

38. Ovunc Kocabas, Tolga Soyata, Jean-Philippe Couderc, Mehmet Aktas, Jean Xia, and Michael Huang. Assessment of cloud-based health monitoring using homomorphic encryption. In *2013 IEEE 31st International Conference on Computer Design (ICCD)*, pages 443–446. IEEE, 2013.

39. Juhee Kwon and M Eric Johnson. Protecting patient data-the economic perspective of healthcare security. *IEEE Security & Privacy*, 13(5):90–95, 2015.

40. Xiaohuan Li, Xumin Huang, Chunhai Li, Rong Yu, and Lei Shu. Edgecare: leveraging edge computing for collaborative data management in mobile healthcare systems. *IEEE Access*, 7:22011–22025, 2019.

41. Beatriz López, Ferran Torrent-Fontbona, Ramón Viñas, and José Manuel Fernández-Real. Single nucleotide polymorphism relevance learning with random forests for type 2 diabetes risk prediction. *Artificial Intelligence in Medicine*, 85:43–49, 2018.

42. Johan P Mackenbach. Sanitation: pragmatism works. *BMJ*, 334(suppl 1):s17–s17, 2007.

43. Senthilkumar Mohan, Chandrasegar Thirumalai, and Gautam Srivastava. Effective heart disease prediction using hybrid machine learning techniques. *IEEE Access*, 7:81542–81554, 2019.

44. John C Moskop, Catherine A Marco, Gregory Luke Larkin, Joel M Geiderman, and Arthur R Derse. From hippocrates to hipaa: privacy and confidentiality in emergency medicine—part i: conceptual, moral, and legal foundations. *Annals of Emergency Medicine*, 45(1):53–59, 2005.

45. Alagumeenaakshi Muthiah, S Ajitha, Monisha Thangam KS, K Kavitha, Ramalatha Marimuthu, et al. Maternal ehealth monitoring system using lora technology. In *2019 IEEE 10th International Conference on Awareness Science and Technology (iCAST)*, pages 1–4. IEEE, 2019.

46. Surya Nepal, Rajiv Ranjan, and Kim-Kwang Raymond Choo. Trustworthy processing of healthcare big data in hybrid clouds. *IEEE Cloud Computing*, 2(2):78–84, 2015.

47. Hans Conrad Peyer. *Roche: A Company History, 1896–1996*. Editiones Roche, 1996.

48. Balamurugan Ramasamy and Abdul Zubar Hameed. Classification of healthcare data using hybridised fuzzy and convolutional neural network. *Healthcare Technology Letters*, 6(3):59–63, 2019.

49. Rahul Saha, Gulshan Kumar, Mritunjay Kumar Rai, Reji Thomas, and Se-Jung Lim. Privacy ensured $\{e\}$-healthcare for fog-enhanced iot based applications. *IEEE Access*, 7:44536–44543, 2019.

50. Kashif Saleem, Abdelouahid Derhab, and Jalal Al-Muhtadi. Low delay and secure m2m communication mechanism for ehealthcare. In *2014 IEEE 16th International Conference on e-Health Networking, Applications and Services (Healthcom)*, pages 105–110. IEEE, 2014.

51. Kashif Saleem, Khan Zeb, Abdelouhid Derhab, Haider Abbas, Jalal Al-Muhtadi, Mehmet A Orgun, and Amjad Gawanmeh. Survey on cybersecurity issues in wireless mesh networks based ehealthcare. In *2016 IEEE 18th International Conference on e-Health Networking, Applications and Services (Healthcom)*, pages 1–7. IEEE, 2016.

52. Ajmal Sawand, Soufiene Djahel, Zonghua Zhang, and Farid Nait-Abdesselam. Multidisciplinary approaches to achieving efficient and trustworthy ehealth monitoring systems. In *2014 IEEE/CIC International Conference on Communications in China (ICCC)*, pages 187–192. IEEE, 2014.

53. Md Abu Sayeed, Saraju P Mohanty, Elias Kougianos, and Hitten P Zaveri. Neuro-detect: A machine learning-based fast and accurate seizure detection system in the iomt. *IEEE Transactions on Consumer Electronics*, 65(3):359–368, 2019.

54. Eirini C Schiza, Theodoros C Kyprianou, Nicolai Petkov, and Christos N Schizas. Proposal for an ehealth based ecosystem serving national healthcare. *IEEE Journal of Biomedical and Health Informatics*, 23(3):1346–1357, 2018.

55. Eirini C Schiza, Kleanthis C Neokleous, Nikolai Petkov, and Christos N Schizas. A patient centered electronic health: ehealth system development. *Technology and Health Care*, 23(4):509–522, 2015.

56. Sagar Sharma, Keke Chen, and Amit Sheth. Toward practical privacy-preserving analytics for iot and cloud-based healthcare systems. *IEEE Internet Computing*, 22(2):42–51, 2018.

57. Rahul Shende, Shailesh Kamble, and Sandeep Kakde. Health data access in cloud-assisted e-healthcare system. 2016.

58. Benjamin Shickel, Patrick James Tighe, Azra Bihorac, and Parisa Rashidi. Deep ehr: a survey of recent advances in deep learning techniques for electronic health record (ehr) analysis. *IEEE Journal of Biomedical and Health Informatics*, 22(5):1589–1604, 2017.

59. Omid Rajabi Shishvan, Daphney-Stavroula Zois, and Tolga Soyata. Machine intelligence in healthcare and medical cyber physical systems: A survey. *IEEE Access*, 6:46419–46494, 2018.

60. Sunil Kumar Srivastava. Adoption of electronic health records: a roadmap for india. *Healthcare Informatics Research*, 22(4):261–269, 2016.

61. Jon Stokes. *Inside the Machine: An illustrated Introduction to Microprocessors and Computer Architecture*. No starch press, 2007.

62. Marco Viceconti, Peter Hunter, and Rod Hose. Big data, big knowledge: big data for personalized healthcare. *IEEE Journal of Biomedical and Health Informatics*, 19(4):1209–1215, 2015.

63. Lauren Vogel. Plan needed to capitalize on robots, ai in health care, 2017.

64. Samuel Fosso Wamba, Abhijith Anand, and Lemuria Carter. A literature review of rfid-enabled healthcare applications and issues. *International Journal of Information Management*, 33(5):875–891, 2013.

65. Guoming Wang, Rongxing Lu, and Yong Liang Guan. Achieve privacy-preserving priority classification on patient health data in remote ehealthcare system. *IEEE Access*, 7:33565–33576, 2019.

66. Shuai Wang, Jing Wang, Xiao Wang, Tianyu Qiu, Yong Yuan, Liwei Ouyang, Yuanyuan Guo, and Fei-Yue Wang. Blockchain-powered parallel healthcare systems based on the acp approach. *IEEE Transactions on Computational Social Systems*, 5(4):942–950, 2018.

67. Mark Wehde. Healthcare 4.0. *IEEE Engineering Management Review*, 47(3):24–28, 2019.

68. Michael Worboys. Vaccines: conquering untreatable diseases. *BMJ*, 334(suppl 1):s19–s19, 2007.

69. Chang Xu, Jiachen Wang, Liehuang Zhu, Chuan Zhang, and Kashif Sharif. Ppmr: A privacy-preserving online medical service recommendation scheme in ehealthcare system. *IEEE Internet of Things Journal*, 6(3):5665–5673, 2019.

70. Ken Qing Yang. Secure and efficient data replay in distributed ehealthcare information system. In *2010 International Conference on Information Society*, pages 634–639. IEEE, 2010.

71. Ying Yu, Min Li, Liangliang Liu, Yaohang Li, and Jianxin Wang. Clinical big data and deep learning: Applications, challenges, and future outlooks. *Big Data Mining and Analytics*, 2(4):288–305, 2019.

72. Lorenzo Zaffiri, Jared Gardner, and Luis H Toledo-Pereyra. History of antibiotics. from salvarsan to cephalosporins. *Journal of Investigative Surgery*, 25(2):67–77, 2012.

73. Khan Zeb, Kashif Saleem, Jalal Al Muhtadi, and Christoph Thuemmler. U-prove based security framework for mobile device authentication in ehealth networks. In *2016 IEEE 18th International Conference on e-Health Networking, Applications and Services (Healthcom)*, pages 1–6. IEEE, 2016.

74. Yinghui Zhang, Dong Zheng, and Robert H Deng. Security and privacy in smart health: Efficient policy-hiding attribute-based access control. *IEEE Internet of Things Journal*, 5(3):2130–2145, 2018.

75. Yandong Zheng and Rongxing Lu. An efficient and privacy-preserving k-nn query scheme for ehealthcare data. In *2018 IEEE International Conference on Internet of Things (iThings) and IEEE Green Computing and Communications (GreenCom) and IEEE Cyber, Physical and Social Computing (CPSCom) and IEEE Smart Data (Smart-Data)*, pages 358–365. IEEE, 2018.

76. Szinay, D., Jones, A., Chadborn, T., Brown, J. and Naughton, F., 2020. Influences on the uptake of and engagement with health and well-being smartphone apps: systematic review. *Journal of medical Internet research*, 22(5), p. e17572.

4 Classification of Lung Diseases Using Machine Learning Techniques

Sudipto Bhattacharjee
Department of Computer Science and Engineering,
University of Calcutta, Kolkata, India

Banani Saha
Department of Computer Science and Engineering,
University of Calcutta, Kolkata, India

Sudipto Saha
Division of Bioinformatics, Bose Institute, Kolkata, India

CONTENTS

4.1 INTRODUCTION

Lung is a vital organ of the human body and lung diseases are a leading cause of death and morbidity in the world [53]. Lung diseases can be broadly classified into four categories–obstructive, restrictive, infectious and lung cancer. *Obstructive* diseases are characterized by airflow limitation, inflammation of airways and hyper-responsiveness (*e.g.* asthma, Chronic Obstructive Pulmonary Disease–COPD). *Restrictive* diseases, also known as non-obstructive diseases, are caused by impar-ity of the lungs to expand properly which results in lung volume reduction (*e.g.* Interstitial Lung Disease–ILD). *Infectious* diseases are caused by inhalation of exter-nal disease-causing organisms like viruses or bacteria (*e.g.* pneumonia, COVID-19). *Lung cancer* is characterized by malignant tumors, and there are different stages of it. These lung diseases are complex and are hard to diagnose due to the similarity of symptoms and existence of comorbidities. Diagnosis and further decision making is often performed based on the patient data. With the plethora of patient data available digitally, as electronic medical records (EMR), different machine learning (ML) models are trained on such data for efficient decision making.

Patient data can be of three kinds–clinical, radiological and genomic. *Clinical data* consists of demographic information, history, medication and investigations. Demographic information relevant to ML applications mainly include age, sex, ethnicity and occupation. Patient history includes smoking profile, comorbidities and allergies. Some examples of investigations include basic examinations (such as blood pressure, SaO_2 and body temperature), blood tests, pulmonary function tests (PFT) and echocardiography. *Radiological data* include lung imagery such as chest X-ray (CXR) and computed tomography (CT) scans. Popular *genomic investigations* include single nucleotide polymorphism (SNP), gene expression and DNA methylation. SNP refers to the substitution of a single nucleotide with another at any position in the DNA sequence. Gene expression investigations are performed using methods such as microarray analysis [23] and RNA-seq [67]. DNA methylation refers to the addition of methyl groups (CH_3) at cytosine (C) nucleotides of DNA in *CpG* sites - the "$5' - C - phosphate - G - 3'$" sequence of nucleotides. Techniques such as Bisulfite Sequencing (BS-seq) [15] and BeadChip Array analysis [3] are used for DNA methylation investigations.

Most ML-based studies obtain patient data from public databases for training and validation purposes. A number of public repositories of clinical data are available, such as *MIMIC III*, that contains a compilation of demographics, vital signs, medication, laboratory test and mortality data of patients admitted in intensive care units [30]. *COPDGene* provides both clinical and genomic data of COPD patients [50]. In comparison to clinical datasets, public radiological and genomic datasets are more abundantly available. *MIMIC-CXR* and *ChestXRay14* are two popular datasets of CXR images with 14 radiological annotations that can be used for supervised learning purposes [29, 65]. *PadChest* is a recently published dataset of CXR images with annotations of 19 differential diagnoses [6]. *The Cancer Imaging Archive (TCIA)* is a comprehensive repository of cancer radiology with various modalities [11]. The *MedGIFT* and *Computed Tomography Emphysema Database* are CT repositories of interstitial lung disease (ILD) and emphysema, respectively [17, 55]. The *Gene Expression Omnibus (GEO)* is a popular repository of genomic data such as gene expression, methylation, genomic variations and protein profiling [20]. *ArrayExpress* is another genomic data repository containing RNA-seq, microarray and methylation data [2]. *Clinvar* is an archive of genomic variations associated with different phenotypes of clinical significance [33]. *dbSNP* is another repository of genomic polymorphisms [51]. *The Cancer Genome Atlas Program (TCGA)* is a comprehensive repository of gene expression, genomic variation, DNA methylation and clinical data of 32 cancer types including lung cancer [27]. *PulmonDB* is a curated gene expression repository of COPD and lung fibrosis [61]. Thus, several public databases containing clinical, radiological and genomic data are available for ML-based analysis.

This chapter discusses the use of ML for different clinical decision making applications that analyze clinical, radiological and genomic data of lung disease patients.

4.2 MACHINE LEARNING ALGORITHMS

Machine learning (ML) is a collection of methods to perform predictive tasks on the basis of a training dataset. It attempts to construct a mathematical model that generalize the given training data. On the basis of the underlying training data, ML algorithms can be categorized as supervised and unsupervised. In *supervised learning*, class labels are known and the task is to train a model that maps the input features to the class label. In *unsupervised learning*, the class labels are unknown and the task is to either capture patterns within the data (*e.g.* clustering) or compute a dimensionally reduced latent representation of the data (*e.g.* auto-encoders). In terms of the mathematical models, ML algorithms can be categorized as probabilistic, non-probabilistic, tree-based and neural networks. The *probabilistic algorithms* use Bayes theorem to construct a probability model of the underlying training data for prediction of the conditional probability of a sample belonging to a class given a set of independent features. The Naive Bayes algorithm is a popular example of probabilistic algorithms [75]. The *non-probabilistic algorithms* do not construct a probability model but attempts to divide the feature space. Support Vector Machine (SVM) is a popular non-probabilistic algorithm that attempts to construct a hyperplane which maximizes the gap between the classes [13]. The general idea of *tree based algorithms* involves one or more decision trees whose nodes represent a set of samples. Initially, all the samples are represented by the root node. The samples are divided by recursively splitting the nodes on the basis of a single feature until the leaves are pure (*i.e.* all samples are of same class). The feature selection for splitting the node is done using different metrics, such as Gini impurity and information gain. Random forest (RF) is an ensemble algorithm that trains a number of decision trees, each with a random subset of samples and features, and final output is obtained by voting or averaging each decision tree outputs [5]. A *neural network* is a circuit of artificial neurons organized in an architecture containing an input layer, one or more hidden layers and an output layer. Neurons in each layer are connected to the next with weighted edges and the training involves tuning the edge weights such that it minimizes the output error. In recent times, deep learning algorithms, such as Convolutional Neural Network (CNN), have emerged as an efficient method for image analysis due to its ability of automatic feature extraction [35]. The convolutional layers of CNNs use moving convolutional kernels with shared weights that capture the correlation between spatially adjacent pixels and are able to extract high level features. The medical data generally contains a large number of features that may lead to overfitting which can be avoided by using cross validation and feature selection methods, such as Anova and chi-squared test. The application of ML in analyzing four categories of lung diseases is shown in Figure 4.1.

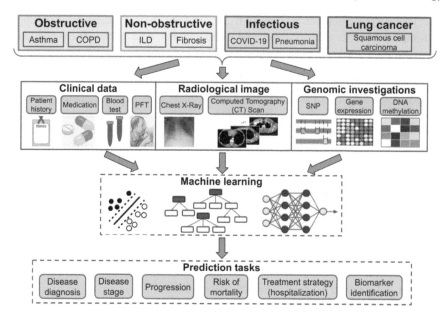

Figure 4.1 Schematic diagram showing the role of ML in analyzing patient data of lung diseases.

4.3 ANALYSIS OF CLINICAL DATA

The clinical data is numerical and represented as either categorical or continuous values. Demographic features, such as sex, ethnicity and occupation, are represented as categorical whereas age is continuous. Comorbidities, allergies and family history are mainly represented as categorical (*yes/no*). Medication, smoking profile, surgery and hospitalization history features are a combination of continuous and categorical values. The features associated with investigations are continuous. ML-based analysis of clinical data is quite straightforward with minimal preprocessing. A missing value imputation step may be required followed by feature selection. The clinical data is one of the first informations available when a patient visits a healthcare facility. Thus, ML can be used for an early diagnosis using such data as features. ML applications to analyze clinical data of lung diseases is shown in Table 4.1.

Asthma and COPD are the most common obstructive lung diseases. Asthma affects patients of all age groups. It is reversible and can be reduced by using bronchodilators. COPD, on the other hand, is irreversible and prevalently caused by smoking. A neural network based asthma prediction study of in adults using PFT results along with symptoms and biochemical findings as features achieved 98% accuracy [59]. In another study, replacing the PFT features with comorbidities and medication data allowed gradient boosted models to achieve area under receiver operator characteristic curve (AUROC) of 0.71, 0.88 and 0.85 in the prediction of exacerbation, emergency department (ED) visit and hospitalization of asthma patients respectively [74]. Another study trained eXtreme Gradient Boosting (XG-Boost) models with features of only clinical history and demographic profile from *Pediatric Big Data (PBD)* resource [60], and achieved accuracy of 81% in predicting persistence of childhood asthma [4]. Here, children were labelled as *"persistent"*, if an asthma diagnosis in ages between 2 and 5 years was followed up by another diagnosis in ages between 5 and 10 years, and *"transient"* otherwise. There was class

Table 4.1

Applications of ML to analyze clinical data of lung diseases

Disease	Function	Features	Sample size	ML algo.^{a}	Optimal metrics
ARDS [73]	Disease prediction	Demographic, vital signs, medication, lab. values	2743 patients	LgR	AUC = 0.81
Asthma [59]	Asthma vs non-asthma classification	Symptoms, biochemical findings, PFT	566 patients	SVM, ANN	Acc = 0.98
Asthma [74]	Prediction of admission; ED visit; exacerbation	Demographic, comorbidity, lab. values medications	60,302 patients	LgR, RF, GBDT	AUC = 0.85; AUC = 0.88; AUC = 0.71
Asthma [4]	Disease-persistence prediction in children	Demographic, clinical history, site of care, insurance geo-location	9934 patients	NB, RF, LgR, KNN, XGB	Acc = 81% Prec = 95% Sens = 82% AUC = 0.86
COPD [49]	Mild vs Severe classification	Blood exam, comorbidity, basic exam	410 patients	CART, ID3, C5.0	Acc = 80% AUC = 0.80
COPD [56]	Patient classification into triaging categories	Patient profile symptoms comorbidity vital signs	2501 patients	NB, ET, SVM, KNN, LgR, GBRF	Acc = 89% Sens = 90% Spec = 99%
COVID-19 [21]	Prediction of respiratory failure in hospitalized patients	signs and symptoms, lab. values, comorbidity	198 patients; 1068 data points	GBDT	Acc = 84% AUC = 0.84
ILD [34]	ILD vs healthy classification	PFT	323 patients	SVM	^{b}Acc = 76% Sens = 62% Spec = 89%

(Continued on next page)

Table 4.1 (Continued)

Disease	Function	Features	Sample size	ML algo.[a]	Optimal metrics
PH [36]	Disease prediction	Patient profile, echocardio-graphy	90 patients	LgR, RF, RFR, SVM, C5.0	Acc = 83% Sens = 89% Spec = 67% AUC = 0.87
Pneu-monia [7]	Post-operative pneumonia prediction	Demographic pre-, post-, intra-operative, lab. values, medication	591 patients	SVM, LgR, RF, MLP, GBDT XGB	AUC = 0.79 Sens = 53% Spec = 78%

[a] The optimal algorithm is underlined. [b] Balanced accuracy Abbreviations: *Acc* Accuracy, *ANN* Artificial Neural Network, *ARDS* Acute Respiratory Distress Syndrome, *AUC* Area Under receiver operator characteristic Curve, *C5.0* C5.0 classifier, *CART* Classification and Regression Tree, *COPD* Chronic Obstructive Pulmonary Disease, *ED* Emergency Department, *ET* Extra Tree, *GBDT* Gradient Boosted Decision Tree, *GBRF* Gradient Boosted Random Forest, *ID3* Iterative Dichotomiser 3, *ILD* Interstitial Lung Disease, *KNN* k-Nearest Neighbor, *LgR* Logistic Regression, *ML algo.* Machine Learning algorithm, *MLP* Multi-Layer Perceptron, *NB* Naive Bayes, *Prec* Precision, *PFT* Pulmonary Function Test, *PH* Pulmonary Hypertension, *RF* Random Forest, *RFR* Random Forest Regression, *Sens* Sensitivity, *Spec* Specificity, *SVM* Support Vector Machine, *XGB* eXtreme Gradient Boosting

imbalance in the dataset which was dealt with techniques such as random under sampling, edited nearest neighbors (ENN), repeated ENN and Tomek links [68, 57, 58]. A COPD risk prediction study developed a novel C5.0 decision tree model to classify patients as high and low risk of mortality using 28 features from blood tests, basic examination and comorbidities and achieved 80% accuracy [49]. The C5.0 algorithm uses information gain ratio for node splitting and allows boosting [31]. Another study for predicting treatment strategy of COPD by classifying patients into four triaging categories – "*Ok*," "*Plan*" (continue normal treatment for 1-2 days), "*Doc*" (call the doctor) and "*ER*" (emergency room) [56]. Seven models were developed for this multi-class classification, where logistic regression achieved optimal accuracy of 89%.

Interstitial lung disease (ILD) is a common non-obstructive disease often clinically diagnosed with PFT. A study for prediction of ILD with PFT features trained SVM model and achieved AUROC of 0.85 [34]. Pulmonary hypertension (PH) is another non-obstructive disease characterized by high PAP. ML models for diagnosis of PH with echocardiography features achieved 83% accuracy (AUROC = 0.89) [36]. Here, the class labels (positive or negative) of every patient were computed by comparing the PAP values. Then, SVM and logistic regression were used for direct classification and random forest regression was used for modelling of the PAP values followed by classification on the basis of a predefined PAP threshold. Acute respiratory distress syndrome (ARDS) is associated with respiratory failure and often leads to intensive care unit admission and ventilation. A ML-based study for ARDS diagnosis among hospitalized patients using LR model achieved AUROC of 0.81 with clinical features obtained within a six hour window [73]. Here, continuous values were quantized and represented as one-hot vectors while multiple measurements within the time window were represented as a vector of summary statistics. Another study that predicted respiratory failure among hospitalized COVID-19 patients using gradient boosted decision tree model, trained with clinical data, achieved 83% accuracy (AUROC = 0.84) [21]. Pneumonia is an infectious disease that often occurs as a side-effect of any surgical procedure. Generally, pneumonia is diagnosed with imagery data. Yet, few recent studies have emerged where ML applications have used clinical features. One such study compared six ML models for prediction of pneumonia as a postoperative complication among patients who underwent liver transplant [7]. The feature set included pre-operative, intra-operative and post-operative factors including comorbidities, laboratory tests and medication. Then, a subset of 14 features, out of 142, were selected using recursive feature elimination (RFE) method [14].

ML-based clinical data analysis for lung disease classification is typically prone to overfitting. Gradient-boosted decision trees and random forest models are very popular as they exploit the advantages of boosting and ensemble learning which allow them to overcome the overfitting problem. Neural network models are not very common in clinical data analysis. Yet, when they are used, they tend to achieve good performance [59].

4.4   ANALYSIS OF RADIOLOGICAL DATA

Image-based prediction often requires good feature extraction methods in addition to the classification methods. The use of CNN is a popular approach for medical image analysis tasks as it requires no explicit feature extraction. In contrast to CNN, a second approach involves extraction of texture and morphological properties of image using image processing (IP) methods. Such features can be, then, applied to a multitude of ML algorithms to perform prediction tasks. Texture properties include the spatial arrangement information (*e.g.* contrast, entropy, homogeneity) that can be utilized for image segmentation whereas morphological properties include information related to the shapes in an image (*e.g.* eccentricity, elongatedness, dispersion). A hybrid approach for image-based classification tasks is also available that combine deep learning models for feature extraction and other ML models for final predictions. The features, known as *bottleneck features*, are obtained as a vector from the output of the final convolutional and pooling layer. Being a numerical vector, such features can be applied to any ML algorithm. It performs a better feature extraction as compared to IP methods, whereas the ability to use any ML algorithm for further prediction tasks reduces the complexity and training time. Several ML applications to analyze radiological image data of lung diseases are shown in Table 4.2.

Table 4.2

Applications of ML to analyze radiological images of lung diseases

Disease	Function	Input	Feature extraction	Sample size	ML algo.[a]	Optimal metrics
COPD [24]	Disease prediction	CT	NA	596 patients	3D-CNN	Acc = 89% Sens = 88% Spec = 94% AUC = 0.94
COVID-19 [25]	Disease prediction	CXR	[b]Extract T&M features	558 images	XGB, KNN, NB, CART	Acc = 80% AUC = 0.87
COVID-19 [28]	Multiclass classification	CXR	NA	6432 images	CNN	Acc = 98%
COVID-19 [62]	Disease prediction	CXR	CNN-extract bottleneck features	1102 images	RF, AB, DT, SVM	Acc = 99% Sens = 99% Spec = 99% AUC = 0.99
Fibrosis [10]	Disease stage prediction	CT	CNN-identify radiological findings	105 patients	RF	Acc = 81% Sens = 79% Spec = 67%
Lung cancer [54]	Benign vs malignant tumor classification	CT	NA	244,527 images, 1010 patients	CNN, ANN, AE+ CNN	Acc = 84% Sens = 84% Spec = 84%
Lung cancer [52]	Benign vs malignant tumor classification	CT	[b]Extract texture features	15750 images	KNN, SVM, MLP, RF, SGD	Acc = 88.55%
Lung cancer [70]	Disease stage prediction	CT	[b]Extract texture, wavelet, features	145 patients	RF	[c]AUC = 0.85
Lung cancer [69]	Treatment response prediction	Time series of CT	NA	759 images, 268 patients	CNN + RNN	AUC = 0.74
Pneumonia [41]	Disease prediction	CXR	NA	3322 images	CNN	AUC = 0.97 Sens = 99% Prec = 84%

[a] The optimal algorithm is underlined. [b] Using image processing methods. [c] Microaverage of AUC values.

Abbreviations: *AB* AdaBoost, *Acc* Accuracy, *AE* Auto-Encoder, *ANN* Artificial Neural Network, *AUC* Area Under receiver operator characteristic Curve, *CART* Classification and Regression Tree, *CNN* Convolutional Neural Network, *COPD* Chronic Obstructive Pulmonary Disease, *CT* Computed Tomography, *CXR* Chest X-Ray, *DT* Decision Tree, *KNN* k-Nearest Neighbor, *ML algo.* Machine Learning algorithm, *MLP* Multi-Layer Perceptron, *NB* Naive Bayes, *Prec* Precision, *RF* Random Forest, *RNN* Recurrent Neural Network, *Sens* Sensitivity, *SGD* Stochastic Gradient Descent, *Spec* Specificity, *SVM* Support Vector Machine, *T&M* Texture and Morphological, *XGB* eXtreme Gradient Boosting

For infectious diseases, like pneumonia and COVID-19, the radiology is characterized by lesions which are observed as haziness and consolidation. An XGBoost model for diagnosis of COVID-19 with 26 texture and morphological features extracted from only 558 CXR images in two datasets [63, 12] achieved accuracy of 79.5% [25]. Another study used the CNN-based approach on a dataset with over 6000 input CXR images [47] and achieved a significantly higher accuracy of 98% for multiclass classification between healthy individuals, COVID-19 and non-COVID-19 pneumonia patients [28]. A further increase in accuracy was achieved by another study that applied the hybrid approach where CNN was used to extract bottleneck features from CXR images of three datasets [63, 12, 44], followed by classification using SVM [62]. For pneumonia prediction, the use of Depth-wise Separable Convolutions (DWSC), instead of conventional convolutions, in a CNN achieved an AUROC of 0.97 [41]. In the DWSC, the convolutions are performed independently for each channel of input images [9]. For prediction of COPD – an obstructive disease – with CT images, a study developed a CNN architecture using 3D convolutional layers [24]. The 3D-convolutions were able to extract volumetric features from every cross-section of CT images and showed better accuracy as compared to the 2D convolution approach. Another study combined both deep learning and other ML models using only 105 CT scan images from *MedGIFT* [17], *Lung Tissue Research Consortium (LTRC)* [45] and *Inselspital ILD* [16] databases for fibrosis stage prediction [10]. Here, CNN was used for identification of radiological pathologies such as reticulation, honeycombing for every lung segment. Then, these pathologies were applied to a RF model as features for diagnosis of different stages of fibrosis – a non-obstructive disease.

Lung cancer is also a prime disease candidate for radiological image based prediction tasks using ML. The three-dimensional malignant tumors are better viewed as nodules in the cross-sectional images of CT scan. A study comparing CNN with conventional neural networks for classification between benign and malignant tumors for diagnosis of cancer showed that CNN outperformed other architectures [54]. CNN often requires a large sample size for efficient training. To perform the same tumor-classification task with a comparatively less number of images, another study extracted 13 texture features from CT images of a public dataset [26] using image processing methods [52]. Then, this small feature set was given as input to seven ML algorithms and obtained good prediction accuracy. Only diagnosis of cancer is not enough to decide treatment strategy, it is also important to predict the stage of cancer (e.g. stage I-IV). The stages can be binarized as "early" (I & II) and "late" (III & IV). An application used the *"Pyradiomics"* Python library to extract features of CT images from *TCIA* dataset [11] and used them as input to a RF model for prediction of lung cancer stages [70]. AUROC of 0.85 was achieved for both 4-class and binarized stage prediction. ML models can also be developed for treatment-response prediction of lung cancer patients that often require a time series of CT images for each patient. Treatment-response is generally defined in terms of survival, progression and recurrence. A study performed such prediction using CT scan data from pre-treatment and three follow-up stages in which the architecture included four CNNs to extract features from CT images at respective time-points [69].

The features were then fed into a Recurrent Neural Network (RNN) which is an ideal deep learning architecture for longitudinal data analysis due to amalgamation of several time-points.

The deep learning models achieve best accuracy with larger sample sizes. Explicit feature extraction followed by classification using classical ML algorithms have also showed good performance with smaller sample sizes. In recent times, there have been cases where the hybrid approach achieved high accuracy [62]. Also, the use of 3D convolutional layers is an innovation that can be adopted for CT analysis of all lung diseases.

4.5   ANALYSIS OF GENOMIC DATA

Genomic investigations for lung diseases are performed on biological samples like lung tissue, pleural fluid, nasal mucosa. The SNP data of each patient is represented as a vector $(s_1, s_2, s_3, \cdots, s_n)$, where each s_i $(for\ all\ i \in [1,n])$ is a categorical value denoting the genotype encoding of a SNP at a specific location in the DNA sequence. Genotype encoding is used to categorically label the SNPs as *(i) homozygous AA* - no substitution in both alleles, *(ii) heterozygous AB* – substitution in one allele, and *(iii) homozygous BB* – substitution in both alleles. Gene expression and DNA methylation datasets are represented as matrices M_{ge} and M_{dm} where each row represents a sample. For M_{ge} and M_{dm} the columns represent genes and *CpG* sites, respectively. Each element $M_{ge}[i, j]$ denotes the expression level of the jth gene for i^{th} sample, whereas each element $M_{dm}[i, j]$ denotes the methylation level represented as β-values [19]. Each row represents the feature vector of a sample that is given as input to a ML model. A standardization step, such as converting expression and β-values to z-scores, is often performed as part of preprocessing. Genomic investigations often generate a large number of features (SNPs, genes or *CpG* sites) which causes ML models to overfit in case of small sample sizes. A good feature selection plays an important role in avoiding the overfitting problem. Also, an efficient feature ranking model can aid in identification of disease-specific genomic biomarkers. Different ML applications to analyze various genomics data of lung diseases are shown in Table 4.3.

For asthma prediction, a study ranked a set of over one million SNPs from *openSNP* database [46] using RF method on the basis of out-of-bag (OOB) error. Then, a subset of 310 top ranked SNPs, selected using RFE method, were used to train a SVM model that obtained accuracy of 62.17% [22]. Another asthma diagnosis study based on DNA methylation data of three tissue types (airway epithelial cells, nasal epithelial cells and peripheral blood mononuclear cells – from *GEO* datasets) using RF model showed a comparatively better performance [37]. Here, potential methylation biomarkers identification was also performed with RF model on the basis of feature importance. For COPD diagnosis, a XGBoost model trained with only nine SNP features achieved 80% accuracy (AUROC = 0.83) [42]. Another study used logistic regression to formulate a novel numerical index based on gene expression values to quantify the COPD-risk and also performed multiclass classification between non-smokers, smokers and COPD patients based on gene expression data with

RF model [43]. The dataset used in this analysis is available in *ArrayExpress* [2]. For diagnosis of infectious disease, such as COVID-19, ML models with incremental feature selection (IFS) achieved 93.8% accuracy with 168 gene features from gene expression data [8, 38]. Here, Boruta feature filtering was used to initially select 604

Table 4.3

Applications of ML to analyze genomic investigation data of lung diseases

Disease	Function	Features	Sample size	ML algo.[a]	Optimal metrics
Asthma [22]	Disease prediction	SNP	128 samples	KNN, SVM	Acc = 62% Prec = 65% Sens = 69%
Asthma [37]	Disease prediction; methylation biomarker identification	DNA methylation	339 samples	RF	[b]AUC = 0.87; 0.99; 0.97
COPD [42]	Disease prediction	SNP	633 patients	LgR, XGB, SVM, DT, KNN, MLP	AUC = 0.83 Acc = 80%
COPD [43]	Disease prediction	GE	204 samples	RF	Avg. true rate = 0.61
COVID-19 [8]	Disease prediction; biomarker identification	GE	195 patients	SVM, KNN, RF, DT	Acc = 94% MCC = 0.92
COVID-19 [64]	Disease severity prediction	SNP	673 samples	RF, ANN	Acc = 90%
Lung cancer [1]	Disease prediction; relapse prediction	GE	1178 patients	XGB	[c]Acc = 0.997, 0.995, 0.967; [c]Acc = 0.794, 0.632
Lung cancer [72]	Disease subtype prediction	GE	150 samples	RF, SVM	Acc = 97% Sens = 100% Spec = 93%
Lung cancer [66]	Disease subtype prediction	DNA methylation	919 samples	AE + LgR	[d]Prec = 95% [d]Sens = 95% Spec = 93%
Lung cancer [71]	Disease-free survival prediction	DNA methylation	610 samples	Ensemble of SVM EN, Cox regression	AUC = 0.77

(*Continued on next page*)

Table 4.3 (Continued)

Disease	Function	Features	Sample size	ML algo.[a]	Optimal metrics
PH [39]	Disease prediction; biomarker identification	GE	140 samples	LgR, SVM, <u>AdaBoost</u>, SVM, KNN	Acc = 95% AUC = 0.99

[a] The optimal algorithm is underlined. [b] AUC for three different tissue samples. [c] Accuracy of three datasets for diagnosis and two datasets for relapse prediction of lung cancer. [d] Microaverage of precision and sensitivity values.

Abbreviations: *Acc* Accuracy, *AE* Auto-Encoder, *ANN* Artificial Neural Network, *AUC* Area Under receiver operator characteristic Curve, *COPD* Chronic Obstructive Pulmonary Disease, *DT* Decision Tree, *EN* Elastic Net, *GE* Gene Expression, *KNN* k-Nearest Neighbor, *LgR* Logistic Regression, *MCC* Matthew's Correlation Coefficient, *ML algo.* Machine Learnng algorithm, *MLP* Multi-Layer Perceptron, *Prec* Precision, *PH* Pulmonary Hypertension, *RF* Random Forest, *Sens* Sensitivity, *SNP* Single Nucleotide Polymorphism, *Spec* Specificity, *SVM* Support Vector Machine; *XGB* eXtreme Gradient Boosting

out of 15k genes in the original dataset and the selected genes were ranked using Max-Relevance and Min-Redundancy (mRMR) algorithm [32, 48]. Another study used frequency of SNPs in haplotype block as features to develop RF models for COVID-19 severity prediction and achieved 90% accuracy [64]. The use of haplotype blocks instead of all the SNPs as features is known to be an important dimensionality reduction technique. A recent study developed a novel, ensemble feature ranking algorithm (*EnRank*) that combined four methods (chi-squared, *t*-test, ridge regression and Lasso) to rank genes associated with pulmonary hypertension using gene expression data [39]. Then, the top ranked genes were used with different ML algorithms for the diagnosis.

Genomic data analysis with ML provides valuable contributions in prediction of lung cancer. Apart from the diagnosis itself, classification of cancer subtypes and disease-free survival prediction based on genomic data is also an important application of ML that helps in accurate therapeutic decision-making. XGBoost models for lung cancer diagnosis with gene expression data achieved an accuracy of over 99% whereas prediction of a relapse event achieved accuracy of 80% [1]. Relapse is the event of re-manifestation of cancer after surgery. A study for classification between two lung cancer subtypes (adenocarcinoma and squamous cell carcinoma) compared Monte-Carlo feature selection (MCFS) with incremental feature selection (IFS) and showed that the IFS-selected genes from gene expression data obtained the optimal result with accuracy of 96.7% using SVM model [38, 72, 18]. Another approach for cancer subtype classification used an auto-encoder on *TCGA* data for dimensionality reduction of DNA methylation dataset and used the latent representation as input features to a logistic regression model [66]. In a disease-free survival prediction study of post-surgery lung cancer patients from *TCGA* dataset, an ensemble classifier (SVM, Cox regression, elastic net) used the arithmetic mean of β-values across methylation-correlated blocks (MCB) as features (instead of directly using β-values of *CpG* sites) and achieved an AUROC of 0.773 [71]. Here, the strongly correlated *CpG* sites in the genome constitute a MCB and it provides a more informed and smaller feature set.

Tree based methods such as RF and XGBoost are widely used ML algorithms for lung diseases classification with genomic data. These are also appropriate for biomarker identification tasks due to their inherent property of node-splitting based on highly informative features. There are also cases where SVM models have outperformed the tree-based methods in classification tasks [8, 72]. Logistic regression is another algorithm that proved to perform well for both biomarker identification and classification tasks [39, 66].

4.6 FUTURE SCOPE

The researchers are making continuous progress to develop efficient ML models to analyze medical data with the goal of making clinical predictions. There are a few investigations, such as Fibre-Optic Bronchoscopy (FOB) and Positron Emission Tomography (PET), which are referred by pulmonologists to aid their decision making, but are not quite commonly used for ML-based predictions. In terms of training CNN models with CT scan images, the use of 3D convolutional kernels can be an efficient choice as it is able to extract volumetric features of lungs. Furthermore, the use of activation heat-maps of the convolutional kernels can provide visual interpretations of the working of CNN as shown by Lujan-Garcia *et al.* (2020) [41]. Doctors often perform diagnosis based on multimodal patient data. So, the different modalities of data must be integrated for a more accurate ML-based prediction. Radiogenomics is an emerging field that attempts to integrate radiological and genomics data [40]. So, there are opportunities for researchers to integrate all data modalities (clinical, radiological and genomic) together to train ML models for better modelling of lung diseases. Identification of genomic biomarkers based on feature selection methods can also provide important hypotheses for further medical research on clinical diagnosis of lung diseases.

REFERENCES

1. Rana Dhia'a Abdu-Aljabar and Osama A Awad. A Comparative analysis study of lung cancer detection and relapse prediction using XGBoost classifier. *IOP Conference Series: Materials Science and Engineering*, 1076(1):12048, 2021.

2. Awais Athar, Anja Füllgrabe, Nancy George, Haider Iqbal, Laura Huerta, Ahmed Ali, Catherine Snow, Nuno A Fonseca, Robert Petryszak, Irene Papatheodorou, Ugis Sarkans, and Alvis Brazma. ArrayExpress update – from bulk to single-cell expression data. *Nucleic Acids Research*, 47(D1):D711–D715, Jan 2019.

3. Marina Bibikova, Zhenwu Lin, Lixin Zhou, Eugene Chudin, Eliza Wickham Garcia, Bonnie Wu, Dennis Doucet, Neal J Thomas, Yunhua Wang, Ekkehard Vollmer, Torsten

Goldmann, Carola Seifart, Wei Jiang, David L Barker, Mark S Chee, Joanna Floros, and Jian-Bing Fan. High-throughput DNA methylation profiling using universal bead arrays. *Genome Research*, 16(3):383–393, Mar 2006.

4. Saurav Bose, Chén C Kenyon, and Aaron J Masino. Personalized prediction of early childhood asthma persistence: A machine learning approach. *PLOS ONE*, 16(3):e0247784, Mar 2021.

5. Leo Breiman. Random Forests. *Machine Learning*, 45(1):5–32, 2001.

6. Aurelia Bustos, Antonio Pertusa, Jose-Maria Salinas, and Maria de la Iglesia-Vayá. Pad-Chest: A large chest x-ray image dataset with multi-label annotated reports. *Medical Image Analysis*, 66:101797, 2020.

7. Chaojin Chen, Dong Yang, Shilong Gao, Yihan Zhang, Liubing Chen, Bohan Wang, Zihan Mo, Yang Yang, Ziqing Hei, and Shaoli Zhou. Development and performance assessment of novel machine learning models to predict pneumonia after liver transplantation. *Respiratory Research*, 22(1):94, 2021.

8. Lei Chen, Zhandong Li, Tao Zeng, Yu-Hang Zhang, KaiYan Feng, Tao Huang, and Yu-Dong Cai. Identifying COVID-19-Specific Transcriptomic Biomarkers with Machine Learning Methods. *BioMed Research International*, 2021:9939134, 2021.

9. F Chollet. Xception: Deep Learning with Depthwise Separable Convolutions. In *2017 IEEE Conference on Computer Vision and Pattern Recognition (CVPR)*, pages 1800–1807, 2017.

10. Andreas Christe, Alan A Peters, Dionysios Drakopoulos, Johannes T Heverhagen, Thomas Geiser, Thomai Stathopoulou, Stergios Christodoulidis, Marios Anthimopoulos, Stavroula G Mougiakakou, and Lukas Ebner. Computer-Aided Diagnosis of Pulmonary Fibrosis Using Deep Learning and CT Images. *Investigative Radiology*, 54(10), 2019.

11. Kenneth Clark, Bruce Vendt, Kirk Smith, John Freymann, Justin Kirby, Paul Koppel, Stephen Moore, Stanley Phillips, David Maffitt, Michael Pringle, Lawrence Tarbox, and Fred Prior. The Cancer Imaging Archive (TCIA): Maintaining and Operating a Public Information Repository. *Journal of Digital Imaging*, 26(6):1045–1057, 2013.

12. Joseph Paul Cohen, Paul Morrison, and Lan Dao. COVID-19 Image Data Collection, 2020.

13. Corinna Cortes and Vladimir Vapnik. Support-vector networks. *Machine Learning*, 20(3):273–297, 1995.

14. Burcu F Darst, Kristen C Malecki, and Corinne D Engelman. Using recursive feature elimination in random forest to account for correlated variables in high dimensional data. *BMC Genetics*, 19(1):65, 2018.

15. Russell P Darst, Carolina E Pardo, Lingbao Ai, Kevin D Brown, and Michael P Kladde. Bisulfite Sequencing of DNA. *Current Protocols in Molecular Biology*, 91(1):7.9.1–7.9.17, Jul 2010.

16. Inselspital ILD database. Database of Interstitial Lung Diseases - Full Text View - https://clinicaltrials.gov/ct2/show/NCT00267800.

17. Adrien Depeursinge, Alejandro Vargas, Alexandra Platon, Antoine Geissbuhler, Pierre-Alexandre Poletti, and Henning Müller. Building a reference multimedia database for interstitial lung diseases. *Computerized Medical Imaging and Graphics*, 36(3):227–238, 2012.

18. Michal Dramiński, Alvaro Rada-Iglesias, Stefan Enroth, Claes Wadelius, Jacek Koronacki, and Jan Komorowski. Monte Carlo feature selection for supervised classification. *Bioinformatics*, 24(1):110–117, Jan 2008.

19. Pan Du, Xiao Zhang, Chiang-Ching Huang, Nadereh Jafari, Warren A Kibbe, Lifang Hou, and Simon M Lin. Comparison of Beta-value and M-value methods for quantifying methylation levels by microarray analysis. *BMC Bioinformatics*, 11(1):587, 2010.

20. Ron Edgar, Michael Domrachev, and Alex E Lash. Gene Expression Omnibus: NCBI gene expression and hybridization array data repository. *Nucleic Acids Research*, 30(1):207–210, Jan 2002.

21. Davide Ferrari, Jovana Milic, Roberto Tonelli, Francesco Ghinelli, Marianna Meschiari, Sara Volpi, Matteo Faltoni, Giacomo Franceschi, Vittorio Iadisernia, Dina Yaacoub, Giacomo Ciusa, Erica Bacca, Carlotta Rogati, Marco Tutone, Giulia Burastero, Alessandro Raimondi, Marianna Menozzi, Erica Franceschini, Gianluca Cuomo, Luca Corradi, Gabriella Orlando, Antonella Santoro, Margherita Digaetano, Cinzia Puzzolante, Federica Carli, Vanni Borghi, Andrea Bedini, Riccardo Fantini, Luca Tabbí, Ivana Castaniere, Stefano Busani, Enrico Clini, Massimo Girardis, Mario Sarti, Andrea Cossarizza, Cristina Mussini, Federica Mandreoli, Paolo Missier, and Giovanni Guaraldi. Machine learning in predicting respiratory failure in patients with COVID-19 pneumonia—Challenges, strengths, and opportunities in a global health emergency. *PLOS ONE*, 15(11):e0239172, Nov 2020.

22. Joverlyn Gaudillo, Jae Joseph Russell Rodriguez, Allen Nazareno, Lei Rigi Baltazar, Julianne Vilela, Rommel Bulalacao, Mario Domingo, and Jason Albia. Machine learning approach to single nucleotide polymorphism-based asthma prediction. *PLOS ONE*, 14(12):e0225574, Dec 2019.

23. Rajeshwar Govindarajan, Jeyapradha Duraiyan, Karunakaran Kaliyappan, and Murugesan Palanisamy. Microarray and its applications. *Journal of Pharmacy And Bioallied Sciences*, 4(6):310–312, Aug 2012.

24. Thao Thi Ho, Taewoo Kim, Woo Jin Kim, Chang Hyun Lee, Kum Ju Chae, So Hyeon Bak, Sung Ok Kwon, Gong Yong Jin, Eun-Kee Park, and Sanghun Choi. A 3D-CNN model with CT-based parametric response mapping for classifying COPD subjects. *Scientific Reports*, 11(1):34, 2021.

25. Lal Hussain, Tony Nguyen, Haifang Li, Adeel A Abbasi, Kashif J Lone, Zirun Zhao, Mahnoor Zaib, Anne Chen, and Tim Q Duong. Machine-learning classification of texture features of portable chest X-ray accurately classifies COVID-19 lung infection. *BioMedical Engineering OnLine*, 19(1):88, 2020.

26. Samuel G Armato III, Lubomir M Hadjiiski, Georgia D Tourassi, Karen Drukker, Maryellen L Giger, Feng Li, George Redmond, Keyvan Farahani, Justin S Kirby, and Laurence P Clarke. Special Section Guest Editorial: LUNGx Challenge for computerized lung nodule classification: reflections and lessons learned. *Journal of Medical Imaging*, 2(2):1–5, Jun 2015.

27. National Cancer Institute. The Cancer Genome Atlas Program - https://www.cancer. gov/tcga.

28. Rachna Jain, Meenu Gupta, Soham Taneja, and D Jude Hemanth. Deep learning based detection and analysis of COVID-19 on chest X-ray images. *Applied Intelligence*, 51(3):1690–1700, 2021.

29. Alistair E W Johnson, Tom J Pollard, Seth J Berkowitz, Nathaniel R Greenbaum, Matthew P Lungren, Chih-ying Deng, Roger G Mark, and Steven Horng. MIMIC-CXR, a de-identified publicly available database of chest radiographs with free-text reports. *Scientific Data*, 6(1):317, 2019.

30. Alistair E W Johnson, Tom J Pollard, Lu Shen, Li-wei H Lehman, Mengling Feng, Mohammad Ghassemi, Benjamin Moody, Peter Szolovits, Leo Anthony Celi, and Roger G Mark. MIMIC-III, a freely accessible critical care database. *Scientific Data*, 3(1):160035, 2016.

31. Max Kuhn, Kjell Johnson, and Others. *Applied Predictive Modeling*, volume 26. Springer, 2013.

32. Miron B Kursa and Witold R Rudnicki. Feature Selection with the Boruta Package. *Journal of Statistical Software; Vol 1, Issue 11 (2010)*, Sep 2010.

33. Melissa J Landrum, Shanmuga Chitipiralla, Garth R Brown, Chao Chen, Baoshan Gu, Jennifer Hart, Douglas Hoffman, Wonhee Jang, Kuljeet Kaur, Chunlei Liu, Vitaly Lyoshin, Zenith Maddipatla, Rama Maiti, Joseph Mitchell, Nuala O'Leary, George R Riley, Wenyao Shi, George Zhou, Valerie Schneider, Donna Maglott, J Bradley Holmes, and Brandi L Kattman. ClinVar: improvements to accessing data. *Nucleic Acids Research*, 48(D1):D835–D844, Jan 2020.

34. Nhat-Nam Le-Dong, Thong Hua-Huy, Huyen-Mi Nguyen-Ngoc, and Anh-Tuan Dinh-Xuan. Applying machine learning and pulmonary function data to detect interstitial lung disease in systemic sclerosis. *European Respiratory Journal*, 50(suppl 61):OA3438, Sep 2017.

35. Yann LeCun, Yoshua Bengio, and Geoffrey Hinton. Deep learning. *Nature*, 521(7553): 436–444, 2015.

36. Andreas Leha, Kristian Hellenkamp, Bernhard Unsöld, Sitali Mushemi-Blake, Ajay M Shah, Gerd Hasenfu, and Tim Seidler. A machine learning approach for the prediction of pulmonary hypertension. *PLOS ONE*, 14(10):e0224453, Oct 2019.

37. Ping-I Lin, Huan Shu, and Tesfaye B Mersha. Comparing DNA methylation profiles across different tissues associated with the diagnosis of pediatric asthma. *Scientific Reports*, 10(1):151, 2020.

38. Huan Liu and Rudy Setiono. Incremental Feature Selection. *Applied Intelligence*, 9(3):217–230, 1998.

39. Xiangju Liu, Yu Zhang, Chunli Fu, Ruochi Zhang, and Fengfeng Zhou. EnRank: An Ensemble Method to Detect Pulmonary Hypertension Biomarkers Based on Feature Selection and Machine Learning Models, 2021.

40. Roberto Lo Gullo, Isaac Daimiel, Elizabeth A Morris, and Katja Pinker. Combining molecular and imaging metrics in cancer: radiogenomics. *Insights into Imaging*, 11(1):1, 2020.

41. Juan E Luján-García, Cornelio Yáñez-Márquez, Yenny Villuendas-Rey, and Oscar Camacho-Nieto. A Transfer Learning Method for Pneumonia Classification and Visualization, 2020.

42. Xia Ma, Yanping Wu, Ling Zhang, Weilan Yuan, Li Yan, Sha Fan, Yunzhi Lian, Xia Zhu, Junhui Gao, Jiangman Zhao, Ping Zhang, Hui Tang, and Weihua Jia. Comparison and development of machine learning tools for the prediction of chronic obstructive pulmonary disease in the Chinese population. *Journal of Translational Medicine*, 18(1):146, 2020.

43. Kazushi Matsumura and Shigeaki Ito. Novel biomarker genes which distinguish between smokers and chronic obstructive pulmonary disease patients with machine learning approach. *BMC Pulmonary Medicine*, 20(1):29, 2020.

44. Paul Mooney. Chest X-Ray Images (Pneumonia) - Kaggle - https://www.kaggle.com/paultimothymooney/chest-xray-pneumonia.

45. NIH. Lung Tissue Research Consortium (LTRC) - NHLBI - https://www.nhlbi.nih.gov/science/lung-tissue-research-consortium-ltrc.

46. openSNP. openSNP - https://opensnp.org/.

47. Prashant Patel. Chest X-ray (Covid-19 & Pneumonia) - Kaggle - https://www.kaggle.com/prashant268/chest-xray-covid19-pneumonia.

48. Hanchuan Peng, Fuhui Long, and C Ding. Feature selection based on mutual information criteria of max-dependency, max-relevance, and min-redundancy. *IEEE Transactions on Pattern Analysis and Machine Intelligence*, 27(8):1226–1238, 2005.

49. Junfeng Peng, Chuan Chen, Mi Zhou, Xiaohua Xie, Yuqi Zhou, and Ching-Hsing Luo. A Machine-learning Approach to Forecast Aggravation Risk in Patients with Acute Exacerbation of Chronic Obstructive Pulmonary Disease with Clinical Indicators. *Scientific Reports*, 10(1):3118, 2020.

50. Elizabeth A Regan, John E Hokanson, James R Murphy, Barry Make, David A Lynch, Terri H Beaty, Douglas Curran-Everett, Edwin K Silverman, and James D Crapo. Genetic Epidemiology of COPD (COPDGene) Study Design. *COPD: Journal of Chronic Obstructive Pulmonary Disease*, 7(1):32–43, Feb 2011.

51. S T Sherry, M.-H. Ward, M Kholodov, J Baker, L Phan, E M Smigielski, and K Sirotkin. dbSNP: the NCBI database of genetic variation. *Nucleic Acids Research*, 29(1):308–311, Jan 2001.

52. Gur Amrit Pal Singh and P K Gupta. Performance analysis of various machine learning-based approaches for detection and classification of lung cancer in humans. *Neural Computing and Applications*, 31(10):6863–6877, 2019.

53. Forum of International Respiratory Societies. *The Global Impact of Respiratory Disease*. European Respiratory Society, Sheffield, second edition, 2017.

54. QingZeng Song, Lei Zhao, XingKe Luo, and XueChen Dou. Using Deep Learning for Classification of Lung Nodules on Computed Tomography Images. *Journal of Healthcare Engineering*, 2017:8314740, 2017.

55. L Sorensen, S B Shaker, and M de Bruijne. Quantitative Analysis of Pulmonary Emphysema Using Local Binary Patterns. *IEEE Transactions on Medical Imaging*, 29(2):559–569, Feb 2010.

56. Sumanth Swaminathan, Klajdi Qirko, Ted Smith, Ethan Corcoran, Nicholas G Wysham, Gaurav Bazaz, George Kappel, and Anthony N Gerber. A machine learning approach to triaging patients with chronic obstructive pulmonary disease. *PloS one*, 12(11):e0188532–e0188532, Nov 2017.

57. Ivan Tomek. An Experiment with the Edited Nearest-Neighbor Rule. *IEEE Transactions on Systems, Man, and Cybernetics*, SMC-6(6):448–452, 1976.

58. Ivan Tomek. Two Modifications of CNN. *IEEE Transactions on Systems, Man, and Cybernetics*, SMC-6(11):769–772, 1976.

59. Katsuyuki Tomita, Ryota Nagao, Hirokazu Touge, Tomoyuki Ikeuchi, Hiroyuki Sano, Akira Yamasaki, and Yuji Tohda. Deep learning facilitates the diagnosis of adult asthma. *Allergology International*, 68(4):456–461, 2019.

60. Drexel University. Pediatric Big Data Project - Drexel Urban Health Collaborative - https://drexel.edu/uhc/research/projects/pediatric-big-data-project/.

61. Ana B Villaseñor-Altamirano, Marco Moretto, Mariel Maldonado, Alejandra Zayas-Del Moral, Adrián Munguia-Reyes, Yair Romero, Jair. S Garca-Sotelo, Luis A Aguilar, Oscar Aldana-Assad, Kristof Engelen, Moisés Selman, Julio Collado-Vides, Yalbi I Balderas-Martnez, and Alejandra Medina-Rivera. PulmonDB: a curated lung disease gene expression database. *Scientific Reports*, 10(1):514, 2020.

62. Dingding Wang, Jiaqing Mo, Gang Zhou, Liang Xu, and Yajun Liu. An efficient mixture of deep and machine learning models for COVID-19 diagnosis in chest X-ray images. *PLOS ONE*, 15(11):e0242535, Nov 2020.

63. Linda Wang and Alexander Wong. COVID-Net: A Tailored Deep Convolutional Neural Network Design for Detection of COVID-19 Cases from Chest X-Ray Images, 2020.

64. Ryan Yixiang Wang, Tim Qinsong Guo, Leo Guanhua Li, Julia Yutian Jiao, and Lena Yiqi Wang. Predictions of COVID-19 Infection Severity Based on Co-associations between the SNPs of Co-morbid Diseases and COVID-19 through Machine Learning of Genetic Data. In *2020 IEEE 8th International Conference on Computer Science and Network Technology (ICCSNT)*, pages 92–96, 2020.

65. X Wang, Y Peng, L Lu, Z Lu, M Bagheri, and R M Summers. ChestX-Ray8: Hospital-Scale Chest X-Ray Database and Benchmarks on Weakly-Supervised Classification and Localization of Common Thorax Diseases. In *2017 IEEE Conference on Computer Vision and Pattern Recognition (CVPR)*, pages 3462–3471, Jul 2017.

66. Z Wang and Y Wang. Exploring DNA Methylation Data of Lung Cancer Samples with Variational Autoencoders. In *2018 IEEE International Conference on Bioinformatics and Biomedicine (BIBM)*, pages 1286–1289, 2018.

67. Zhong Wang, Mark Gerstein, and Michael Snyder. RNA-Seq: a revolutionary tool for transcriptomics. *Nature Reviews Genetics*, 10(1):57–63, 2009.

68. D L Wilson. Asymptotic Properties of Nearest Neighbor Rules Using Edited Data. *IEEE Transactions on Systems, Man, and Cybernetics*, SMC-2(3):408–421, 1972.

69. Yiwen Xu, Ahmed Hosny, Roman Zeleznik, Chintan Parmar, Thibaud Coroller, Idalid Franco, Raymond H Mak, and Hugo J W L Aerts. Deep Learning Predicts Lung Cancer Treatment Response from Serial Medical Imaging. *Clinical Cancer Research*, 25(11):3266 LP – 3275, Jun 2019.

70. Lingming Yu, Guangyu Tao, Lei Zhu, Gang Wang, Ziming Li, Jianding Ye, and Qunhui Chen. Prediction of pathologic stage in non-small cell lung cancer using machine learning algorithm based on CT image feature analysis. *BMC Cancer*, 19(1):464, 2019.

71. Xin Yu, Qian Yang, Dong Wang, Zhaoyang Li, Nianhang Chen, and De-Xin Kong. Predicting lung adenocarcinoma disease progression using methylation-correlated blocks and ensemble machine learning classifiers. *PeerJ*, 9:e10884, 2021.

72. Fei Yuan, Lin Lu, and Quan Zou. Analysis of gene expression profiles of lung cancer subtypes with machine learning algorithms. *Biochimica et Biophysica Acta (BBA) - Molecular Basis of Disease*, 1866(8):165822, 2020.

73. Daniel Zeiberg, Tejas Prahlad, Brahmajee K Nallamothu, Theodore J Iwashyna, Jenna Wiens, and Michael W Sjoding. Machine learning for patient risk stratification for acute respiratory distress syndrome. *PLOS ONE*, 14(3):e0214465, Mar 2019.

74. Joe G Zein, Chao-Ping Wu, Amy H Attaway, Peng Zhang, and Aziz Nazha. Novel Machine Learning Can Predict Acute Asthma Exacerbation. *Chest*, 159(5):1747–1757, 2021.

75. Harry Zhang. The Optimality of Naïve Bayes. In *Proceedings of the Seventeenth International Florida Artificial Intelligence Research Society Conference (FLAIRS 2004)*, pages 562–567, Florida, USA, 2004. AAAI Press.

5 Multi Objective Bacterial Foraging Optimization: A Survey

R. Vasundhara Devi
Department of Computer Science, Perunthalaivar
Kamarajar Arts College, Puducherry, India

S. Siva Sathya
Department of Computer Science, Pondicherry University,
Puducherry, India

CONTENTS

5.1 INTRODUCTION

For the past two decades, optimization algorithms have been applied to all fields such as engineering, banking and industrial applications. Also physical, chemical and biological sciences problems are being solved using meta-heuristic algorithms. These algorithms include evolutionary algorithms and swarm intelligence (SI) algorithms. SI, one of the artificial intelligence techniques uses a meta-heuristic approach to solve optimization problems. Evolutionary algorithm is categorized as Genetic

DOI: 10.1201/9781003246688-5

algorithm, Evolutionary programming and evolutionary strategies. These algorithms are also combined with the traditional approaches such as back tracking, dynamic programming, simulated annealing, Tabu search and niching, etc. to enhance its performance. Lot of researchers from various disciplines are attracted towards SI. Bonabeau defined SI as "The emergent collective intelligence of groups of simple agents" [4]. Self-organization and division of labour are the fundamental concepts of SI algorithms. SI algorithms are those meta-heuristic techniques simulated from the collective behavior of decentralized and self-organised natural organisms and their systems such as flock of birds/fish in Particle swarm optimization [24], group of monkeys [55], collection of bacteria [41], herd of Grey wolves [35] to the Whales [34] that have reasonable advantage over conventional algorithms. Holland [21] proposed an outstanding algorithm based upon the survival of the fittest and the evolution in nature, namely, genetic algorithm and for decades, it has been applied to the real-world optimization problems. Then came the evolutionary programming [18], Genetic programming [28] and evolutionary strategies [2] made a well-known impact in the field of evolutionary algorithms. Next level of algorithms framed wad SI algorithms such as particle swarm optimization [42], ant colony optimization [15], Cuckoo search [49], Bat algorithm [50], Monkey algorithm [55], Bacterial foraging optimization [40], Grey wolf optimization [35], whale optimization algorithm [34], ant lion optimization algorithm [33], grasshopper optimization algorithm [46] to the latest Biofilm algorithm [11].

Bacterial foraging optimization [40] has been found to be a very effective foraging algorithm for single objective optimization problems and number of multi-objective versions of BFO were proposed recently. This chapter begins with a lucid outline of classical BFOA and reports an in-depth study of existing multi-objective bacterial foraging algorithms and various techniques proposed to solve multi objective optimization (MOO) problems. Generally, in SI algorithms [3, 14], the individuals or the candidate solutions are modified with the help of objective function values. As swarm is a collection of interacting agents or individuals, the fitness values make the individuals of the algorithm to converge towards the better optimal non-dominated solutions after performing the exploitation and exploration in the problem space. There are few more categories of algorithms such as physics based, chemistry based, natural processes-based algorithms each specialized for domain specific real-world problems which satisfy the no free lunch theorem as shown in Figure 5.1. Among this list, Figure 5.2 shows a peek at where the bacterial foraging optimization algorithm stands. The rest of the chapter is organized as follows: Section 2 provides the details about the multi-objective optimization, Section 3 discusses the basics of BFOA and Section 4 describes the existing multi objective bacterial foraging optimization algorithms. Section 5 analyses the existing methods and future research paths in the area of multi-objective BFOA. Section 6 concludes the chapter.

5.2 BACTERIAL FORAGING OPTIMIZATION

One of the interesting SI algorithms is the bacterial foraging optimization algorithm simulated by Passino [40] inspired by the foraging behavior of Esterichia Coli

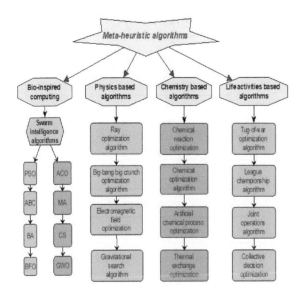

Figure 5.1 A glimpse of few meta-heuristic algorithms.

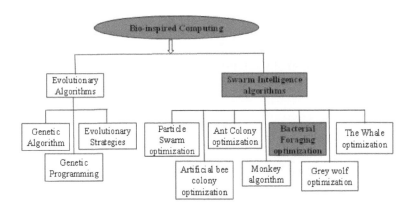

Figure 5.2 Position of bacterial foraging optimization in the hierarchy.

bacteria in the intestine. In this foraging procedure, bacteria moves or tumbles with its two flagella in chemotactic steps of defined step lengths to friendly environments. There are three loops for elimination and dispersal, Reproduction and Chemotactic steps in BFO. Bacteria using flagella either rotate or swim with the help of the chemotactic step size. In the reproduction, the half of the least fit bacteria duplicate leaving the other half to die. In the elimination and dispersal loop, the environment plays a major role in eliminating the unfit bacteria. There are variations in the BFO

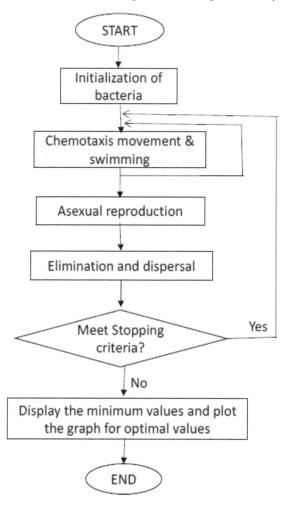

Figure 5.3 Flow chart of the bacterial foraging optimization algorithm.

and hybrids of BFO for multiple objective optimization. The flow chart of the basic bacterial foraging optimization is given in Figure 5.3.

The E.Coli bacteria has 8–10 left-handed helix-shaped flagella grown on its body. These flagella are the locomotive organs of E.Coli which can move and rotate at high speeds [30]. After the rotation, the flagella stop for a moment to change the direction of rotation carefully [16]. When the E.Coli bacteria moves in the human intestine and sense unfavorable conditions, the bacteria alter their random walk to a biased one. The bacteria is said to move with chemotactic steps towards chemical signals. The helix configuration of the flagella of the bacteria help the bacteria to propel forward during their swim. When the flagella rotate in clockwise direction, the bacteria destabilize and move in various directions during their tumble movement randomly. Bacteria moves with various combinations of swim and tumble movements and explore their living surfaces. These combinations also depend on the gradients of the environments. When the bacteria move to the food-rich environment, they swim more compared to tumbling to have a good grasp of their food. In a neutral environment conditions, they alternatively swim and tumble in the process of searching for a food-rich environment. When bacteria senses dangerous or noxious environment, they swim faster to avoid or run away from there to find a better environment [41].

5.3 MULTI OBJECTIVE OPTIMIZATION

Optimization algorithms are categorized into single objective optimization and multi-objective optimization algorithms. Famous single objective optimization algorithm include PSO [24], ACO [15], ABC [20], MA [55], GWO [35], Whale optimization algorithm [34] to the recent Biofilm algorithms [11] which has single objective to solve the optimization problems.

Examples of multi objective optimization algorithms include SPEA2 [56], NS-GAII [10], MOPSO [7], MOACO [6], MOBFO [37], MOGWO [36], etc and are used to solve the optimization problems with two objectives and at the maximum of three objectives. Most of the real-time optimization problems fall in this category such as transmission emission, interdisciplinary problems, etc. New category of optimization algorithm is the many objective optimization algorithms that can solve 3–15 objectives. For example, NSGAIII [22], PSO for many objectives [5], etc.

Given below is the MOO problem formulated in terms of a maximization problem without loss of generality:

$$Maximize : F\left(\underset{x}{\rightarrow}\right) = f_1\left(\underset{x}{\rightarrow}\right), f_2\left(\underset{x}{\rightarrow}\right), ..., f_k\left(\underset{x}{\rightarrow}\right) \tag{5.1}$$

Subject to:

$$g_i(x) \geq 0, i = 1, 2, 3, ..., c \tag{5.2}$$

$$h_i\left(\underset{x}{\rightarrow}\right) = 0, i = 1, 2, ..., e \tag{5.3}$$

$$L_i \leq\leq x_i \leq U_i, i = 1, 2, 3, ..., s \tag{5.4}$$

where s – number of variables, k – number of objectives, c – number of inequality constraints and e – number of equality constraints. Lower and upper boundaries are Li and Ui of the multi objective problem space of the respective problem.

5.4 MULTI OBJECTIVE BACTERIAL FORAGING OPTIMIZATION APPROACHES

In this review, multi-objective bacterial foraging optimization methods available till date have been collected and analysed. There are quite a number of work conducted to enable BFO to solve multi objective problems. As shown in Figure 5.4, BFO

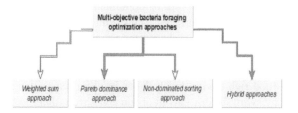

Figure 5.4 Techniques used to solve MOBFO algorithms.

is modified into several multi objective versions using the methods discussed below:
1. Weighted sum approach, 2. Pareto-dominance approach, 3. Non-dominated sorting
approach and 4. Hybrid approaches with BFO.

The bacterial foraging optimization algorithm due to its efficiency attracted the
researchers to develop its multi objective BFO in their own flavors. As there was no
such consolidation and discussion about their application as far as our knowledge
in the literature, this chapter consolidates the available literature and analyzes them
in terms of methods used, advantages and disadvantages of the methods, how the
algorithms are used in the respective applications and so on.

5.4.1 WEIGHTED SUM APPROACH

As per [51], in weighted sum approach, the multiple objective functions are combined into a composite, single scalar solution. The format of weighted sum approach
is given in (1)

$$wF(x) = w_1 f_1(x) + w_2 f_2(x) + ... + w_n f_n(x) \qquad (5.5)$$

The addition of these weights chosen should provide a positive value of 1.

$$\sum_{i=0}^{n} w_i = 1, w_i \varepsilon (0,1) \qquad (5.6)$$

where

$$w = (w_1, w_2, ..., w_n) \qquad (5.7)$$

are the weights provided to the corresponding n objectives. In this approach, the
MOO problem is converted into a single objective problem. In weighted sum approach, the objectives are provided with weight values based upon the importance
or the priority of the objective in the corresponding real-world problems. In the case
of [47], the BFO algorithm is modified to solve the multiple objectives of the single phase transformer design. Transformers are the electrical equipment used in the
transmission and distribution of electricity in a stable and reliable manner. They
help in the successful operation of electrical utilities. The transformers should be
designed with high quality, highly efficient transformers with a minimal cost. This
single-phase transformer design needs to minimize the cost while improving the performance. As there are two objectives that need to be simultaneously solved, the

weighted sum method is used. The first objective is the cost involved represented as f1, and the second objective is the efficiency represented as f2. The weighting values associated with cost (Rs) and efficiency (%) are w1 and w2, respectively. The solutions obtained should minimize the cost and the objective values are represented as

$$f[f_1, f_2] = f(w_1, w_2) = w_1 f_1 + w_2 (1/f_2) \tag{5.8}$$

In this work, [47], without altering the solution, the objective function is divided by a positive number, followed by dividing (3) by w1, w2/w1 is redefined as w.

In a real vector optimization problem, all the objective functions cannot be solved by a single solution. To bring a balance among all the objectives, there will be a set of solutions known as Pareto set. In this weighting method, the decision maker interacts with the system and provide weights as per the priority of the objectives. These choice of weights provide a pareto set of optimal solutions and the decision maker (DM) chooses the best feasible and compromising solution from the Pareto set. The results of cost and efficiency are compared to the multi objective PSO. The results of weighted sum BFA indicate that the weighted sum approach followed is equally competitive to the MOPSO algorithm. In [26], multi objective BFOA is applied in power factor correction in Switched mode power supplies, as they have to optimize multiple objectives. The calculation of multiple objectives is carried out with the help of weighted sum approach. These power factor corrections are needed to solve the poor power factors due to the non-linearities in the switching devices. The fitness functions (FF) considered for optimization are the error criteria, namely, ITSE (Integral Time Square Error) , IAE (Integral absolute error) ISE (Integral square error) and ITAE (Integral Time Absolute Error). The optimal solutions are to be minimized while improving the dynamic performance. Also, efficiency has to be maximized and power factor has to attain a value close to one. As this work uses weighted sum approach, the error criteria are assigned weights. The criterion ISE, ITSE, IAE, ITAE are represented by F1, F2, F3 and F4 and their weights are a1, a2, a3, a4, respectively. Therefore, the FF is calculated using (4) as given below:

$$FF = a1 * F1 + a2 * F2 + a3 * F3 + a4 * F4 \tag{5.9}$$

This approach gives the best optimal parameters for the PI controller of the simulated converter in terms of settling time, rise time, percentage of total harmonic distortion, peak overshoot, and percentage of efficiency. When the load, line and reference voltage are varied dynamically, this multi objective BFOA with weighted sum approach provide an improved performance.

In [19], an investigation is made about the non-gaussian stochastic engine effect over the performance of the meta-heuristics algorithms using bacterial foraging algorithm. Four stochastic engines or random generators such as Weibull, Gamma, Gaussian distribution and a chaotic mechanism are utilized simultaneously in bacterial foraging algorithm. The BFA technique utilizing all these above-mentioned distributions is mentioned by G-BFA, γ-BFA and W-BFA, respectively, whereas the BFA using chaotic generator is termed as Ch-BFA. These developed methods are actually applied to the real-world MOO problem, namely, resin bonded sand mould problem. The objectives of this resin bonded sand mould problem are the maximization of permeability, compression, tensile and shear strengths with constraints. These multi objectives are solved with the help of the weighted sum approach. In this work, 53 solutions are obtained with different weights for each of these four BFA variants. The performance metric HVI (hyper volume indicator) is used to measure the nature of these solution sets. Another novel metric AER (Average Explorative Rate) metric for the measuring the search region on average covered by the computational method at each iteration. Among the four variants, W-BFA performs better than the other variants in terms of their approximate pareto frontiers obtained. In [17], an improved BFOA (IBFA) is considered to solve the STHTS (short-term hydro thermal scheduling) problem. STHTS problem has to minimize the gas emissions and provide efficient hydro thermal scheduling which is short term. As the basic BFO algorithm has convergence problems when used for large-scale and complex problems, an IBFA is proposed to solve hydro-thermal generation system. The objective functions of this work are to minimize operating cost of thermal generating units and the gaseous emissions of NO2, co2 and SO2 over a scheduling period. The constraints include operational, control, thermal and hydro constraints. A test for hydro-thermal generation is employed as the case study to validate the IBFA performance.

STHTS problem is considered as the problem with two objectives, F1 is the cost of thermal units Nk, F2 is the NO2 emissions, F3 is the SO2 emissions and F4 is the CO2 emissions. By the weighted-sum method, the a single solution is obtained using:

$$min \sum_{K=1}^{O} w_k F_k P_{gi} \qquad (5.10)$$

where O is the number of objective functions and wk is the weight assigned to the kth objective. Fk is the total cost values of kth objective. The consolidated weight value is equal to one. Run-length parameter is the unique parameter which has hold on local and global search ability of BFOA by providing a balance among exploration and exploitation capabilities. As the run length unit of the BFOA is always constant when applied to small scale problems, the constant run length unit will not provide feasible results when applied to large-scale problems with high dimensions. Therefore, to handle the large-scale problems, a decreasing dynamic function is introduced to perform the swim walk. The new run length parameter defined in this work is calculated with the help of maximum number of chemotactic steps and two predefined parameters C(Nc) and C(1).

In the entire scheduling period, the hourly load demand for each time interval is taken as the parameter. Two case studies are followed to optimize the objectives. In the first case, the four objective functions are considered individually and these fuel costs of NO2, SO2 and CO2 are optimized to provide minimal values using single objective optimization technique. In the second case, the optimization of fuel cost NO2, SO2 and CO2 emissions are performed simultaneously as a multi objective problem with various weighting combinations, i.e weighted sum approach.

In [29], a multi objective BFOA algorithm is proposed to design and control harmonic filters in a practical system. The objective functions used for this selection, planning and control of passive filters in the system are the minimization of the filter installation cost, generation cost and power system losses. This methodology is chosen to minimize voltage distortion and harmonic current in the point of couplings. The multi objective optimization of selection of passive filter is carried out by framing a FF integrating all the objectives of the problem. The constraints considered include harmonic voltage and current distortion limits, voltage magnitude limits, voltage angles above 0.95 and the passive filter after its installation, the system should not be over or under compensated. In this work, weighted sum approach is used and two case studies are performed in IEEE 5 bus system. The results of these two case studies achieve current and voltage harmonic reduction, power factor improvement and resonance damp out in the network. The performance of the optimal solution proved to be good compared to the conventional shunt passive filter.

5.4.2 PARETO-DOMINANCE APPROACH

As per [8], the mathematical definition of pareto dominance, without loss of generality, for a problem of maximization problem are provided below: Definition 1: Pareto Dominance Two vectors such as

$$\underset{x}{\rightarrow} = (x_1, x_2, ..., x_k) \, and \, \underset{y}{\rightarrow} = (y_1, y_2, ..., y_k) \tag{5.11}$$

Vector x dominates vector y (denoted as x) iff:

$$\forall i = \{1, 2, ..., k\}, [f(x_i) \geqslant f(y_i) \wedge [\exists i\varepsilon 1, 2, ..., k : f(x_i)]]] \tag{5.12}$$

Definition 2. Pareto optimality: A solution

$$\underset{x}{\rightarrow} \varepsilon X \tag{5.13}$$

is known Pareto-optimal iff:

$$\nexists \underset{y}{\rightarrow} \varepsilon X || F \underset{y}{\rightarrow} \succ F \underset{x}{\rightarrow} \tag{5.14}$$

Pareto optimal set is the set of all non-dominated solutions of a problem. and it is defined as: Definition 3. Pareto optimal set: Pareto set is the set of all Pareto-optimal solutions and is defined as:

$$P_s := \{x, y\varepsilon X | \exists F(y) \succ F(y)\} \tag{5.15}$$

Pareto optimal front is the set of objective values of solutions in Pareto optimal set and is shown as:

Definition 4. Pareto optimal front: A set of objective function values s for Pareto solutions set:

$$P_f := \{F(x) | x\varepsilon P_s\} \tag{5.16}$$

All the above algorithms were primarily designed for the optimization problems using single objective. Hence, interesting algorithms have evolved for solving MOO problems such as NSGAII [10], SPEA2 [56], PESA2 [9], PAESII [25],

MOPSO [45], MBFO [38], MOMA [13], MOGWO [36], MOGOA [53], MOALO [33] to the recent MoBifi [12]. Although each of these algorithms can solve sets of benchmark problems, they were applied to the interesting real-world problems such as engineering problems, Portfolio optimization, cervix lesion classification, multiple sequence alignment (MSA), parallel cell entropy, De novo drug design, etc. This survey covers the available multi objective bacterial foraging optimization algorithms and their interesting applications. Multi objective optimization of the real-world problems are done with the help of apriori, aposteriori and interactive methods. As per [27], there are three different approaches to solve multiple objectives using optimization algorithms. They are weighted sum approach, altering objective functions, and the pareto ranking approach. Among these three approaches, pareto approach has gained popularity due to less human intervention in decision making. Recently multi objective optimization algorithms that include MOGWO [36], MOALO [33], MOGOA [53] and MoBifi [12] also use the pareto-based approach to solve the benchmark problems. Multi-objective optimization problems may have conflicting constraints. Multi-objective optimization algorithms have gradient-less mechanism along with the local optima avoidance property. This made the optimization algorithms best suited to solve the real-world optimization problems in a cost efficient manner.

Pareto-based approaches is followed in [37] along with the BFO algorithm. The survival of the fittest mechanism helps in maintaining the objectives. The basic idea of this mechanism is that the bacteria with better values will survive. The pareto dominance mechanism is integrated with a health sorting approach. The diversity of the solutions is improved with the help of the unfeasible border solutions on a given probability. Two performance metrics, namely, Diversity metric and Generational distance, are used to evaluate the performance of the algorithm. MBFO algorithm is applied to multi objective benchmark problems.

In [52], multi objective BFO (MOBFO) is proposed and applied to environment-friendly aluminium electrolysis production process. This application has a number of highly coupled and nonlinear parameters. The objectives of this problem are maximization of current efficiency and minimization of production of PFC (perfluorocarbons). A task-oriented framework called parallel cell coordinate system is proposed in this work with the help of the pareto solutions produced by MOBFO. These pareto solutions are arrived using the adaptive foraging strategy (AFS) and Pareto-archived evolution approach (PAEA). These help in balancing the diversity and convergence of the Pareto optimal set in the optimization process. MOBFO is applied to the ZDT1-ZDT4, DTLZ1 and DTLZ2 benchmark functions which yield much better results compared to algorithms namely NSGAII, SPEA2, MOPSO, MOEA/D and MBFO which is further applied to the aluminum electrolysis production process.

In [48], the single objective BFO algorithm is modified to solve multiple objectives called an archive bacterial foraging optimizer for multi objective optimization (MABFO). MABFO proposes four optimization mechanisms such as chemotaxis, reproduction, conjugation and elimination-and-dispersal. MABFO incorporates the pareto dominance concept along with an external archive to save the elite individuals and non-dominated individuals. A crowding distance operator is utilized to improve the diversity of the pareto solutions. This algorithm is tested with ZDT1 – ZDT4 functions and compared with two famous algorithms, NSGAII and SPEA2. Chemotaxis and conjugation are the important behaviours of bacteria. Conjugation is regarded as the mating between bacteria and in this work it is considered to simulate the information exchange mechanism between bacteria in entire Population and archive. A new parameter called conjugation length L is measured by the count of decision variables. Each bacteria b in the population, randomly choose another bacteria bb in the archive and a conjugation point to update the position of the solution. This conjugation mechanism helps the individual bacteria to get the guidance from the superior individual in archive and converge to the global pareto front. To assess the convergence and the spread of the obtained solutions in the pareto front, two performance metrics, namely, generational distance and spread metric are utilized.

In [43], modified bacterial foraging optimization algorithm (MBFOA) is proposed. It is applied to the two objective mechanical design problem with dynamic constraint. This proposed MBFOA is modified to solve multi objective optimization problems namely multi-objective modified bacterial foraging algorithm (MOMB-FOA). The main features of this MOMBFOA are pareto dominance for selection criterion, elite individual saved in external archive and diversity of the optimal solutions maintained using the crowding distance. To solve the constraints, feasibility rules are defined. The constraint handling mechanism changes the criteria for selection used in the tumble-swim movements and sorting performed in the reproduction process of the bacterial swarm. Isolated solutions have higher crowding distance values. The archive is populated with higher crowding distance solutions which simultaneously extends the obtained pareto front. An attractive operator is proposed in this work. The results are compared with two evolutionary algorithms, NSGAII and Differential evolution. A two-set coverage metric among two sets is used as a performance metric for the comparison of pareto fronts of algorithms MOMBFOA, NSGAII and DE.

5.4.3   NON-DOMINATED SORTING APPROACH

Multiple objectives are solved using non-dominated sorting approach along with knowledge sharing or message passing parameters as proposal in NSGAII [10]. In this section, the multi objective BFOA algorithms using non-dominated sorting approaches are discussed.

In [39], BFOA algorithm is upgraded to solve multiple objectives with the inclusion of non-dominated sorting approach called non-denominated sorting and applied to the real-world non-linear constrained environmental/economic dispatch problem of a standard 30-bus 6-Generator test system. The two objectives considered are emission and fuel cost that are conflicting in nature. The constraints associated with this problem are bus voltages, generator real power, power flow of transmission lines, reactive power outputs, POZ (prohibited operating zones) and ramp rate limits. The fuel cost has to be minimized which simultaneously reduce the total generation cost along with the constraints. The emission dispatch has to be minimized which simultaneously minimizes the classical economic dispatch. The power balance constraint is the equality constraint and generation capacity, ramp rate limit, POZ and line flow limits are the inequality constraints. Using this non-denominated sorting bacteria foraging (NSBF), experiments are carried out with the IEEE 30-bus 6-Generator using three different cases of simulations with single or multiple constraints. Performance metrics such as divergence and spacing of the non-dominated solutions are calculated for 30 trial runs which proved to provide a good pareto front. These experimental results are compared with NSGA, NPGA and SPEA algorithms. The previous best positions are stored in memory.

In [1], a variation in the classical BFOA is proposed based upon the chemotaxis movement of bacteria termed as BCMOA (bacterial chemotaxis multi-objective optimization algorithm) and applied to 11 benchmark problems. The characteristics of the new proposed algorithm are the fast non-dominated sorting approach and a simple chemotactic strategy to alter the bacterial positions for better exploration of the search space to find number of optimal solutions.

The features of the BCMOA include initialization, objective evaluation followed by non-domination classification, chemotactic movement for strong bacteria, communication and chemotactic strategy for weak bacteria, step size, diversity preservation and constraint handling. Two goals obtained from the literature for a successful Multi objective optimization algorithm are the global pareto optimal region and the population diversity in the pareto optimal front. There are 11 benchmark functions used: ZDT1 - ZDT4, SCH, FON [10], unconstrained problems KUR ,POL, constraint problems such as SRN, CONSTR and TNK. The benchmark results of BCMOA are compared with that of the NSGAII and non-dominated sorting particle swarm optimization (NSPSO) algorithms. The performance metrics used to evaluate BCMOA performance are convergence metric and diversity metric [10] and generational distance [31].

In [23], an elitist multi objective bacterial foraging evolutionary algorithm is proposed to provide service to resource providers and consumers. It is a multi criteria based scheduling approach based upon Grid computing. The performance of this work is compared with OMOPSO, NSGAII and Grid's conventional mechanisms such as Min-min and Max-min on ALEA 3.0 simulator. Also this proposed work is combined with ALEA 3.0. These comparisons prove that the elitist multi objective BFOA is a competitive candidate for solving the problems in Grid with conflicting multiple objectives. MOBFOA is supposed to provide a schedule which satisfy multiple grid scheduling criteria. In compute intensive grids, grid scheduling is a real-world multi-objective problem. Four scheduling methods are considered in this work. Four objective functions used are Makespan, Job success rate, Cost and flowtime. Crowding distance mechanism is used to select solutions with better ranks. The pareto front are obtained after the non-dominated sorting.

5.4.4 HYBRID APPROACHES WITH BFO

Usually, the optimization problems with multiple objectives can also be solved with the help of combining more than one algorithms. These algorithms with their own salient features when brought together provide better optimal solutions.

In [32], bacterial foraging algorithm is combined with genetic algorithm to carry out the multiple objectives optimization and applied to find MSA using pareto approach. Multiple objectives used are similarity variable maximization, gap penalty minimization and non-gap percentage. This work is measured with MSA methods, namely, Clustal Omega, Muscle,T-Coffee, MAFFT, GA, K-Align, ACO, PSO and ABC.

This algorithm is applied on four benchmark datasets:, BAliBASE 3.0, SABmark 1.65 and Oxbench 1.3. Better results were obtained When compared with the famous methods. MSA is a real-world complex optimization problem. MSA has complexities in the choice of sequences, objective function and its optimization. SOP (Sum of pairs) and TOC (total column score) are the performance measures for analysing the MSA algorithms. In this work, the choice of the sequences is chosen based on the non-dominated solution obtained with the help of crowding distance measure. The objective functions used to solve the MSA problem are maximization of similarity, minimization of variable gap penalty and maximization of non-gap percentage. SOP and the TCS are used as performance measures to analyse the algorithms of multiple sequence alignment.

Within the BFOA [40] algorithm, after the swim and tumble, selection, crossover and mutation are combined. In the swarming phase, to draw the attention of the other bacteria, a penalty function from a non-dominated sorting algorithm is taken to find the fittest bacteria with highest crowding distance and lower social status. The original cost function gets added with the relative lengths of every bacterium. This algorithm is executed to find the best solution with highest crowding distance and lower rank in the selection phase too.

In [54], for active distribution network (ADN), a model with environmental protection and energy protection is proposed. This model has multiple objectives such as minimization of the network loss, difference of peak, valley load and voltage deviation. This is a two stage algorithm, first a set of pareto solutions from the hybrid of PSO with BFO are obtained followed by the optimal schedule model of ADN by the evaluation of pareto solutions with entropy weigh decision-making method before determining the optimal scheduling strategy. For information exchange, a two-value crossover operator is used among sub-populations which update the related particles position to avoid the convergence to local optimal solution. The convergence speed of this algorithm is improved with the help of the adaptive adjusting inertia constant strategy. The objectives of ADN optimization scheduling are the minimization of run loss and maximized utilization of green renewable energy. The objective functions of optimal ADN scheduling are minimized power loss of the system, minimized peak-valley difference and minimum voltage deviation. These objectives have to assure equality constraints of power flow equation constraints and inequality constraints including all the variables within their limits.

This work is the hybrid of PSO and BFO algorithms with the advantages of both along with the elite archiving strategy. Pareto optimal solutions of PSO-BFO algorithm are obtained first followed by its evaluation with the help of the entropy weight decision making method to determine the optimal scheduling strategy.

In [44], a version of Bacterial Foraging Optimization was applied to align the biological sequences to produce non-dominated solutions. It employs multiple objectives namely, non-gap percentage, similarity maximization, gap penalty minimization and conserved blocks. In this work, two algorithms have been proposed: Hybrid Genetic Algorithm with ABC algorithm and Bacterial Foraging Optimization. Next algorithm proposed is MO-BFO (Multi-Objective Bacterial Foraging Optimization Algorithm) was compared with MSA methods MAFFT, Kalign, MUSCLE, Clustal Omega, Ant Colony Optimization (ACO), Genetic Algorithm (GA), Particle Swarm Optimization (PSO), Artificial Bee Colony (ABC) and Hybrid Genetic Algorithm with Artificial Bee Colony (GA-ABC). In this work, the combination of gap penalty, similarity, non-gap percentage and conserved blocks was used as multi objective to obtain a non-dominated optimal alignment. Getting an optimal alignment score by aligning the protein or DNA sequences is the major idea of multiple sequence alignment problem. The MOBFO algorithm uses BAliBASE3.0 benchmark datasets with 6255 protein sequences. This database also contains five reference sets.

5.5 ANALYSIS

The findings from this survey are summarized below.

5.5.1 MULTI OBJECTIVE OPTIMIZATION

The multi objective bacterial foraging algorithms used in this survey for solving the real-world optimization problems uses four approaches. The popular approach is the weighted-sum approach where the multiple objectives along with their weights, converted into a single objective function and [17, 26, 29, 47, 19, 51] use weighted sum technique with slight modification in their algorithms. Pareto-dominance is the important approach where two objective vectors are compared precisely without adding any preference to the problem definition as in [38, 43, 48, 52]. The third multi-objectives solving approach is the non-dominated sorting approaches and its associated elite and archive mechanisms are used to solve multiple objectives of the real world problems as in [1, 23, 39] with significant variations in the approaches. Hybrid approaches using multiple SI algorithms for their unique features and hybridize them to get the best out of them [32, 44, 54].

5.5.2 MODIFICATIONS TO MOBFO

The parameter of BFOA, modified in multi objective BFOA is the run length parameter. In [17], it was a user-defined parameter. The operators included are the crossover operator [32], features added are the crowding distance, elite strategy, archive strategy. Another improvement made to BFOA while solving multiple objectives are non-dominated sorting feature and the elite individuals addition and archive strategy to store the non-dominated individuals [48].

5.5.3 MULTI OBJECTIVE PROBLEMS CONSIDERED

The multi objective bacterial foraging algorithm and their variants are applied to the real-world problems such as environmental economic dispatch problem [39], short-term hydro thermal scheduling [17], three variations of multi objective BFOA are tested with benchmark problems, optimal design of single phase transformer [47], multiple sequence alignment with multi objectives, multi criteria based grid scheduling problem [23], power factor correction using interleaved DC-DC sepic converter [26], parallel cell entropy for aluminum electrolysis production process [52], active distribution network [54], mechanical design problem [43] and for the resin bonded sand mould problems [19].

5.5.4 FUTURE RESEARCH PATHS

As a result of this survey, the suggestions proposed for modifications using multi-objective BFOA are listed below:

1. Other bacteria life activities can be incorporated for new features to the existing algorithms to improve the current exploration and exploitation capabilities.

2. A broader set of benchmark problems could be solved and evaluated using performance metrics.

3. More hybrid approaches can be considered with the new features of the recent SI algorithms and evolutionary algorithms

4. In-depth analysis of many objective algorithms and incorporate the many objectives solving feature to the multi objective BFOA

5.6 CONCLUSION

This chapter presented a survey of various bacterial foraging algorithms which help to solve multiple objectives in real-world optimization problems. After a short introduction about the existing BFO algorithm and its salient features to solve single objective optimization, details about multi objective BFOA are discussed. Existing multi objective BFO algorithms are categorized into four approaches. These approaches are described through the existing literature and are discussed in detail. Different features of each of the four approaches are analyzed (step size, run length parameter, elitism, archive, non-dominated sorting approach, crowding distance operator, etc.) and discussed in detail. Moreover, exploration and exploitation techniques, operators are highlighted from each algorithm and its usefulness is analyzed. Multi objective BFOA can be still enhanced using the features of evolutionary algorithms or SI algorithm. To conclude, multi-objective BFOA is not used with more and more real-world applications which should be encouraged. Therefore, more research work is expected to happen and solve multi objective optimization problems. Even this algorithm can also be improved and applied to solve many objectives real-world optimization problems.

REFERENCES

1. M. Alejandra Guzmán, A. Delgado, and J. De Carvalho. A novel multiobjective optimization algorithm based on bacterial chemotaxis. *Engineering Applications of Artificial Intelligence*, 23(3):292–301.

2. H. Beyer and H. Schwefel. Evolution strategies–a comprehensive introduction. *Natural Computing*, 1:3–52.

3. S Binitha, S Siva Sathya, et al. A survey of bio inspired optimization algorithms. *International Journal of Soft Computing and Engineering*, 2(2):137–151, 2012.

4. Eric Bonabeau and G.T. Marco Dorigo. *Swarm Intelligence: From Natural to Artificial Systems*. Oxford University Press.

5. A. Britto and A. Pozo. I-mopso: A suitable pso algorithm for many-objective optimization. In *Proceedings - Brazilian Symposium on Neural Networks, SBRN*, page 166–171.

6. Jing, Li, Zhang Zhuo-qun, Zhang Li-li, and Shao Kang-jie. Multi-objective ant colony optimization algorithm based on discrete variables. In IOP Conference Series: Earth and Environmental Science, vol. 189, no. 4, p. 042031. IOP Publishing, 2018.

7. CA Coello Coello and Maximino Salazar Lechuga. Mopso: A proposal for multiple objective particle swarm optimization. In *Proceedings of the 2002 Congress on Evolutionary Computation. CEC'02 (Cat. No. 02TH8600)*, volume 2, pages 1051–1056. IEEE, 2002.

8. C.C. Coello. Evolutionary multi-objective optimization: some current research trends and topics that remain to be explored. *Frontiers of Computer Science in China*, 3(1):18–30.

9. D. Corne, N. Jerram, J.D. Knowles, J. Martin, and Oates. Pesa-ii: Region-based selection in evolutionary multiobjective optimization. In *3rd Annual Conference on Genetic and Evolutionary Computation*, page 283–290.

10. K. Deb, A. Pratap, S. Agarwal, and T. Meyarivan. A fast and elitist multiobjective genetic algorithm: Nsga-ii. *IEEE Transactions on Evolutionary Computation*, 6(2):182–197.

11. R. Vasundhara Devi and S. Siva Sathya. Biofilm algorithm for global numerical optimization. *International Journal of Innovative Technology and Exploring Engineering ISSN.*, 8:23–29, 2019.

12. R.V. Devi, S. Siva Sathya, and M.S. Coumar. Multi-objective biofilm algorithm (mobifi) for de novo drug design with special focus to anti-diabetic drugs. *Applied Soft Computing*, 96:106655.

13. R. Vasundhara Devi and S. Siva Sathya. Multi-objective monkey algorithm for drug design. *I.J. Intelligent Systems and Applications*, 3(March):31–41.

14. Tansel Dökeroglu, Ender Sevinç, Tayfun Kucukyilmaz, and Ahmet Cosar. A survey on new generation metaheuristic algorithms. *Computers & Industrial Engineering*, 137, 2019.

15. M. Dorigo, V. Maniezzo, and A. Colorni. Ant system : Optimization by a colony of cooperating agents. *IEEE Transactions on Systems, Man, and Cybernetics*, 26(1):13.

16. M. Eisenbach. Towards understanding the molecular mechanism of sperm chemotaxis. *The Journal of General Physiology*, 124(2):105.

17. I.A. Farhat and M.E. El-Hawary. Multi-objective short-term hydro-thermal scheduling using bacterial foraging algorithm. *IEEE Electrical Power and Energy Conference*, page 176–181.

18. L.J. Fogel, A.J. Owens, and M.J. Walsh. Artificial intelligence through a simulation of evolution. In *Evolutionary Computation: The Fossil Record*.

19. Timothy Ganesan, Pandian Vasant, and Irraivan Elamvazuthi. Non-gaussian random generators in bacteria foraging algorithm for multiobjective optimization. *arXiv preprint arXiv:1605.07364*, 2016.

20. E. Gerhardt and H.M. Gomes. Artificial bee colony (abc) algorithm for engineering optimization problems. In *3rd International Conference on Engineering Optimization*, page 1–11, Brazil. Rio de Janeiro.

21. D.E. Goldberg. *Genetic Algorithms in Search, Optimization, and Machine Learning*. Addison-Wesley Longman Publishing Co.

22. A. Ibrahim, I. Member, M. Vargas Martin, S. Rahnamayan, and K. Deb. EliteNSGA-III: An Improved Evolutionary Many-Objective Optimization Algorithm.

23. M. Kaur. Elitist multi-objective bacterial foraging evolutionary algorithm for multi-criteria based grid scheduling problem by. In *International Conference on Internet of Things and Applications (IOTA*, page 431–436.

24. J. Kennedy and R. Eberhart. Particle swarm optimization. In *Proceedings of the IEEE International Conference on Neural Networks*, volume 4, page 1942–1948.

25. J.D. Knowles and D.W. Corne. Approximating the nondominated front using the pareto archived evolution strategy. *Evolutionary Computation*, 8(2):149–172.

26. C. Komathi. Multi objective bacterial foraging optimization algorithm for power factor correction using interleaved dc-dc sepic converter. In *Trends in Industrial Measurement and Automation (TIMA*, page 7.

27. A. Konak, D.W. Coit, and A.E. Smith. Multi-objective optimization using genetic algorithms: A tutorial. *Reliability Engineering and System Safety*, 91(9):992–1007.

28. J.R. Koza. Genetic programming on the programming of computers by means of natural selection. In *Biosystems*, volume 33.

29. P. Krishnapriya and M.R. Sindhu. Multi-objective optimal design and control of auto-tuned passive filter using bacterial foraging algorithm to improve power quality and to minimise power losses. *IJCTA*, 8(5):2013–2020.

30. S. Kudo, S. Tamura, T. Nakajima, H. Yamano, H. Kusaka, and H. Watanabe. Diagnosis of colorectal tumorous lesions by magnifying endoscopy. *Gastrointestinal Endoscopy*, 44(1):8–14.

31. X. Li. A non-dominated sorting particle swarm optimizer for multiobjective optimization. *Genetic and Evolutionary Computation — GECCO*, 2723:37–48.

32. P. Manikandan and D. Ramyachitra. Bacterial foraging optimization – genetic algorithm for multiple sequence alignment with multi-objectives. *Scientific Reports*, 7(April):1–14.

33. Seyedali Mirjalili, P. Jangir, and S. Saremi. Multi-objective ant lion optimizer : a multi-objective optimization algorithm for solving engineering problems. *Applied Intelligence*, 46:79–95.

34. Seyedali Mirjalili and A. Lewis. The whale optimization algorithm. *Advances in Engineering Software*, 95:51–67.

35. Seyedali Mirjalili, S. Mohammad, and A. Lewis. Grey wolf optimizer. *Advances in Engineering Software*, 69:46–61.

36. Seyedali Mirjalili, S. Saremi, and S. Mohammad. Multi-objective grey wolf optimizer: A novel algorithm for multi-criterion optimization. *Expert Systems with Applications*, 47:106–119.

37. B. Niu, H. Wang, L. Tan, and J. Xu. Multi-objective optimization using bfo algorithm. *Neuro Computing*, 116:582–587.

38. B. Niu, H. Wang, J. Wang, and L. Tan. Multi-objective bacterial foraging optimization. *Neurocomputing*, 116:336–345.

39. V.R. Pandi, B.K. Panigrahi, W.C. Hong, and R. Sharma. A multiobjective bacterial foraging algorithm to solve the environmental economic dispatch problem. *Energy Sources Part B-Economics Planning and Policy*, 9(3):236–247.

40. Kevin M. Passino. Bacterial foraging optimization. *International Journal of Swarm Intelligence Research*, 1(1):1–16.

41. K.M. Passino. Biomimicry of bacterial foraging for distributed optimization and control. *Control Systems, IEEE*, 22(3):52–67.

42. R. Poli, J. Kennedy, and T. Blackwell. Particle swarm optimization an overview. *Swarm Intelligence*, 1:33–57.

43. E.A. Portilla-flores. Optimization of a mechanical design problem with the modified bacterial foraging algorithm. In *Congreso Argentino de Ciencias de La Computación*, page 171–180.

44. R.R. Rani and D. Ramyachitra. Multiple sequence alignment using multi-objective based bacterial foraging optimization algorithm. *BioSystems*, 150:177–189.

45. M. Reyes-sierra and C.a C. Coello. Multi-objective particle swarm optimizers : A survey. *Departamento de Ingenieria Electrica Sección de Computación*, 2(March):48.

46. S. Saremi, S. Mirjalili, and A. Lewis. Grasshopper optimisation algorithm : Theory and application. *Advances in Engineering Software*, 105:30–47.

47. S. Subramanian and S. Padma. Bacterial foraging algorithm based multiobjective optimal design of single phase transformer. *Journal of Computer Science and Engineering*, 6(2):2–7.

48. C. Yang and J. Ji. Multiobjective bacterial foraging optimization using archive strategy. In *ICPRAM*, page 185–192.

49. X.-S. Yang and S. Deb. Engineering optimisation by cuckoo search. *International Journal of Mathematical Modelling and Numerical Optimisation*, 1(4):330–343.

50. Xin She Yang. A new metaheuristic bat-inspired algorithm. *Studies in Computational Intelligence*, 284:65–74.

51. X.S. Yang, M. Karamanoglu, and X.S. He. Multi-objective flower algorithm for optimization. *Procedia Computer Science*, 18:861–868.

52. J. Yi, D. Huang, S. Fu, H. He, and T. Li. Multi-objective bacterial foraging optimization algorithm based on parallel cell entropy for aluminum electrolysis production process. *IEEE Transactions on Industrial Electronics*, 63(4):2488–2500.

53. S. Zahra, M. Seyedali, M. Shahrzad, S. Hossam, and I. Aljarah. Grasshopper optimization algorithm for multi-objective optimization problems. *Applied Intelligence*, 48:805–820.

54. F. Zhao, J. Si, and J. Wang. Research on optimal schedule strategy for active distribution network using particle swarm optimization combined with bacterial foraging algorithm. *International Journal of Electrical Power & Energy Systems*, 78:637–646.

55. R. Zhao. Monkey algorithm for global numerical optimization. *Journal of Uncertain Systems*, 2(3):165–176.

56. E. Zitzler, M. Laumanns, and L. Thiele. Spea2: Improving the strength pareto evolutionary algorithm. In *Evolutionary Methods for Design Optimization and Control with Applications to Industrial Problems*, page 95–100.

6 Artificial Intelligence for Biomedical Informatics

Shahid Azim, Samridhi Dev
School of Computer and Systems Sciences,
Jawaharlal Nehru University, Delhi, India

Sushil Kumar and Aditi Sharan
School of Computer and Systems Sciences,
Jawaharlal Nehru University, Delhi, India

CONTENTS

DOI: 10.1201/9781003246688-6

6.1 INTRODUCTION

"Biomedical informatics (BMI) is the inter-disciplinary field that studies and pursues the effective uses of biomedical data, information, and knowledge for scientific inquiry, problem solving, and decision making, driven by efforts to improve human health" [15]. They lay an earliest grounding work for defining the scope and breadth of biomedical informatics and specify core competencies for teaching biomedical informatics as a graduate level course. Biomedical informatics consists of mainly three sub-disciplines like health informatics, bioinformatics and translational informatics. Clinical and public health informatics are the two types of health informatics, which addresses the health concerns at individual and population levels, respectively. These are briefly described below:

Clinical Informatics – here the goal is to improve the health of individual patient by developing intelligent systems for clinical decision support, recommendation, reminders and checklists.

Public Health Informatics – aims at improving the health of population and communities in general. For example, monitoring disease indicators in real-time or near real-time to detect outbreaks of Covid-19.

Bioinformatics – it focuses at a lower biological (molecular, cellular) level. For example DNA and protein sequencing.

Translational Informatics – aims at bridging the gap between discoveries in laboratory and clinical practices. Using genomic and cellular mechanisms to explain clinical phenomena is one of the examples. Characteristics of Biomedical Data is as follows:

- Unstructured
- Heterogeneous
- Multimodality
- Vastness/Massive (Volume)
- Generated at very fast speed (Velocity)
- Specialized Discipline–Reliance on domain experts

The varied type (Figure 6.1) and characteristics of biomedical data and the need for utilizing it effectively for better health-care delivery requires capability that greatly exceeds human cognitive capacity and to support the delivery of better healthcare, Artificial intelligence (AI) is envisaged to have an important and supportive role in human cognition. In healthcare, AI has the potential to enhance patient and clinical team outcomes at lower cost, and influence public health.

6.1.0.0.1 Challenges to the advancement and application of AI tools

Challenges with respect to AI Techniques and Reliance

- Data Quality, Availability and Access
- Lack of Labeled and Benchmarked Data
- Demand for High Performance/ Errorless Systems
- Poor Evaluation Measures

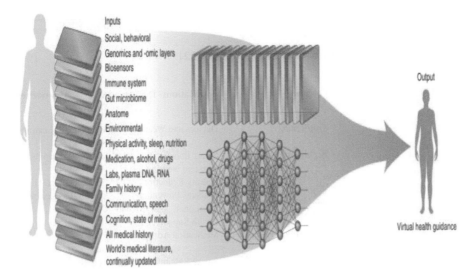

Figure 6.1 The virtual medical coach model using multimodal data inputs and algorithms to deliver individualized guidance [31].

- De-Identification and Privacy Concerns
- Transparency and Reproducibility
- Explainability and Interpretability
- Fairness and Bias
- Other ethical concerns like accountability

Challenges with respect to Adoption and Integration in Clinical Settings

- Workflow integration
- Availability at point of care
- Integration and inter-operability between systems
- Education of patients and clinicians

6.2 RECENT ADVANCES IN MACHINE LEARNING AND AI IN HEALTH

After a long AI-Winter, AI began its resurgence to prominence due to the success of machine learning esp. deep learning techniques as well as significant increases in computational power using GPUs, TPUs and modern computational and storage infrastructure, in the form of cloud computing. These advances have fueled the growth in the domain of natural language processing and computer vision. Table 6.1 summarizes some of the applications of AI tools and techniques in health.

One of the important applications of AI Tools and Techniques is Natural language processing (NLP). It has become a fundamental tool for biomedical informatics (BMI) research, with uses ranging from mining new protein–protein interactions

Table 6.1

Examples of AI applications in healthcare [22]

User Group	Category	Applications/ Devices	AI Approaches/ Techniques
	Risk/Benefit	Devices and Wearables	Machine Learning, NLP
Patients and families	Health Monitoring	Smart Phones and tablet apps and websites	Speech Recognition and Chatbots
	Disease Prevention and Management	Obesity Reduction	Conversational AI
		Diabetes Prevention	Speech Recognition
		Emotional and Mental Health Support	Chatbots
	Medication Management	Medication Adherence	Robotic Home Tele Health
	Rehabilitation	Stroke Rehabilitation using apps and robots	Robotics
Clinical Care Teams	Early Detection, Prevention and Diagnostic Tools	Imagingg for cardiac arrhythmia	Machine Learning
		Early Cancer Detection	Computer Vision
	Surgical Procedure	Robotic Surgery	Robotics
		AI-Supported Surgical Road Map	Machine Learning
	Precision Medicine	Personalized Chemotherapy Treatment	Information Retrieval and NLP
	Patient Safety	Early Detection of Sepsis	Machine Learning
Public Health Program Managers	Identification of Individuals at risk	Suicide risk identification using social media	Deep learning (convolutional and recurrent neural networks)
	Population Health	Eldercare Monitoring	Ambient AI Sensors

from scientific literature to recommending medications in clinical care including classification and analysis of genes [28]. One of the common concerns in biomedical informatics esp. clinical informatics is – how to create clinical decision support systems or medical search engines that are capable of supporting both experts (e.g., physicians) and novices (e.g., patients) (e.g., patients and their next-of-kin) tackling complex tasks(e.g., searching for diagnosis, searching for a treatment). The underlying technology behind a search system is an information retrieval (IR) system. Current success in information retrieval is attributed to success in NLP, esp. document representation strategies; Transformer based language models like BERT and Knowledge Graphs. From now onwards we will limit our discussion to information retrieval (discussed in Section 6.3), named entity recognition, NER (in Section 6.4) and knowledge graphs (in Section 6.5).

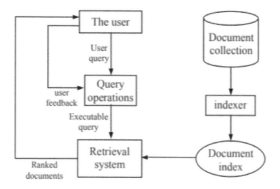

Figure 6.2   Typical information retrieval system [20].

6.3   INFORMATION RETRIEVAL (IR) SYSTEMS

Information retrieval (Figure 6.2) involves representing, searching and modifying large amounts of electronic text and other human-language data. To obtain information, the user must submit a natural-language query that identifies the data needed. After that, the IR system would return the proper records for the relevant information. In traditional IR architecture, Gerard Salton and his colleagues presented the vector space model, which is based on Luhn's similarity criterion [29]. "The more two representations overlap in given elements and distribution, the more probable they are to represent identical information," according to Luhn's similarity criterion. Data representation, storage and retrieval are all research areas in IR from a practical perspective.

6.3.1   DOCUMENT COLLECTION AND REPRESENTATION

Document collection entails selecting documents and other items from various web resources to satisfy user requests, such as data, statistics, photos, maps, trademarks, sounds, and so on. The content of a document can be depicted by a collection of terms, such as words, phrases, or other units. Every phrase will be allocated a weight, signifying the importance of the document. The following are some examples of how terms can be depicted.

6.3.1.0.1   Bag of words

The bag of words view ignores the specific ordering of terms in a text, but it visualizes the number of times each phrase appears. Only the number of times each term appears is taken into consideration. In Figure 6.3, the word arrangement in the phrases like "John likes movies" and "Mary likes movies too" can simply be considered that the word "John" occurred 1 time, word "likes" occurred 2 times, and so on. Documents and queries both can be represented in the bag of words view. The cosine

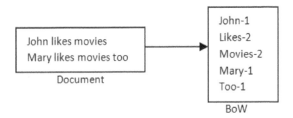

Figure 6.3 Bag of words (BoW).

similarity between a query vector and a document vector can be used to calculate the document's score for a given query.

6.3.1.0.2   Term document matrix

The term-document matrix could be seen by viewing a collection of N documents as a collection of vectors. It is an $M \times N$ matrix, with M denoting the number of documents and N denoting the total number of phrases that each document corresponds to.

6.3.2   TERM WEIGHTING AND DOCUMENT SCORING

The frequency of a word in a given document is defined as the count of its occurrences in the document. Each word in a document is given a weight based on its frequency. Based on the weight of t in d, we would choose to compute a similarity score between a query term t and a document d. The easiest method is to set the weight equal to the number of times word t emerged in document d. The set of weights for document d is calculated by retaining the documents in the Bag of Word View. This scheme of weighting is known as term frequency denoted by $Tf_{(t,d)}$, for the term t in the document d. The Tf-Idf weighting and the BM25 weighting systems are the most often utilized term weighting schemes.

6.3.2.0.1   Tf-Idf weighting

The term frequency denoted by Tf and the inverse document frequency denoted by Idf are combined to form Tf-Idf. Term frequency tells us count of a term appearing in a document. We commonly use log 10 of the word frequency instead of the actual occurrence. We can't calculate the frequency of phrases that appear 0 times in a document because we're using the log function for weighting. As a result, we usually add 1 to the count, altering the function to:

$$Tf_{(t,d)} = \log(count(t,d) + 1) \tag{6.1}$$

The document frequency dft of a term t is the count of documents containing that term. When trying to score documents for the term, it is preferable to utilize a document-level statistic rather than a collection-wide statistic. The inverse document frequency (Idf) of a term t is calculated by dividing the total number of documents in a collection denoted by N, and the count of documents in which term t emerges df_t.

$$Idf_t = \log \frac{N}{df_t} \qquad (6.2)$$

Then the product of term frequency $Tf_{(t,d)}$, and Idf_t is the $Tf\text{-}Idf$ value for word t in document d:

$$Tf\text{-}Idf_{(t,d)} = Tf_{(t,d)} \times Idf_t \qquad (6.3)$$

The cosine similarity function can be used to determine how similar the query and document vectors are. The cosine similarity metric is used to determine how similar papers are regardless of their size.

$$Score(q,d) = \sum_{t \in q} \frac{Tf\text{-}Idf_{(t,d)}}{|d|} \qquad (6.4)$$

6.3.2.0.2 BM25 weighting scheme

BM25 is a retrieval function that aligns a set of documents by providing them a ranking based on the query phrases that exist in each document. A parameter k is included in BM25 to achieve a equilibrium between term frequency and Idf, and a parameter b is added to determine the relevance of document length normalization. The BM25 score for a document d can be calculated using a given query vector q:

$$Score(d,q) = \sum_{t \in q} \log \left(\frac{N}{df_t} \right) \frac{Tf_{(t,d)}}{k \left(1 - b + \left(\frac{|d|}{|d|_{avg}} \right) \right) + Tf_{(t,d)}} \qquad (6.5)$$

6.3.3 INVERTED INDEXING

The inverted index maintains all of the terms that appeared in the document set, as well as references to all of those terms' occurrences across all documents. Inverted indexes, which can be considered as a list of keywords and links to relevant pages, are generally used to provide quick, concurrent query processing. It is divided into two sections: a dictionary and postings. The dictionary is a list of terms, each of which leads to a list of postings for that keyword. A posts list is a list of document IDs connected with each term, which can also include information such as term frequency or exact position of terms in the document.

6.3.4 EVALUATION OF IR SYSTEMS

In an information retrieval system, evaluation is a very important and time-consuming activity. The performance of ranked retrieval systems can be assessed by assuming that every document retrieved by the model is either relevant or not relevant to our needs. Following metrics can be utilized to evaluate the system.

6.3.4.0.1 Precision

Precision (P) defines the retrieved documents that are relevant

$$Precision = \frac{relevant\ items\ obtained}{obtained\ items} \tag{6.6}$$

6.3.4.0.2 Recall

Recall (R) defines the relevant documents that are fetched

$$Recall = \frac{relevant\ items\ obtained}{relevant\ items} \tag{6.7}$$

6.3.4.0.3 Mean Average Precision

The precision for each locale in the ranking where a relevant item is retrieved is averaged to get the average precision [36].

$$MAP = \frac{1}{n(Re)} = \sum Rek \sum_{i=1}^{k} \frac{Rei}{k} \tag{6.8}$$

where n(Re) is the total relevant items.

6.3.4.0.4 Precision at k

P@k is the precision defined at a cut-off point named k. This metric quantifies how many items in the top-K results were relevant.

$$P@k = \frac{true\ positives@k}{true\ positives@k + (false\ positives@k)} \tag{6.9}$$

6.3.5 PROBLEMS WITH TRADITIONAL IR MODELS

Boolean models are unable to incorporate the weighting of terms and expect boolean queries only which makes it suitable for highly skilled users only. Also, there is no notion of ranking of search results. Probabilistic models and vector space models suffer from a problem: that they work only if it finds an exact match of words between the query and document (Vocabulary Mismatch Problem). In other words, they are unable to capture the semantics between the query and the documents; instead it focuses only on matching the keywords between query and documents, which is a serious drawback for an intelligent system. Also, the text representation in these models especially Vector Space Model is highly sparse and has high dimensionality which makes it computationally intensive.

6.3.6 QUERY EXPANSION

In order to deal with vocabulary mismatch problem of IR, Query expansion(QE) has been used by the researchers to improve retrieval efficiency. In most general way, query expansion involves selection of some source for providing expansion terms, selecting the appropriate terms for expansion based on similarity between the original and the candidate expansion terms, reformulating the query by weighting/ reweighting of original and expanded terms. Three main source options for selecting QE terms: (i) manually extracted knowledge resources inclusive of thesauri, dictionaries, and ontologies; (ii) the documents needed in the retrieval process; (iii) external text clusters and resources. Based on these sources query expansion can be categorized as knowledge based or corpus based. In biomedical domain, because of the availability of biomedical knowledge bases like Mesh, UMLS, Gene-ontology etc., mostly knowledge based query expansion has been performed by the researchers.

6.3.7 RECENT ADVANCEMENTS IN IR

Recent advancement in deep learning-based techniques is able to overcome these problems to some extent. These advancements are discussed below-

6.3.7.0.1 Advancement in Text Representation Technique

The first step towards capturing the semantic relationship between query and document is the "notion of similarity" between the words of query and the documents in the corpus. The idea that a word is characterized by the company it keeps popularized by Firth, J. R. 1957 [7], gave rise to vector semantics where the semantic representations is learned from the distribution of word neighbors/context words in a large corpora. Two words are considered similar if they are surrounded by same context words in large corpora. And hence have similar vector representations. The vectors learned this way are called word embeddings. The problem that word embedding technique tries to solve is "how to come up with the vector representation of words such that two words having similar meaning should have similar vector representations?" For example vector representation of "cat" and "dog" should be closer/less distant than that of "cat" and "car". Popular word embeddings are discussed below.

Word2Vec Word Vectors were popularized by [23], as they proposed an efficient way of estimating word representations in vector space. They proposed two new model architectures of their neural language model: continuous bag-of-words (CBOW) and skip-gram as shown in Figure 6.4. They also provide the implementation of their model in a software package called "Word2Vec". CBOW predicts the current word given the context words while Skip-Gram predicts the surrounding words given the current word as illustrated below.

The word2vec methods can be trained efficiently at a very fast speed, and easily available online with code and pre-trained embeddings and scalable for billion words. These word embeddings (short and dense) were able to capture many syntactic and semantic relationship between two words. For example one could easily show algebraically that if you add word vectors of "King" and "Woman" and subtract word vector of "Man" resulting vector will be most similar to the vector representation of the word "Queen".

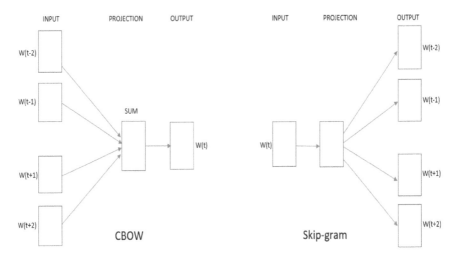

Figure 6.4   CBOW and skip-gram model [24].

Glove Apart from word2vec, Glove is most widely used static embedding model [26], short for Global Vectors, as the model focusses on acquiring global corpus statistics. It utilizes the fact that count-based approaches can be effective in capturing global information while still maintaining the relevant linear substructures found in prediction-based methods such as word2vec. It's based on probability ratios of word co-occurrence.

Sentence and Document Embeddings (Doc2Vec) Le and Mikolov [16] proposes paragraph vector, an unsupervised approach for getting fixed-length feature representations out of variable-length texts like sentences, paragraphs, and documents as shown in Figure 6.5. They show that when it comes to broader text representation, paragraph vectors outperform bag-of-words models and other strategies.

Contextualized Word Vectors (Language Modeling Based) Word2vec embeddings are static embeddings, which means that the method learns one fixed vector representation called embedding for each word in the vocabulary. Static vectors for the word "Bank" are same in the context of "River Bank" and "Financial Bank". Contextualized word vectors are able to differentiate between these two contexts. Replacing static vectors with contextualized word representations has led to significant improvements on virtually every NLP task. Contextualized word vectors can be obtained from transformer-based models like BERT. BERT [5] provides a masked language modeling (MLM) approach, in which some tokens in the input sequence are randomly masked, and the goal is to anticipate these masked positions using the corrupted sequence as input. During pre-training, BERT uses a Transformer encoder to

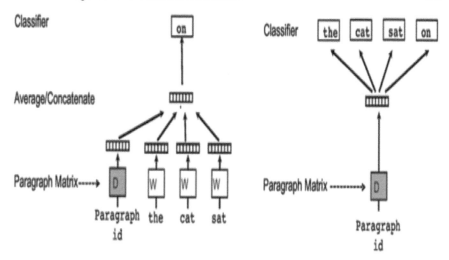

Figure 6.5   Doc2Vec [16].

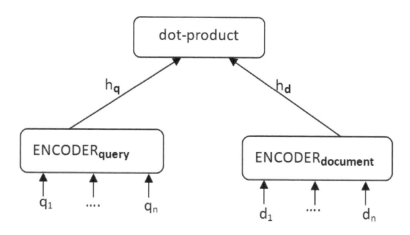

Figure 6.6   BERT bi-encoder for computation of relevance of a document to a query.

focus on bi-directional contexts. Figure 6.6 below depicts how query and document can be represented by encoder of BERT model and later matched for similarity using dot-product [13].

6.3.8   ADVANCEMENT IN SEMANTIC MATCHING

Semantic matching techniques aim to compare two sentences to determine if they have a similar meaning. For example, the questions "Where do you sleep?" and "Where are you sleeping?" have almost the same words in them so we can say that

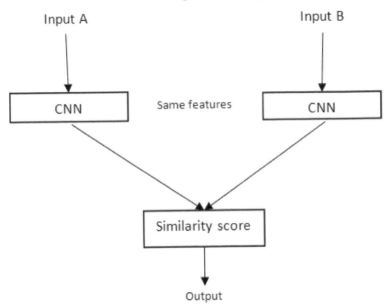

Figure 6.7 Siamese model architecture.

they are asking the same question. But if you consider another question "where are you sleeping?" this also looks similar to the last question but has an entirely different meaning. Even in some cases, the words may totally not match but the questions are the same such as "How aged are you?" and "What is your age?" are exactly two same questions but have so common words. So in process of semantic matching, we train a network that returns a high similarity score when the questions are similar and a low similarity score when the questions are different.

6.3.8.0.1   Models for semantic matching

Siamese Neural Networks A Siamese neural network (Figure 6.7), also called a twin neural network, is a kind of artificial neural network made up of two or more identical sub-networks with the same topology and weights. One of the output vectors is typically pre-computed, which serves as a benchmark against which the other output vector is assessed. In a Siamese network, we keep the fundamental network for acquiring entity features while passing the two entities we want to compare through it. At the end of a common network, we obtain a vectored representation of our input, which we can then use to test or quantify their similarity.

MNet-Sim It's a multilayered semantic similarity network approach that compares sentences using a variety of similarity measures. Phrasal Overlap, Cosine Similarity, Euclidean distance, Jaccard similarity and Word mover's distance are currently the dimensional parameters employed in this model. In this model, the Cosine Similarity score between sentence vector pairs relate to the upper-most layer. In the intermediate layer, the phrasal overlap network will appear. The bottom-most layer, as well as any expanding layers, may be linked to other variables.

Sentence Transformers Sentence transformers are a group of tools for creating dense vector representations of sentences or paragraphs. These models are fine-tuned transformer networks for semantic textual similarity. There are several models available with efficient accuracy but in those models, the semantic textual similarity is considered as a regression task. This means whenever we need to calculate the similarity score between two sentences, we need to pass them together into the model and the model outputs the numerical score between them. For example, you need to search over say 1000 documents, you would need to perform 1000 separate inference computations, it is not possible to compute the embeddings separately. To overcome this problem sentence transformers are used.

Multi-Perspective Convolutional Neural Networks Multi-Perspective Convolutional Neural Networks were proposed by [9]. It's a paradigm for examining sentences that employ a variety of viewpoints. They first utilize a convolutional neural network to model each sentence, extracting features at many levels of granularity and employing multiple pooling, and then compare our sentence representations at various granularities using multiple similarity measures. In the MPCNN model, the two sentences are processed independently of each other and have no interaction until the full-connected layer, which causes a loss of useful information.

6.3.9 ADVANCEMENT IN QUERY EXPANSION

6.3.9.0.1 *Deep Learning-based Query Expansion*

Various deep learning-based algorithms for learning high-quality vector representations of terms from vast amounts of unstructured text data have recently been developed. Because they are learned from hundreds of millions of words, deep learning vectors could be a source of query expansion. Artificial neural networks, deep belief networks, word2vec, and other deep learning models are currently in use. Deep learning vectors are useful for overcoming over-expansion or the initial query, as well as lowering the negative impact of additional terms. Using a deep belief network, the model [11] extracts relevant phrases to expand a query and removes irrelevant terms from the query. The query expansion method based on pseudo correlation feedback with deep learning [18] where documents were trained using Word2vec tool and have better performance than the traditional pseudo-correlation feedback algorithm on query expansion. In order to minimize the effects of insufficient training data, a siamese neural network architecture was utilized to pick optimal expansion terms by learning latent properties included within word embeddings that are responsible for term efficacy or ineffectiveness when employed for query expansion [12].

6.3.9.0.2 Hybrid Methods

Knowledge-based methods are used in conjunction with corpus-based methods to improve the effectiveness of the query expansion approach. To create expansion characteristics, the corpus-based method relies on term frequency, but it ignores meaning and term dependency. The semantic technique, on the other hand, is context-dependent, but it suffers from the restricted number of terms and relationships in the context. The most promising solution is a hybrid approach that combines a statistical method for generating expansion terms with a semantic method for ensuring the correct order of phrases in their context.

Knowledge-based relevance feedback (KBRF) [25] is a hybrid framework for automatic query expansion that enhances the PRF process by applying external knowledge to measure the semantic relatedness of the query to the top-ranked articles and re-ranking the most relevant documents higher in respect to the user information demand. Rather than utilizing precise matching methodologies, it analyses the semantic relatedness of the top-ranked documents of the initial query, combining the benefits of pseudo relevance feedback and knowledge-based relevance feedback.

6.3.9.0.3 Query Expansion Example

We are presenting here the results of query expansion performed on the Covid-19 - related information retrieval dataset: CORD-19.

CORD-19 [32] is a benchmarked IR database released by TREC. The dataset consists of over 500,000 scientific papers regarding COVID-19, SARS-CoV-2, and similar viruses, including over 200,000 with full text, along with 50 covid related queries with their relevance judgment. This publicly available dataset is made available to the world's researchers so that they may use current advancements in natural language processing and other AI approaches to create new insights in support of the ongoing fight against this deadly disease.

We have proposed a model for improving retrieval efficiency for retrieving covid related documents based on the user's query. The proposed query expansion uses the domain-specific resources and pre-trained vectors for expanding the query. The proposed expansion catches the semantic relationship between documents and queries. To catch this semantic relationship, we performed query expansion by extracting named entities from the query and utilized these entities to extract similar/related terms from the MeSH ontology and used pre-trained bio-specific word2vec model. The results were compared with unexpanded queries. The experiments were performed on TREC-COVID dataset which contains fifty queries and relevant queries wise documents for the information retrieval task. Standard parameters were used for evaluation. Following are the results of our experiments as shown in Table 6.2.

Following is an example of query terms obtained for expansion using our proposed model-

- **Original Query** – coronavirus origin
- **Expanded Query** – coronavirus origin origins originated originating origination originate proto-bulgariansorgin h1a-m82 palaeoamericansoriginthe

Table 6.2
Shows the overall performance

Precision@K	Un-expanded Query	Expanded Query
P@5	21.20%	42.80%
P@10	18.20%	43.00%
P@15	18.27%	41.47%
P@20	17.90%	39.90%
P@30	16.40%	38.27%
P@100	12.60%	31.14%

2019 Novel Coronavirus 2019 Novel Coronavirus Disease 2019 Novel Coronavirus Disease Testing 2019 Novel Coronavirus Infection 2019 Novel Coronavirus Testing Amyotrophy, Thenar, Of Carpal Origin CNS Origin Vertigo Central Nervous System Origin Vertigo Congenital Amaurosis of Retinal Origin Delirium of Mixed Origin.

6.4 NAMED ENTITY RECOGNITION

Named entity recognition (NER) is a sub-task of the information extraction in which we try to classify the words in free text into pre-defined categories like people, places, organizations, events, etc.

6.4.1 NER APPLICATIONS

- Named entity recognition is extensively used in the biomedical domain such as DNA identification, gene identification, drug name identification, disease name identification, etc. from biomedical texts.
- Named entity recognition can be combined in the information retrieval model to optimize the retrieval process and the question-answering model to extract the relevant information.
- Named entity recognition is extensively used in social media domains such as opinion mining, text summarization, and finding the most relevant information.

6.4.2 NER TECHNIQUES

"NER consists of three different problems – the recognition of a named entity in text, the assignment of a class to this entity (gene, protein, drug, etc.), and the selection of a preferred term for naming the object in case that synonyms exist" [17]. Earlier traditional NER models were based on sequence labeling, such models are Hidden

Markov Models (HMM) and Conditional Random Fields (CRF). Performing hand-crafted feature extraction is a difficult and time-consuming task in natural language processing, many pieces of research show deep learning models do automated feature extraction which is better than the handcrafted feature extraction technique and less time-consuming, so if features extracted by the deep learning models are applied to traditional sequence labeling models it may show considerable good performance. Cho & Lee [4] proposed "contextual long short-term memory networks with CRF (CLSTM)" model for named entity recognition. They are incorporated n-gram with the BI-LSTM and CRF. BI-LSTM with CRF encoding performed better in Named Entity Recognition than the RNN model [2]. To handle the out of vocabulary problem character embedding can be incorporated with the deep learning models. And to capture the relationship between entities CRF encoding layer can be added at the output layer. Character embedding showed improvement in text mining in natural language processing and can also handle the inconsistency of words [1].

6.4.2.0.1 Problems with Traditional Methods of Named Entity Recognition

Traditional methods were heavily dependent on the quality of handcrafted features. In natural language processing extracting features from the text data is a hectic and time-consuming task. If we are unable to feed proper features to the model, it might not generalize well.

6.4.2.0.2 Hybridizing of Deep Learning with Traditional Model

The main advantage of deep learning technique is that it has overcome a lot of manual feature extraction and vectorization technique of input words. Earlier static representation of text data like Bag of Words, one-hot encoding, etc. was used which is independent of the context. With the coming of deep learning there is a major shift from the static representation of the text to context based vector representation.

Considering the popular deep learning architectures BI-LSTM and CNN, it was observed that BI-LSTM performed well for natural language processing-related tasks. In addition to word embedding, character embedding is found useful for handling out of vocabulary (OOV) problems. The context of the biomedical domain might have small variations like spelling, suffixes, prefixes, hyphenations, etc. Character embedding can also help deal with these variations. To extract the character embedding-based feature extraction CNN can be utilized.

We used CORD-19 dataset described earlier have performed four experiments using Bi-LSTM, Bi-LSTM-CRF, Bi-LSTM-CNN and Bi-LSTM-CNN-CRF.

6.4.3 DEEPLEARNING-BASED NER MODEL

We used CORD-19 dataset described earlier and performed four experiments using Bi-LSTM, Bi-LSTM-CRF, Bi-LSTM-CNN and Bi-LSTM-CNN-CRF. We concatenate the CNN-based character embedding and glove pre-trained embedding. The concatenated embedding was fed as the input to the Bi-LSTM. In this way, we modified the input representation to the Bi-LSTM. On other hand, we observed that the

CRF encoding model shows good performance for the traditional named entity extraction. However, major limitation being that, it required a lot of handcrafted feature design. With the emergence of the deep learning model, the output of the deep learning model may be utilized to provide proper features to traditional sequence labeling models like CRF. It has been observed that the CRF decoding-based output is more accurate than the general probabilistic classifier like Softmax. So, we fed the output of the Bi-LSTM to the CRF layer for predicting the tag of each term. The architecture of our proposed model is illustrated in Figure 6.8, and the results obtained are shown in Table 6.3.

Our experimental results shows that combining traditional models like CRF with deep learning based models like Bi-LSTM and CNN along with character embeddings gives better result for biomedical NER as compared to simpler Bi-LSTM-based deep learning models.

6.5   KNOWLEDGE GRAPHS

A knowledge graph is a type of data modeling that uses a graph to describe related data, with nodes representing data entities and edges indicating relationships between them. Facts are stored in knowledge graphs as SPO (subject, predicate, and object) triples, with subjects and objects representing knowledge entities and predicates representing knowledge relations. Biomedical ontologies have become increasingly significant in recent years for characterizing existing biological information as knowledge graphs. Traditionally, graphs with interconnected biological components have been employed to depict complex biological systems.

In knowledge graphs, exploring all associated paths is less scalable and time expensive. Knowledge graph embedding models are used since this method relies on path discovery. Embeddings work by learning low-rank vector representations of network nodes and edges that keep the graph's structural integrity. These embedding models are frequently utilized in bioinformatics tasks such as DNA mapping, genomic study and protein sequencing because of analytical skills such as learning clusters and similarity measurements.

6.5.1   KNOWLEDGE GRAPH EMBEDDING MODELS

Knowledge graph entities and relations are represented in low-rank representations by KGE models. Random embeddings are assigned to all entities and relations at first. They're then updated utilizing a multiphase learning technique that takes into account the true and corrupted triplets' matching embeddings. The embeddings are then processed to generate scores for all of the triplets using model-dependent scoring functions, followed by the computation of training loss using model-dependent loss functions such as gradient descent algorithm variants, with the goal of maximizing true triplet scores while minimizing corrupted triplet scores.

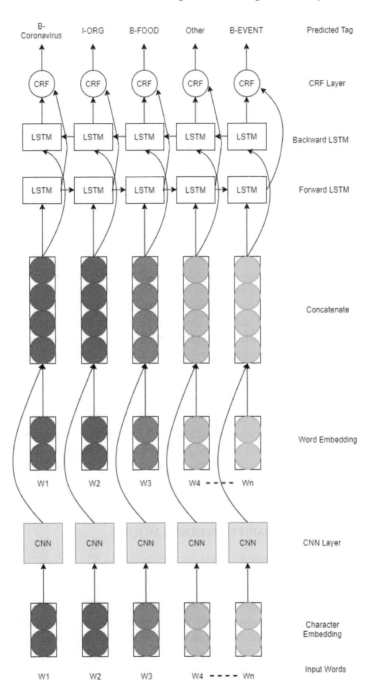

Figure 6.8   Architecture for NER (Bi-LSTM CRF with character embedding).

Table 6.3
Shows the results of NER models

Models	Precision	Recall	F1-score
BILSTM	0.75756	0.72035	0.73848
BILSTM-CRF	0.77162	0.79610	0.78367
BILSTM-CNN	0.76432	0.73097	0.74728
BILSTM-CNN-CRF	0.79513	0.80386	0.79950

Table 6.4
Databases for knowledge graph

Database Name	Format	Coverage
UNIPROT	Structured and unstructured	Proteins
Drug bank	Structured and unstructured	Drugs
Gene ontology	Structured	Genes
CTD	Structured and unstructured	Chemicals
ChEMBL	Structured and unstructured	Chemicals
HPA	Structured and unstructured	Proteins
STRING	Structured	Proteins
BIOGRID	Structured	Proteins
InAct	Structured	Proteins
InterPro	Structured	Proteins
PharmaGKB	Structured	Drugs
TTD	Structured	Drugs
Supertarget	Structured	Drugs

6.5.2 EXISTING BIOMEDICAL KNOWLEDGE GRAPH

Table 6.4 shows a popular biological knowledge graph in terms of format and coverage. These databases contain biological entities related to protein, genes, chemicals and drugs.

6.5.2.0.1 Contribution of knowledge graphs for enhancement in bioinformatics

Predicting Drug Target Interactions The study of drug targets has grown in popularity as a means of elucidating the mechanisms of current drugs' functions as well as their potential off-target effects. Knowing possible clinically significant targets is also essential in the process of rational drug development. With this knowledge, potential drugs targeting specific proteins can be designed to accomplish the desired therapeutic effects. DTI prediction on a large scale and with high accuracy can help speed up the development of novel treatments. To date, a number of DTI prediction algorithms have been proposed. [24] propose a novel computational approach for predicting drug-target proteins considering the problem as the task of link prediction.

Drug discovery and repurposing The process of identifying potential new medicines is known as drug discovery. Knowledge graphs have proven to be effective systems for storing and retrieving data. Due to their capacity to model complex data structures, knowledge graphs can reflect an increase in profitable research and development in bioinformatics. The process of finding a new possible use for a medicine is known as drug repurposing. Drug reuse has become a significant complement to the standard strategy due to the high costs connected with the discovery and development of new treatments. Various successful attempts have been made for drug discovery and repurposing. One of the earliest notable attempts to prioritize drugs for repurposing was the work of developing Hetionet [10], SemaTyP is a programme that mines published biological literature to find potential medication candidates for disorders. [30] [35] used knowledge graph for drug repurposing [21]. To prioritise medication repurposing candidates, semantic features in a knowledge graph were used. For the situations, a simple disjoint cross-validation approach was proposed for validating drug-drug interaction predictions [4].

Clinical Uses Knowledge graphs due to their commendable properties are emerging as an impactful tool for clinical processes and question answering systems. Several knowledge graphs have been proposed from a different perspective for medical health classification [27] and different diseases like stroke [3], [8], osteoarthritis [19], [33], PubMed knowledge graph (PKG) has also been constructed for PubMed abstracts [34], Covid-19 knowledge graph [6].

6.5.3   STEPS FOR CONSTRUCTING BIOMEDICAL KNOWLEDGE GRAPHS

- **Dataset Selection** The process for developing a database entail acquiring relevant content, such as journal articles, abstracts, or web-based information and having curators scan it to find sentences that suggest a relationship.
- **Data Curation / Information Extraction** Curation involves the annotation of data. The curation process can be done in either manual or semi-automated ways where initially sentences from the text are extracted in an automated manner. Information extraction is a process of finding essential and relevant entities and their semantic properties from the sentences of a given corpus. Entity recognition, relation extraction and coreference resolution are the three subtasks. Entity recognition entails determining the most appropriate entity type label for a given entity. Relation extraction entails determining the best relation type label between two entities, while coreference resolution entails grouping spans that refer to the same entity.
- **Knowledge Graph Visualization** In this step, we find triples from extracted facts and these triples will make up the knowledge base. Here we

Table 6.5
Extracted entity pairs

Entity1	Entity2
bcr abl fusion aberrant	tyrosine kinase
REcently we	poor IFN alpha treatment
regulatory factors	binding domain
They	such IFNs
impaired T expression data	predominately T cells immune cells

formalize our data model using standards like RDF Schema and OWL. Alternative techniques may be used to generate graphs once the information has been retrieved and saved in the form of triples.

6.5.4 AN EXAMPLE OF A KNOWLEDGE GRAPH

For the construction of the biomedical knowledge graph, we have used GENIA biomedical-event dataset [14] which is used to identify certain medical conditions in a person. GENIA dataset consists of 8000 sentences related to biomedical events extracted from MEDLINE abstracts. Table 6.5 shows the extracted entity pairs from the dataset. Table 6.6 shows the relationships between the entities and their count. Figure 6.9 shows the generated sub-knowledge graph containing 10 entities and 7 relationships.

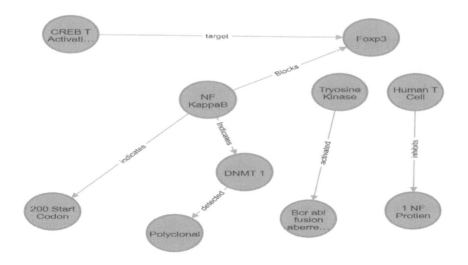

Figure 6.9 Sub-graph of constructed knowledge graph.

Table 6.6
Entity relationships

Relationship type	Count	Relationship type	Count
suggest	203	observed	36
show	144	conclude	35
demonstrate	136	demonstrated	118
was	116	showed	108
indicate	105	shown	102
investigate	93	revealed	85
induced	76	identified	75
are	74	inhibited	71
examined	66	indicated	52
used	50	induces	47
resulted in	46	were	45
report	44	had	43
suggests	42	associated with	42
contains	42	inhibits	41
appears	38	increased	37
observed	36	conclude	35
inhibited by	35	detected in	34
requires	34	studied	33
required for	32	analyzed	32
reported	32	decreased	32
suggested	30	known	30
provide	30	blocked by	27
detected	26	expressed	26
blocked	25	expressed in	25
activated	24	found	141

6.6 FUTURE DIRECTIONS

6.6.0.0.1 Pre-trained word vectors

For heterogeneous textual data, word vectors can be regarded as a latent semantic representation. Vectors for semantically related keywords lie in close proximity to each other in the space of word embeddings, which aids to improve the quality of information retrieval systems. In this aspect, there is a shortfall of standards and pre-trained vectors in the biomedical context, resulting in unsatisfactory results.

6.6.0.0.2 Character level embeddings

Recently popular models use a unique vector to represent each word and ignore the internal structure of words. In certain situations, word embeddings are insufficient to capture all words such as out-of-vocabulary words, different written forms of the same entity, misspelled words. Character embeddings should be utilized to build vector representations of words based on their character-level structure to identify such terms, as it has the potential to substantially enhance biomedical information retrieval systems.

6.6.0.0.3   Domain-specific pre-processing tools

Biomedical text is not the same as ordinary text. By focusing on building preprocessing tools specifically for the biomedical text from the ground up, we may contribute to the development of an effective biomedical IR system.

6.6.0.0.4   Ontology integration

Current common models use a single ontology for information retrieval since it makes the process quicker to execute; however, maintaining global consistency with a single large ontology is difficult. As a result, integration of various ontologies may be recommended to solve this problem with the present search scenario, making information retrieval more efficient.

6.6.0.0.5   Entity tags enhancement

For the biomedical domain, named entity recognition is concerned with determining the boundaries of domain-specific terminology such as diseases, procedures, and substances. The availability of annotated corpora is primarily limited to English; however, high-quality annotated corpora for other languages are also required. Furthermore, in the biomedical domain, the main issue is that POS taggers' accuracy drops when dealing with unknown words; therefore, emphasis should be placed on improving the set of entity tags.

6.6.0.0.6   Knowledge graph for information retrieval

The usage of a knowledge graph within the information retrieval system can assist enhance the comprehension of a user's purpose in a query that is more likely to be answered directly.

6.6.0.0.7   A hybrid approach for query expansion

Knowledge-based approaches are used in conjunction with corpus-based methods to improve the efficacy of the query expansion strategy. To produce expansion features, the corpus-based technique relies on term frequency, but it ignores meaning and term dependence. The semantic approach, on the other hand, is context-dependent, but it suffers from the restricted number of terms and relationships in the context. The most viable approach is a hybrid approach that combines a statistical method for generating expansion terms with a semantic method for ensuring the right order of phrases and their context.

REFERENCES

1. Titipat Achakulvisut and Daniel E Acuna. Pubmed parser, 2015.

2. Raghavendra Chalapathy, Ehsan Zare Borzeshi, and Massimo Piccardi. Bidirectional lstm-crf for clinical concept extraction. *arXiv preprint arXiv:1611.08373*, 2016.

3. Binjie Cheng, Jin Zhang, Hong Liu, Meiling Cai, and Ying Wang. Research on medical knowledge graph for stroke. *Journal of Healthcare Engineering*, 2021, 2021.

4. Hyejin Cho and Hyunju Lee. Biomedical named entity recognition using deep neural networks with contextual information. *BMC Bioinformatics*, 20(1):1–11, 2019.

5. Jacob Devlin, Ming-Wei Chang, Kenton Lee, and Kristina Toutanova. Bert: Pre-training of deep bidirectional transformers for language understanding. *arXiv preprint arXiv:1810.04805*, 2018.

6. Daniel Domingo-Fernández, Shounak Baksi, Bruce Schultz, Yojana Gadiya, Reagon Karki, Tamara Raschka, Christian Ebeling, Martin Hofmann-Apitius, and Alpha Tom Kodamullil. Covid-19 knowledge graph: a computable, multi-modal, cause-and-effect knowledge model of covid-19 pathophysiology. *Bioinformatics*, 37(9):1332–1334, 2021.

7. John R Firth. A synopsis of linguistic theory, 1930-1955. *Studies in Linguistic Analysis*, 1957.

8. SM Shamimul Hasan, Donna Rivera, Xiao-Cheng Wu, Eric B Durbin, J Blair Christian, and Georgia Tourassi. Knowledge graph-enabled cancer data analytics. *IEEE Journal of Biomedical and Health Informatics*, 24(7):1952–1967, 2020.

9. Hua He, Kevin Gimpel, and Jimmy Lin. Multi-perspective sentence similarity modeling with convolutional neural networks. In *Proceedings of the 2015 Conference on Empirical Methods in Natural Language Processing*, pages 1576–1586, 2015.

10. Daniel Scott Himmelstein, Antoine Lizee, Christine Hessler, Leo Brueggeman, Sabrina L Chen, Dexter Hadley, Ari Green, Pouya Khankhanian, and Sergio E Baranzini. Systematic integration of biomedical knowledge prioritizes drugs for repurposing. *Elife*, 6:e26726, 2017.

11. Qing Huang, Yang Yang, and Ming Cheng. Deep learning the semantics of change sequences for query expansion. *Software: Practice and Experience*, 49(11):1600–1617, 2019.

12. Ayyoob Imani, Amir Vakili, Ali Montazer, and Azadeh Shakery. Deep neural networks for query expansion using word embeddings. In *European Conference on Information Retrieval*, pages 203–210. Springer, 2019.

13. Vladimir Karpukhin, Barlas Oguz, Sewon Min, Patrick Lewis, Ledell Wu, Sergey Edunov, Danqi Chen, and Wen-tau Yih. Dense passage retrieval for open-domain question answering. pages 6769–6781. Association for Computational Linguistics.

14. J-D Kim, Tomoko Ohta, Yuka Tateisi, and Jun'ichi Tsujii. Genia corpus—a semantically annotated corpus for bio-textmining. *Bioinformatics*, 19(suppl_1):i180–i182, 2003.

15. Casimir A Kulikowski, Edward H Shortliffe, Leanne M Currie, Peter L Elkin, Lawrence E Hunter, Todd R Johnson, Ira J Kalet, Leslie A Lenert, Mark A Musen, Judy G Ozbolt, et al. Amia board white paper: definition of biomedical informatics and specification of core competencies for graduate education in the discipline. *Journal of the American Medical Informatics Association*, 19(6):931–938, 2012.

16. Quoc Le and Tomas Mikolov. Distributed representations of sentences and documents. In *International Conference on Machine Learning*, pages 1188–1196. PMLR, 2014.

17. Ulf Leser and Jörg Hakenberg. What makes a gene name? named entity recognition in the biomedical literature. *Briefings in Bioinformatics*, 6(4):357–369, 2005.

18. Weijiang Li, Sheng Wang, and Zhengtao Yu. Deep learning and semantic concept spaceare used in query expansion. *Automatic Control and Computer Sciences*, 52(3):175–183, 2018.

19. Xin Li, Haoyang Liu, Xu Zhao, Guigang Zhang, and Chunxiao Xing. Automatic approach for constructing a knowledge graph of knee osteoarthritis in chinese. *Health Information Science and Systems*, 8(1):1–8, 2020.

20. Bing Liu. *Web Data Mining Exploring Hyperlinks, Contents, and Usage Data*. Springer, 2011.

21. Tareq B Malas, Wytze J Vlietstra, Roman Kudrin, Sergey Starikov, Mohammed Charrout, Marco Roos, Dorien JM Peters, Jan A Kors, Rein Vos, Peter AC't Hoen, et al. Drug prioritization using the semantic properties of a knowledge graph. *Scientific reports*, 9(1):1–10, 2019.

22. Michael Matheny, S Thadaney Israni, Mahnoor Ahmed, and Danielle Whicher. Artificial intelligence in health care: the hope, the hype, the promise, the peril. *NAM Special Publication. Washington, DC: National Academy of Medicine*, page 154, 2019.

23. Tomas Mikolov, Kai Chen, Greg Corrado, and Jeffrey Dean. Efficient estimation of word representations in vector space. *arXiv preprint arXiv:1301.3781*, 2013.

24. Sameh K Mohamed, Vit Novaek, and Aayah Nounu. Discovering protein drug targets using knowledge graph.

25. Jamal Abdul Nasir, Iraklis Varlamis, and Samreen Ishfaq. A knowledge-based semantic framework for query expansion. *Information Processing & Management*, 56(5):1605–1617, 2019.

26. Jeffrey Pennington, Richard Socher, and Christopher D Manning. Glove: Global vectors for word representation. In *Proceedings of the 2014 Conference on Empirical Methods in Natural Language Processing (EMNLP)*, pages 1532–1543, 2014.

27. Thuan Pham, Xiaohui Tao, Ji Zhang, and Jianming Yong. Constructing a knowledge-based heterogeneous information graph for medical health status classification. *Health Information Science and Systems*, 8(1):1–14, 2020.

28. Ranjeet Kumar Rout, SK Sarif Hassan, Sanchit Sindhwani, Hari Mohan Pandey, and Saiyed Umer. Intelligent classification and analysis of essential genes using quantitative methods. *ACM Transactions on Multimedia Computing, Communications, and Applications (TOMM)*, 16(1s):1–21, 2020.

29. Gerard Salton, Anita Wong, and Chung-Shu Yang. A vector space model for automatic indexing. *Communications of the ACM*, 18(11):613–620, 1975.

30. Shengtian Sang, Zhihao Yang, Lei Wang, Xiaoxia Liu, Hongfei Lin, and Jian Wang. Sematyp: a knowledge graph based literature mining method for drug discovery. *BMC Bioinformatics*, 19(1):1–11, 2018.

31. Eric J Topol. High-performance medicine: the convergence of human and artificial intelligence. *Nature Medicine*, 25(1):44–56, 2019.

32. Lucy Lu Wang, Kyle Lo, Yoganand Chandrasekhar, Russell Reas, Jiangjiang Yang, Darrin Eide, Kathryn Funk, Rodney Kinney, Ziyang Liu, William Merrill, et al. Cord-19: The covid-19 open research dataset. *ArXiv*, 2020.

33. Xiaolei Xiu, Qing Qian, and Sizhu Wu. Construction of a digestive system tumor knowledge graph based on chinese electronic medical records: development and usability study. *JMIR Medical Informatics*, 8(10):e18287, 2020.

34. Ming Ming Yang, Jun Wang, Li Dong, Yan Teng, Ping Liu, Jiao Jie Fan, Xu Hui Yu, et al. Lack of association of c3 gene with uveitis: additional insights into the genetic profile of uveitis regarding complement pathway genes. *Scientific Reports*, 7(1):1–8, 2017.

35. Yongjun Zhu, Chao Che, Bo Jin, Ningrui Zhang, Chang Su, and Fei Wang. Knowledge-driven drug repurposing using a comprehensive drug knowledge graph. *Health Informatics Journal*, 26(4):2737–2750, 2020.

36. Keneilwe Zuva and Tranos Zuva. Evaluation of information retrieval systems. *Journal of Computer Science & Information Technology*, 4(3):35, 2012.

7 A Novel Approach for Feature Selection Using Artificial Neural Networks and Particle Swarm Optimization

Venkata Maha Lakshmi N and Ranjeet Kumar Rout
Computer Science and Engineering,
National Institute of Technology, Srinagar, Hazratbal,
Jammu and Kashmir, India

CONTENTS

7.1 INTRODUCTION

In the digital era, massive data is available due to advances in technology. The reliable and efficient processing of these enormous data poses challenges for the users [8]. Useful information can be extracted by processing this data. It is impractical to retrieve information manually from such a vast volume of data. This has led to the development of various data mining procedures to extract information automatically

DOI: 10.1201/9781003246688-7

from the stored data. Furthermore, the machine learning methods proved to be useful in automatically providing patterns among the mined data. Classification algorithms in data mining and machine learning help us to find patterns among datasets and provide information automatically. Classification algorithms categorize the elements in the data set as per the labels associated with them. Typically many features will be introduced in the data set while using classification algorithms. However, the efficiency of classification algorithms will be affected due to the irrelevant, noisy and repeating data from these data sets. Hence, features selection procedures were developed to determine that subset, which is requisite to reach the target [11]. Future selection procedures not only allow us to identify these subsets but also improve the efficiency of algorithms, optimize the running time complexity, and simplify the structure of learned classifiers by removing the irrelevant, noisy, and repeating data from these data sets. The difficulty of a feature selection algorithm is directly proportional to the number of features. The feature selection algorithms are generally categorized into two categories [16, 29, 6, 13]. The first one is Filter algorithms, which are used before the classification algorithm as it carefully selects the feature subset and removes the features from the subset, which are not important. The second one is the Wrapper algorithm, which illustrates three parameters, the first being search strategy followed by performance criteria and, finally, the learning algorithm. The learning algorithm identifies this subset using two parameters, the first is the training data, followed by the performance of the subset. Hence, wrapper algorithms give good results when compared to filter algorithms. Wrapper algorithms typically use exhaustive search methods when the number of features is more and to get an optimal solution. An exhaustive search method cannot be used in all situations, as its time complexity is very high. Different search algorithms are tried to obtain the feature selection but failed as most of them used high memory, very expensive, and classifier architecture complexity. A good criterion for the selection of an efficient search technique that is global is to see how effectively it addresses the issue of feature selection. Evolutionary computing (EC) techniques are highly recognized by their ability to search globally. Many (EC) algorithms like a genetic algorithm (GA) [37, 26, 7], Genetic programming (GP) [24], and Ant Colony Optimization (ACO) [31, 23], Firefly algorithm and others as discussed [38, 32, 5, 4, 28]) have been introduced to feature selection problems recently. Algorithmic specific parameters, in addition to the standard controlling parameters, are used in feature selection by the EC algorithms and also play a significant role in the efficiency of the machine learning models. Out of all search algorithms, Particle Swarm Optimization (PSO) algorithm proved to be good globally as it can search large spaces, less expensive, easy to implement and use only a few parameters. PSO is the one which is based on Swarm Intelligence (SI) algorithm, and it requires common optimization parameters only. (Like solutions in the population and stopping criteria). We've adapted the PSO technique to reduce the strain of setting parameter values while selecting the features.

In our proposed work, the classification is carried out using Artificial Neural networks (ANN), while the Binary PSO is proposed and implemented to select the subset of optimal features. The main proposed models are:

1. A binary Particle Swarm Optimization algorithm (FS-BPSO) is developed to select the subset of optimal features.

2. The selected subset of features are used to train an ANN model for classification.

7.2 RELATED WORKS

The predicament of feature selection is tackled using various approaches. We also studied various existing methods for feature selection to identify the limitations and to extend them to different applications. In [36], Xu et al. came up with an image classification technique that had better accuracy. This technique was based on the discriminative L2 regularization for implementing sparse representation. Nie et al. [25] used joint L2,1-Norms minimization for feature selection. In Bahassine et al. [3], the author's recommended a novel feature selection strategy for text classification in Arabic by implementing the chi-square method. Lai et al. [19] for contemplating feature extraction used robust discriminate regression that reduced the dimensionality of the data set. The major drawback of this type of regression is that it cannot provide the sparse projections for feature selection. L. Peterson et al. [20] developed a dimensionality reduction methodology based on k-means and PCA for DNA microarray-based prostate cancer detection and used ANN as a classifier. C.D.A Vanitha et al. [34] selected a set of most relevant features using mutual information techniques and used the SVM algorithm for classification. H.T Huynh et al. [12] used the single value decomposition (SVD) technique along with a feed-forward artificial neural network for dimensionality reduction to classify DNA microarray. In recent times, swarm intelligence techniques have been widely used by many researchers to select the subset of optimal features. For example, Garro [10] introduced a novel approach for classifying microarray data by implementing ANN and ABC algorithms. Thawkar et al. [33] suggested a feature selection technique for the classification of digital mammograms using the biogeography-based optimization algorithm. In [2], Mohan Allah et al. compared their gene selection algorithm based on binary teaching-learning based optimization with the conventional genetic algorithm. The paper deals with gene selection for breast cancer diagnosis. Sayed et al. [14] proposed a new meta-heuristic method for feature selection based on the crow search algorithm and achieved good results for benchmark datasets. Agarwal et al. [1] created a novel feature selection strategy for the classification of cervical cancer CT images by implementing the Artificial Bee Colony (ABC) and the KNN algorithm. In [22], Mazini et al. put forth a new technique for network-based intrusion detection. This technique included a hybrid model in which the relevant genes were selected using the artificial bee colony technique and classified using the AdaBoost algorithm.

7.3 PROPOSED MODEL

The proposed model consists of two phases. The first phase is confined to selecting the optimal feature subset. Instead of using traditional feature selection approaches,

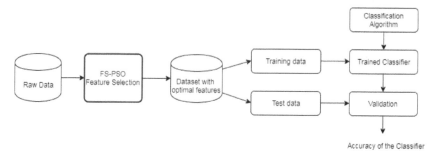

Figure 7.1 Flow of work for the proposed model for feature selection.

we opted to use the Binary Particle Swarm Intelligence algorithm, as described in [15]. This results in producing the best subset of features. The second stage encompasses training a neural network on the previously selected features, with constantly adjusting the weights. The values for the synaptic weights are varied according to their impact on the classification rate. Figure 7.1 represents the model for the proposed work.

7.3.1 FEATURE SELECTION

The step of feature selection enables the user to find the best subset of features that has the least number of features. This method is essential and appropriate for the classification of the dataset by the classifier. The number of features in the selected subset 'S' is necessarily less than the original number of features'N.' The sequence followed while selecting the feature subset is shown in Figure 7.2.

7.3.2 SELECTION OF THE FEATURE SUBSET USING PSO

PSO derived from the simulation of birds' social behaviour in a flock [18]. Through PSO, each particle travels at speed in the search space and transitions through its own travelling memories and the flying experience of its companion. Figure 7.3 illustrates the working of the PSO. We developed a method which is a binary revised version of the particle swan optimization algorithm and called it FS-BPSO. Its job is to select the most appropriatefeature subset for the concerned problem at hand. The significant difference between the binary variant of particle swarm optimization and the original variant is regarding the particle movement within the search space. For FS-BPSO, it is binary; hence, the position vectors of particles are allotted only binary value 0 or 1. The FS-BPSO is represented in Algorithm 1. Just like in PSO; the initial steps are almost similar. The first one being the selection of a population for a given number of particles. Every particle in a multi-dimensional search space provides a viable solution. For instance, consider a seven-dimensional space and a

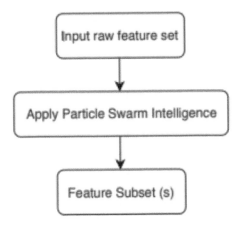

Figure 7.2   Sequence of steps in choosing the subset of features.

solution in which alternate features are selected, e.g., 1010101. Particles will update themselves through iterations by tracking the best value and the global best value. The best value is also referred to as the best individual, is the best position of each particle that has been found so far. $p^{Best_i} = (p^{Best_i^1}, p^{Best_i^2}.....p^{Best_i^n})$ is the best individual of the ith particle. The global best is the best solution until a particular instant, found throughout the swarm. $g^{Best_i} = (g^{Best_i^1}, g^{Best_i^2}.....g^{Best_i^n})$ The individual values for each particle are compared with the g^{Best}, and if they are equal, then the number of features is compared between the two. Then replace the g^{Best} with the particle which produces a smaller number of features. Particle movement in a swarm is affected by position and velocity vectors. The position vector of any particle, say i^{th} particle is given by $Y_i = [y_i^1, y_i^2, ..., y_i^n]$ and its velocity vector $Vl_i = [Vl_i^1, Vl_i^2, ..., Vl_i^n]$ where $y_i^k \, \varepsilon \, 0,1$, $k=1,2,...,m$ (where m is the number of features) in $i=1,2,...m$ (m represents the number of particles). The velocity is adjusted according to code of the inner loop (17-23) in the pseudo code (Algorithm 1). This contains inertia, cognitive and social component. Every particle's velocity is regulated by inertia component. The inertia component regulates the velocity of each particle. The cognitive variable with the cognitive coefficient c1 will reflect the particle memory. This facilitates in projecting particles to their best-known individual positions. Finally, the social component with the social coefficient c2 helps to guide the particles to the swarm's best-known global position. In the case of c1> c2, the search behavior is inclines to the individual particle limits, otherwise if c1<c2, the search behavior is inclined to the global swarm limit. Adjustment of position is based on the code of the fifth if-else section (24-28) in the pseudo code (Algorithm 1) and is explained in Sigmoid function.

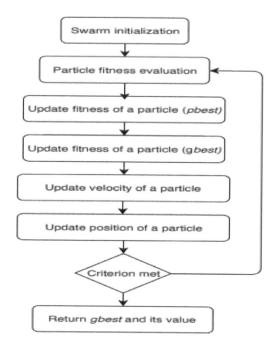

Figure 7.3   Flow of PSO for feature selection.

7.3.3   SIGMOID FUNCTION

To represent the particle velocity vl_{ik}^{new} within the range of 0 and 1, the sigmoid function is applied, represented in Eq. (1).

$$Sigmoid(vl_{ik}^{new}) = \frac{1}{1 + e^{-vl_{ik}^{new}}} \tag{7.1}$$

where e is the traditional logarithm's base. The resulting value is then compared to the randomly generated value uniformly distributed between 0 and 1 (U(0, 1) function). A decision about the particle X_{ik}'s new position is now probabilistic. This implies the greater the likelihood of vl_{ik}^{new} the greater the Sigmoid function's value. The possibility of the value 1 allocated to the x_{ik}^{new} is therefore increasing. If vl_{ik}^{new} rises, the sigmoid vl_{ik}^{new} function will be limited to 1. For example, if $vl_{ik}^{new} > 5$, X_{ik}'s probability is nearly 1, but not absolutely 1. Therefore, for $vl_{ik}^{new}=5$, the $X_{ik}=1$ probability is 0.998, and the $X_{ik}=0$ probability is 0.002.

7.3.4   FITNESS FUNCTION

The fitness function aims to evaluate every particle's efficiency. The fitness function input is a particle that selects features labelled with 1. This is followed by developing the classifier stemming from the selected features. The output of a fitness function

---

Algorithm 1 Particle Swarm Optimisation Algorithm

Require: c1,c2,vlmin,vlmax,ω,maximum-no-of-Iterations

Ensure: Global best particle, g^{Best} (feature subset) and its fitness value fitness
    while *maximum no of Iterations* **do**
        for *i* = 1 *to number of particles* **do**
            fitness = *fitness function of a particle*
            fsp = *features selected in a particle*
            if fitness(Y_i) > fitness(P_i^{Best}) **then**
                $P_i^{Best} = Y_i$
            end if
            if fitness(Y_i) > fitness(g^{Best}) **then**
                $g_i^{Best} = Y_i$
            end if
            if fitness(Y_i) = fitness(g^{Best}) and fsp(Y_i) ¡ (fsp(g^{Best})) **then**
                $g_i^{Best} = Y_i$
            end if
        end for
        for *i* = 1 *to number of particles* **do**
            for *k* = 1 *to number of features* **do**
                $vl_{ik}^{new} = \omega \times vl_{ik}^{old} + c_1 \times r_1 \times (p^{Best_{ik}^{old}} - y_{ik}^{old}) + c_2 \times r_2 \times (g^{Best_{ik}^{old}} - y_{ik}^{old})$
                if $vl_{ik}^{new} > vl_{max}$ **then**
                    $vl_{ik}^{new} = vl_{max}$
                end if
                if $vl_{ik}^{new} < vl_{min}$ **then**
                    $vl_{ik}^{new} = vl_{min}$
                end if
                if *sigmoid*($vl_{ik}^{new} > U(0,1)$ **then**
                    $y_{ik}^{new} = 1$
                else
                    $y_{ik}^{new} = 0$
                end if
            end for
        end for
    end while

---

is classifier's classification accuracy, developed with the ANN algorithm by taking a subset of features, identified by a particular particle. Classification accuracy is the most widely used metric to assess the classification performance. The equation represented as Eq. 7.2 has been used by us to find the classification performance to compute the accuracy of the artificial neural network.

$$Accuracy = \frac{number\ of\ correctly\ classified\ instances}{number\ of\ instances} \tag{7.2}$$

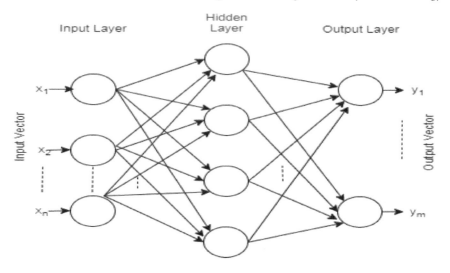

Figure 7.4   Schematic representation of an artificial neural network.

7.4   ARTIFICIAL NEURAL NETWORK (ANN)

Artificial Neural Network (ANN) is an exemple of modern computing systems that are modeled to replicate the functioning of the human brain for data analysis and information processing [27]. It is useful in solving problems like classification, forecasting, regression, optimization, etc. [9], which are difficult for humans due to the large dataset size. In addition to handling the huge volumes of data, the complex structure of this data is also seen as a trending and potential field that requires immediate attention from researchers in the field of problem-solving. ANN are being designed to solve such problems [21, 30]. Due to its multiple layers, this neural network performs better self-learning and hence delivers better outcomes [17]. The number of input and output layers ina multi-layer neural network is constant, i.e., one, while they can "n" the number of neurons depending on the requirement of the problem. However, the number of hidden layers or intermediate layers ranges from 0 to n. The presence of a considerable number of hidden layers develops a non-linear mapping between the input and the output layer [35]. Such a network is illustrated in Figure 7.4. This network also has a few characteristic features:

It is a fully connected network, with each neuron of the current layer (say k^{th} layer) being connected to every other neuron of the succeeding layer (say $k+1^{th}$ layer). The flow of information is unidirectional. No two neurons that belongs to the same layer is connected to each other. The number of nodes in the input and output layers depends on the input and output objects. The number of neurons in the hidden layer is depends on the complexity of the problem and the size of the data set. If the hidden layer contains less number of neurons, the network may not have enough freedom to form a representation. The network may become over trained if too many neurons are used. Hence an optimal design is needed for the number of neurons in

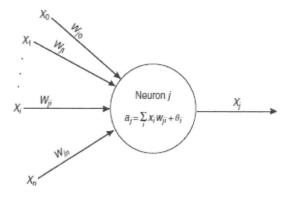

Figure 7.5   Basic processing element (neuron).

the hidden layer. Additionally, it is considered another two parameters called weight and bias. Every neuron has a quantity called weight associated with it. It represents the strength of the connection. Bias serves as a constant that makes the model fit for the given data. The net input of every neuron is measured by the sum of all its connecting input values multiplied by their corresponding connection weights and fed forward to become an input to the neurons in the succeeding layer of the network through a non-linear activation function or transfer function. (See Figure 7.5). In the case of a multi-layer perceptron (MLP), each neuron calculates its own output according to (Eq.3 and Eq.4). During the training phase, the perceptron is fed with the training data repeatedly.

$$o(w,x,\theta) = \sum_{i=1}^{N_L} x_i w_i + \theta \tag{7.3}$$

$$y_j = F_j(O_j(w_j,x,\theta)) \tag{7.4}$$

where N_L is the number of neurons, w represents the synaptic weights, θ represents bias, and F can be any transfer function from amongst the basic transfer functions, (linear, sigmoid, piecewise, hyperbolic, etc.) The output produced is a measure of our ANN's efficiency with respect to the mean square error (MSE) given by Eq.(5).

$$MSE = \frac{1}{n} \sum_{i=i}^{n} (y_i - \overline{y_i})^2 \tag{7.5}$$

The network weights are then modified until the mean square error (MSE) becomes low, or a fixed number of epochs is reached. The backpropagation (BP) algorithm is one of the most commonly used techniques for training multi-layer perception. The proposed model is based on the methodology of gradient descent.

7.5 EXPERIMENTAL RESULTS AND DISCUSSION

The key objectives of the proposed model are

1. Bring down the number of features that are needed to classify a given data set efficiently.

2. Design a classification model that is better in terms of accuracy from the conventional models used.

3. Foreshorten the training time of the learning model, trained using the smaller chosen subset of features.

With the aforementioned objectives, we performed various experiments on different high-dimensional datasets and obtained improved performance relative to traditional feature selection approaches. All datasets are downloaded from the UCI repository. The initial data distribution between the training and testing phases for each dataset is described below.

In this work, 80% of the dataset was used for training 20% of the dataset was used to test the efficiency of the learned model. For the training and testing of the proposed model with the following data sets, three separate cases were created. In the first case, the model was equipped with the datasets' traditional features. But in the nextexample, we implemented our FS-BPSO algorithm in order to select the optimal subset of features from the dataset. This was followed by training the models. In the last case, the model was validated using a selected subset of features from FS-BPSO.

1. Membrane dataset: It consists of 5061 protein samples with 464 features. The training set, the testing set, contains 4016 and 1045 samples, respectively.

2. Breast cancer dataset: The data set includes features taken from the Fine Needle Aspiration (FNA) image of a patient's breast mass. The training dataset contains 455 patient records. The testing dataset contains 144 records. The dataset has a total of 569 records. Each record includes 30 features.

3. Heart dataset: This dataset is used to determine the patient's existence of heart disease. It contains 303 patient samples with 14 features. The training set includes 242 samples, and the testing set contains 61 samples.

4. Sawdust: It contains the features of Sawdust from birch, pine, and spruce that were blended in specific ratios. It consists of 54 samples with 1206 features. It was divided into 44 samples for training and 10 samples for the testing dataset.

5. Parkinson's disease dataset: The dataset includes information obtained from Parkinson's disease patients dataset (107 males and 81 females), all aged between 33 and 87, $(65.1A \pm 10.9)$ years. It contains 756 samples with 754 features. The training set contains 605 samples, and the test set contains 151 samples.

The data set is divided into training and test data in the ratio of 80:20 to show the validity of the proposed technique. K-fold cross-validation approach has been used for efficient validation, in which the entire sample space is distributed into the same sized k number of sub-samples. Out of the total k samples, one sample is used for testing the model by placing it as the validation data. The rest of the (k-1) samples are fed to the neural network for training the ANN. The process of cross-validation is implemented in a loop that is run k times, with each sample used for validation once. The result of each implementation is used to generate a single input using either finding the average of the individual results or combining all the results together. This approach has many advantages; firstly, a separate set of data is not needed for validation as the training set in-turn is used for validation. Secondly, each round of implementation generates a single result. Since all the observations are used for training, the network can be trained better, and also for validation, every observation is used just once, which improvises the efficiency of validation. Usually, stratified k-fold cross-validation is chosen for classification problems, where the folds are selected so that each fold contains approximately the same proportions of class labels. 10-fold cross-validation was performed on the Membrane, heart, breast cancer and Parkinson's datasets before running the proposed methodology. This is done by dividing the initial data into 10 equal stratified sections. The classification model is then trained on training data (9 parts) and then evaluated on the test set (1 part). The two fold cross-validation process has been used on the sawdust dataset, where data is distributed into two equal stratified parts. In that first part was used for training set remaining second part for the test set.

After the successful execution of our proposed model, we compare its performance in terms of classification accuracy by varying the number of features used to train the network. Out of the two sets of features, the first set is the complete data set, while the other feature se isa subset of this feature in which the features are selected using the FS-BPSO algorithm. The results obtained have been tabulated in Table 7.5. For the membrane dataset, we obtain 89.5 classification accuracy with the proposed model by taking the original features in contrast to 99.1 with FS-BPSO selected features. We were also able to achieve a reduction of 51.72% in terms of the number of features. The accuracy of the model was found to be 99.12 and achieved 62.5.1% of the reduction in terms of the number of features, which is best when compared to the values 98.24 of initial features for WBC (Diagnostic) dataset. Heart dataset provided 91.8 accuracy with all features and 98.8 with FS-BPSO selected features and achieved 57.1% of the reduction in terms of the number of features. Accuracy value achieved with the sawdust dataset is 90.9 with all features, and 1.0 with FS-BPSO selected features and gained 52.40% of the reduction in terms of the number of features. Finally, Parkinson's dataset gave comparable values of 90.78 and 91.44 respectively and achieved 51.59% of the reduction in terms of the number of features. FS-BPSO based selected features achieved a better result with the sawdust dataset compared to conventional features. The results proved that FS-BPSO hasa high proportion of reduction rate as well as the maximum accuracy of the reduced datasets, as like the original datasets. The graphical representation of the classification accuracies for the models described above is shown in Figure 7.6. The number

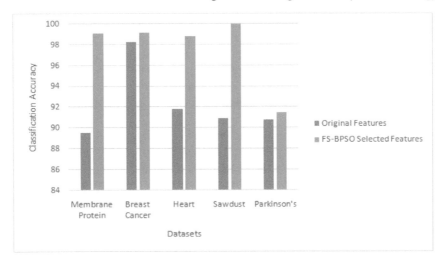

Figure 7.6 Comparison of classification accuracy with FS-BPSO selected features.

Table 7.1

Characteristics of the datasets used for experimentation

Dataset	Number of samples	Number of features	Number of classes
Membrane protein	5061	464	3
Breast cancer	569	32	2
Heart	303	14	2
Sawdust	54	1206	27
Parkinson's	756	754	2

of features also substantially decreased for various datasets with the proposed model, as shown in Table 7.4. The selected subset of features is used to train ANN to make more reliable predictions. So as to determine the accuracy of the model with reduced features, the proposed approach is executed and experimentally investigated.

To select the optimum features using FS-BPSO, some parameters are to be tuned earlier than moving round inside the search space called hyper parameters, which are mentioned in Table 7.1.

The proposed FS-BPSO algorithm is evaluated by computing the classification accuracy using Artificial Neural Network. In Table 7.3: the 5 hyperparameters, which is probably to be tuned earlier than training an ANN are presented. The neural network used here is a three-layered, feed-forward neural network that incorporates learning from backpropagation. So as to determine the accuracy of the model with reduced features, the proposed approach is executed and experimentally investigated.

Table 7.2

Initialization of learning parameters in PSO

Parameter	Value
Social coefficient, c1	2
Cognitive coefficient, c2	2
Minimum velocity, Vmin	-5
Maximum velocity, Vmax	5
Inertia weight, w	0.5
Number of particles	10
Number of iterations	100

Table 7.3

Artificial neural network hyperparameters

Parameter	Value
Number of layers	Three (input, output, hidden)
Activation function	ReLu
Number of nodes in hidden layer	250
Learning rate	0.0001
Maximum number of iterations	2000

Table 7.4

Proposed methodology reduction rate

Datasets	Number of features	Reduction rate %
Membrane protein	464	51.72
Breast cancer	32	57.1
Heart	14	62.5
Sawdust	1206	51.59
Parkinson's	754	52.40

7.5.1 PERFORMANCE EVALUATION

Classification accuracy metric alone does not determine a learning model as the best performing. Several parameters exist for validating a model's performance. We use a confusion matrix to summarize the classifier performance. A confusion matrix reports the four possible outcomes of the comparison between the actual and the predicted class. That is, the actual class value is yes, and the predicted class value is also

Table 7.5
Performance comparison of the proposed methodology using classification accuracy

Datasets	All features	FS-BPSO selected features
Membrane protein	89.5	99.1
Breast cancer	98.24	99.12
Heart	91.8	98.8
Sawdust	90.9	1.00
Parkinson's	90.78	91.4

yes (YY), the actual class value is no and predicted class value is also no (NN), the actual class value is no and predicted class value is yes (NY), the actual class value is yes but predicted class value in no (YN). The elements of the matrix allow the definition of the following performance metrics:

i. Accuracy: It is defined as a ratio of the correctly classified samples, i.e., the sum of all the true-positives and true-negatives to the total number of samples. It can be represented as follows:

$$Accuracy = \frac{YY + NN}{YY + NN + YN + NY} \tag{7.6}$$

ii. Precision: It is represented as the ratio correctly classified positive samples or true-positives (YY) to the sum of correctly classified positive samples (YY) and the false-positives (NY), i.e., the actual value is negative, but the predicted value is positive. It can be represented as follows:

$$Precision(P) = \frac{YY}{YY + NY} \tag{7.7}$$

iii. Recall: It is defined as the product of correctly classified positive samples (YY) and the inverse of the sum of correctly classified positive samples (YY) and the true negative. This represents that the actual result is positive, while the predicted value is negative(YN). It can be represented as follows:

$$Recall(R) = \frac{YY}{YY + YN} \tag{7.8}$$

iv. F-1 Score: It can be described as the harmonic mean between the two previously discussed parameters, i.e., Precision and Recall. It can have values ranging from 0 to 1. 1 represents the best value, while 0 represents the worst.

$$F - 1 Score = \frac{2 \times (R \times P)}{R + P} \tag{7.9}$$

Table 7.6

Performance comparison of the proposed methodology based on precision

Datasets	All features	FS-BPSO selected features
Membrane protein	91	99
Breast cancer	98.5	99.5
Heart	92	
Sawdust	83	1
Parkinson's	91.5	88

Table 7.7

Performance comparison of the proposed methodology based on recall

Datasets	All features	FS-BPSO selected features
Membrane protein	89.6	98.3
Breast cancer	98.5	99
Heart	91.5	
Sawdust	83	1
Parkinson's	83	84.5

Table 7.8

Performance comparison of the proposed methodology based on F-1 score

Datasets	All features	FS-BPSO selected features
Membrane protein	90.3	98.6
Breast cancer	98.5	99
Heart	92	
Sawdust	83	1
Parkinson's	86	87

The values for Precision, Recall and F-1 score are computed during the classification carried out by ANN on the whole data set with all features and a subset of features that are selected using FS-BPSO algorithm. The results for Precision, Recall and f1-score are tabulated in Tables 7.6, 7.7 and 7.8. Amongst all the datasets, the sawdust dataset was found to have the highest vales for the above-mentioned parameters, while implementing the feature selection algorithm as proposed by us. Going back to the results tabulated in Table 7.4, we have illustrated that using our proposed model, the number of features has also substantially decreased for various datasets. Hence, we conclude that using fewer features we can train a model which generates better predictions.

The results speak for themselves as they clearly indicate that FS-BPSO shows a considerable performance improvement from other dimension reduction techniques. By using the proposed methodology, only less than 45% of the total information is used, and experimentally investigated results prove that the accuracy obtained is much better.

7.6 CONCLUSION

The proposed method, which is implemented in this paper, serves as an efficient solution for classifying high-dimensional data. On further analysis, we can conclude that it works reasonably well for low-dimensional data as well. An experimental setup was designed to generate the results. Five benchmark datasets were taken into consideration for the classification problem. More focus has been laid on the feature selection stage of the classification problem, and an improvised version of PSO called the BPSO was used, which takes subsets of features as the input instead of the entire feature space. With the aid FS-BPSO, better results are achieved, in addition to reducing the training feature subset. The outcomes prove that the proposed idea outperforms other preceding works. In the future, we will intend to develop a novel evolutionary computation approach to reduce the number of features with different classification algorithms.

REFERENCES

1. Vartika Agrawal and Satish Chandra. Feature selection using artificial bee colony algorithm for medical image classification. pages 171–176, 08 2015.

2. Mohan Allam and Nandhini Malaiyappan. Optimal feature selection using binary teaching learning based optimization algorithm. *Journal of King Saud University - Computer and Information Sciences*, 12 2018.

3. Said Bahassine, Abdellah Madani, Mohammed Al-Sarem, and Mohamed Kissi. Feature selection using an improved chi-square for arabic text classification. *Journal of King Saud University - Computer and Information Sciences*, 32, 05 2018.

4. Tapasi Bhattacharjee, Ranjeet Kumar Rout, and Santi P Maity. Affine boolean classification in secret image sharing for progressive quality access control. *Journal of Information Security and Applications*, 33:16–29, 2017.

5. Hugh Cartwright. Swarm intelligence. by james kennedy and russell c eberhart with yuhui shi. morgan kaufmann publishers: San francisco, 2001. 43.95. xxvii + 512 pp. isbn 1-55860-595-9. *The Chemical Educator*, 7:123–124, 04 2002.

6. Li-Yeh Chuang, Sheng-Wei Tsai, and Cheng-Hong Yang. Improved binary particle swarm optimization using catfish effect for feature selection. *Expert Systems with Applications*, 38(10):12699–12707, September 2011.

7. Asit Das, Sunanda Das, and Arka Ghosh. Ensemble feature selection using bi-objective genetic algorithm. *Knowledge-Based Systems*, 123, 02 2017.

8. M. Dash and H. Liu. Feature selection for classification. *Intelligent Data Analysis*, 1(3):131–156, May 1997.

9. M.W. Gardner and Stephen Dorling. Artificial neural networks (the multilayer perceptron) – a review of applications in the atmospheric sciences. *Atmospheric Environment*, 32, 08 1998.

10. Beatriz A Garro, Katya Rodriguez, and Roberto Vázquez. Classification of dna microarrays using artificial neural networks and abc algorithm. *Applied Soft Computing*, 38, 10 2015.

11. Isabelle Guyon and André Elisseeff. An introduction of variable and feature selection. *J. Machine Learning Research*, 3:1157 – 1182, 01 2003.

12. Hieu Huynh, Jung-Ja Kim, and Yonggwan Won. Classification study on dna microarray with feedforward neural network trained by singular value decomposition. *International Journal International Journal International Journal International Journal*, 1, 12 2009.

13. Nuhu Ibrahim, H.A. Hamid, Shuzlina Rahman, and Simon Fong. Feature selection methods: Case of filter and wrapper approaches for maximising classification accuracy. *Pertanika Journal of Science and Technology*, 26:329–340, 01 2018.

14. Gehad Ismail Sayed, Aboul Ella Hassanien, and Ahmad Azar. Feature selection via a novel chaotic crow search algorithm. *Neural Computing and Applications*, 31, 01 2019.

15. J. Kennedy and R. Eberhart. Particle swarm optimization. In *Proceedings of ICNN'95 - International Conference on Neural Networks*, volume 4, pages 1942–1948 vol.4, 1995.

16. Ron Kohavi and George John. Wrappers for feature subset selection. *Artificial Intelligence*, 97:273–324, 12 1997.

17. Gregory Krauss. An introduction to neural networks: J.a. anderson (mit press, cambridge, ma, 1995, 672 p., price: Us$ 55.00). *Electroencephalography and Clinical Neurophysiology*, 99:99, 07 1996.

18. Snezana Kustrin, Rosemary Beresford, and Ahmad Yusof. Theoretically-derived molecular descriptors important in human intestinal absorption. *Journal of Pharmaceutical and Biomedical Analysis*, 25:227–37, 06 2001.

19. Zhihui Lai, Dongmei Mo, Wai Wong, Yong xu, Duoqian Miao, and David Zhang. Robust discriminant regression for feature extraction. *IEEE Transactions on Cybernetics*, PP:1–13, 10 2017.

20. Y. Lu, Q. Tian, and M. Sanchez. Hybrid pca and lda analysis of microarray gene expression data, computational intelligence in bioinformatics and computational biology, 2005. cibcb '05. *Proceedings of the 2005 IEEE Symposium on 14-15*, pages 1–6, 01 2005.

21. Federico Marini, Antonio Magri, and Remo Bucci. Multilayer feed-forward artificial neural networks for class modeling. *CHEMOMETRICS AND INTELLIGENT LABORATORY SYSTEMS*, 88:118–124, 08 2007.

22. Mehrnaz Mazini, Babak Shirazi, and Iraj Mahdavi. Anomaly network-based intrusion detection system using a reliable hybrid artificial bee colony and adaboost algorithms. *Journal of King Saud University - Computer and Information Sciences*, 31, 03 2018.

23. Parham Moradi and Mehrdad Rostami. Integration of graph clustering with ant colony optimization for feature selection. *Knowledge-Based Systems*, 84, 04 2015.

24. Kourosh Neshatian, Mengjie Zhang, and Peter Andreae. A filter approach to multiple feature construction for symbolic learning classifiers using genetic programming. *Evolutionary Computation, IEEE Transactions on*, 16:645–661, 10 2012.

25. Feiping Nie, Heng Huang, Xiao Cai, and Chris Ding. Efficient and robust feature selection via joint $\ell 2$, 1-norms minimization. pages 1813–1821, 01 2010.

26. Il-Seok Oh, Jin-Seon Lee, and Byung-Ro Moon. Hybrid genetic algorithms for feature selection. *IEEE Transactions on Pattern Analysis and Machine Intelligence*, 26:1424–37, 12 2004.

27. André Rossi and Maria Angelica Camargo-Brunetto. Protein classification using artificial neural networks with different protein encoding methods. 10 2007.

28. Ranjeet Kumar Rout, Pabitra Pal Choudhury, Sudhakar Sahoo, and Camellia Ray. Partitioning 1-variable boolean functions for various classification of n-variable boolean functions. *International Journal of Computer Mathematics*, 92(10):2066–2090, 2015.

29. Ranjeet Kumar Rout, Pabitra Pal Choudhury, and Sudhakar Sahoo. Classification of boolean functions where affine functions are uniformly distributed. *Journal of Discrete Mathematics*, 2013, 2013.

30. D. Rumelhart, G. Hinton, and Williams RJ. Learning internal representations by error propagation. *Parallel Distributed Processing: Explorations in the Microstructure of Cognition*, 1, 07 1986.

31. Rahul Sivagaminathan and Sreeram Ramakrishnan. A hybrid approach for feature subset selection using neural networks and ant colony optimization. *Expert Systems with Applications*, 33:49–60, 07 2007.

32. Aureli Soria-Frisch. Andries p. engelbrecht (university of pretoria), computational intelligence: An introduction, john wiley & sons ltd., west sussex, england, 2002, isbn 0-470-84870-7. *Applied Soft Computing*, 7:628–629, 03 2007.

33. Shankar Thawkar and Ranjana Ingolikar. Classification of masses in digital mammograms using biogeography-based optimization technique. *Journal of King Saud University - Computer and Information Sciences*, 32, 02 2018.

34. C. Vanitha, D. Devaraj, and iM Venkatesulu. Gene expression data classification using support vector machine and mutual information-based gene selection. *Procedia Computer Science*, 47:13–21, 12 2015.

35. Paul Werbos. Backpropagation through time: what it does and how to do it. *Proceedings of the IEEE*, 78:1550 – 1560, 11 1990.

36. Yong Xu, Zhong Zuofeng, Jian Yang, Jane You, and David Zhang. A new discriminative sparse representation method for robust face recognition via l_2 regularization. *IEEE Transactions on Neural Networks and Learning Systems*, PP:1–10, 06 2016.

37. Jihoon Yang and Vasant Honavar. Feature subset selection using a genetic algorithm. *Intelligent Systems and their Applications, IEEE*, 13:44 – 49, 04 1998.

38. Long Zhang, Linlin Shan, and Jianhua Wang. Optimal feature selection using distance-based discrete firefly algorithm with mutual information criterion. *Neural Computing and Applications*, 28, 09 2017.

8 In Search for the Optimal Preprocessing Technique for Deep Learning-Based Diabetic Retinopathy Stage Classification from Retinal Fundus Images

Nilarun Mukherjee
Department of Computer Science and Information Technology,
Bengal Institute of Technology, Kolkata, India

Souvik Sengupta
Department of Computer Science and Engineering,
Aliah University, Kolkata, India

CONTENTS

DOI: 10.1201/9781003246688-8

161

8.1　INTRODUCTION

Diabetic Retinopathy (DR) is the damage of the micro-vascular system in the retina, due to prolonged hyperglycemia and blockages or clots that are formed due to high level of glucose in the small blood vessels of the retina. This in effect raptures the wall of those weak vessels due to high pressure and leakage of blood on surface of retina which leads to vascular disorder, blurred vision and sometimes complete blindness [18]. DR is one of the most severe microvascular complications in patients with type 2 diabetes mellitus and has become the leading cause of vision loss resulting irreversible blindness among working-aged adults (20–74 years) [18] [12]. A recent hospital-based study conducted by Bhutia et al. [3] on the type 2 diabetic population in north east India have reported 17.4% overall prevalence of diabetic retinopathy, similar to that observed by Rema et al. (17.6%) [17] and Raman et al. (18.1%) [16], in studies done in the southern states of India.

Figure 8.1 depicts the normal retinal components such as blood vessels, optic disc, macula and fovea. It also shows different DR anomalies like microaneurysm, exudates and hemorrhages which are the main pathognomonic signs of DR. DR can be broadly classified into two main stages, namely, non-proliferative (NPDR) and proliferative (PDR), based on its severity of vascular degeneration and other ischemic changes in retina. NPDR is an early stage, which contains at least one microaneurysm or hemorrhage with or without presence of any hard exudates. NPDR is further subdivided into four stages i) mild (presence of MA), ii) moderate (appearance of HM and EX along with MA) and iii) severe (venous beading in at least two quadrants and MA, HM in four retina quadrants), according to Scottish DR grading protocol [21]. Proliferative DR (PDR) is an advanced stage which is characterized by neovascularization, where circulation of blood in vessels experiences lack of oxygen and leads to the growth of new fragile blood vessels, causing vitreous hemorrhages and tractional retinal detachment.

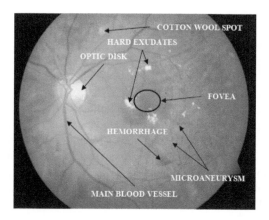

Figure 8.1　Normal components and DR anomalies in the fundus image.

Proliferative DR (PDR) is an advanced stage of DR which is characterized by neovascularization, where circulation of blood in vessels experiences lack of oxygen and leads to the growth of new fragile blood vessels, causing vitreous hemorrhages and tractional retinal detachment. Most clinical practitioners recommend regular screening of diabetic retinopathy using digital fundus photography of the patients, especially those with mild and moderate retinopathy. The five-stage DR severity grading (hereafter, referred as DR-grading task) is a vital activity in DR detection, which requires manual inspection of retinal anatomic features such as optic disc, cup, fovea, macula and vascular structure and DR lesions by ophthalmologists or well-trained technicians. Therefore, grading of large-scale retinal images of ever increasing DR patients has now become a highly exhaustive, time-consuming and expensive task that depends entirely on human skill. Computer Aided Diagnostic (CAD) systems facilitate fast and automated classification of fundus images. This enables ophthalmologists with early screening and diagnosis of DR in reduced time and cost. ML Research in the field of DR are categorized into different classification problems such as DR screening (no-DR vs. DR) and referable-DR (normal to mild-NPDR vs. moderate-NPDR to PDR) classification.

Earlier works in the field of DR detection and gradation are predominantly relied on classical image processing-based handcrafted feature extractions and conventional machine learning models like SVM and Random Forest for DR classification [13], [22]. However, most of the significant contributions in recent times on classification and gradation task of DR are based on deep learning (DL) methods. Instead of handcrafted features, DL approaches inherently rely on Convolutional Neural Networks (CNN) for feature extractions [19], [6], [15]-[11]. This paper investigates the different preprocessing techniques and strategies that are explored in the state-of-the-art DL-based approaches (between 2015 -2020) for automated DR detection tasks, such as DR-screening, referable-DR and DR-grading. The schematic overview of the proposed framework for preprocessing and Deep Convolutional Neural Networks (DCNN) based DR detection is depicted in Figure 8.2.

This study contributes to the need of the missing comparative study on the impact of the different preprocessing strategies in DL-based DR classification tasks. To the best of our knowledge, none of the study till date provides any comprehensive comparative study of the performance and effectiveness of different preprocessing strategies in DL-based DR classification. In this study, we have evaluated performances different preprocessing pipelines formed by combining different contrast, edge enhancement and noise reduction techniques and different intensity normalization techniques, in all three DR classification tasks using a benchmark DL model (ResNet-50) [9] on two publicly available large retinal datasets (Kaggle EyePACS [1] and APTOS [2]).

The rest of the paper is organized as follows – Section 2 provides a brief overview of the different preprocessing techniques and strategies. Section 3 illustrates the experimental setup and implementation details. Section 4 depicts the comparative performance metrics for different combinations of preprocessing techniques on a baseline DL model. Finally, Section 5 concludes this work.

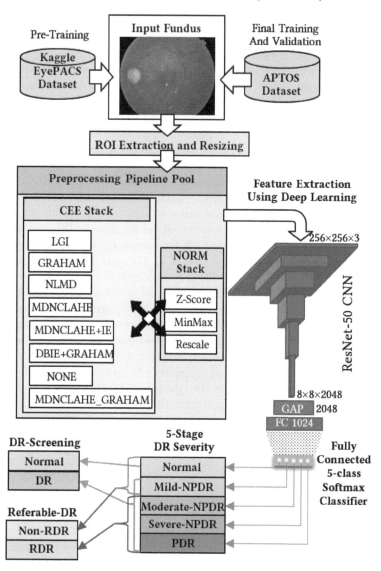

Figure 8.2 Proposed framework for DR classification.

8.2 REVIEW OF PREPROCESSING STRATEGIES

Preprocessing is an integral and crucial part of both conventional image processing for hand-crafted feature extraction, and DL-based approach. It plays an important role in the overall performance of the DR classification models. Since the fundus images in different datasets are captured under different conditions and using different camera settings, they suffer from significant differences in quality. These images possess varying resolutions, non-uniform illuminations, noise and color distortions coming from incorrect focus, and angular positioning of the fundus camera. Therefore, preprocessing techniques are extensively used to unify and enhance the image quality, and sharpen the texture details. This work is restrained to different preprocessing methods that have been employed on the DL-based approaches. The commonly used preprocessing strategies, such as cropping, scale normalization, adaptive thresholding, color space conversions, edge and contrast enhancements, noise reduction and intensity normalization, are discussed below.

In almost all the DL-based DR grading methods, the common preprocessing steps before fitting the input images into the CNN architecture are i) cropping the images around the inner retinal circle to remove the black borders, which contain no information, ii) extracting a square retinal region of interest (ROI) and iii) scale normalizing them by resizing to appropriate dimensions (e.g. 256x256, 448x448 or 512x512 pixels etc.). Vo et al. [19] introduced a novel color space conversion, in which the RGB fundus image are converted into L∗a∗b∗ and $I_1I_2I_3$. A hybrid color space (LGI) is formed by combining the most discriminant channels, from each of the color space i.e. luminance channel L (0-100) from L∗a∗b∗, green channel from RGB and I_1 from $I_1I_2I_3$, which holds most of the chrominance and luminance information and then rescaling the intensity into 0 to 1 range. Doshi et al. [6] proposed a preprocessing strategy which composed of contrast enhancement through Contrast Limited Adaptive Histogram Equalization (CLAHE), and min-max intensity normalization on the extracted green channel, and scaled the images to a fixed resolution of 512x512 pixels.

Another popular preprocessing method ([15] - [11]) for enhancing and normalizing the contrast, and emphasizing the high-frequency components (including blood vessel, the edge of the lesion area, etc.) is the method proposed by Graham et al. [7]. The method is based on linear unsharp masking, where a Gaussian low-pass filtered (*with =ROI radius/30*) average image is subtracted from the ROI image, and the resultant is scaled, and then a constant intensity (*I=128*) image is added to obtain the enhanced output. This method also suppresses the low-frequency information, reduces noise, removes illumination problems and unwanted DC component, and maps the background to gray color.

In another work, Quellec et al. [15] eroded the contrast and edge enhanced ROI images to remove illumination artifacts around the edges. The resulting image is resized and cropped to 448x448 pixels to remove the boundary effects. Wan et al. [20] preprocessed the images by non-local means denoising, and then enhanced edge and contrast using Graham's method and applied z-score intensity normalization of the result. Chen et al. [4] and Lam et al. [10], both used Otsu's thresholding to generate binary ROI mask to extract the circular retinal region through background segmentation. Chen et al. [4] then enhanced and normalized the contrast and enhanced the edges using the method proposed by Graham et al. [7]. On the other hand, Lam et al. [10] normalized the images by subtracting the minimum intensity and dividing by the mean intensity, before enhancement through CLAHE. Orlando et al. [14] normalized the image intensities by subtracting the average image intensities calculated over the entire training set of each Graham's enhanced ROI images. Zhou et al. [22] introduced a distance based illumination equalization technique to minimize the brightness difference between the edge area and center area of the fundus images. Each pixel of the fundus image is weighted based on the distances between their co-ordinates and the fundus centre. The Brightness is balanced by adding the brightness of the original image with the weighted pixel values multiplied by a coefficient found by fitting.

Intensity normalization by z-score normalization [20], [10], [11] is obtained by subtracting the channel-wise mean and then dividing by the channel-wise standard deviation to make them zero mean unit variance. It is the most popular preprocessing method used to standardize the image, and to unify the image illumination, contrast and colour. Many researchers reported that z-score normalization or through mean subtraction has significantly boosted the learning of the DL models and is especially effective for the five-stage DR-grading task [22]. The retinal ROI extraction together with the z-score intensity normalization has been extensively used as a successful preprocessing step for all the three DR classification tasks. The contrast, edge enhancement and noise reduction method based on background image estimation through Gaussian filtering, as proposed by Graham et al. [7], has also been proven as a highly effective and successful preprocessing strategy, which have been adopted by many of the researchers.

From all the above reviewed deep learning models, it has been observed that models' ability to learn both the low and high level features are increased successfully with intuitive and effective preprocessing strategies.

8.3 EXPERIMENTAL SETUP

In this study, we compare and evaluate the efficacy of different preprocessing techniques in CNN-based feature extraction and classification. In Section 2, we have investigated and identified different preprocessing strategies, which have been commonly used by the researchers in DL approaches for the DR gradation and classification tasks with help of baseline DCNN architecture. This section describes the components of the experimental setup used in this work for the comparative analysis and performance evaluation of different preprocessing strategies (Figure 8.1).

8.3.1 DATASETS

The DCNN-based deep learning models have millions of learnable parameters; therefore, they require sufficiently large training set with atleast several thousand of annotated images to achieve effective training, i.e. to successfully learn from the data. In this section, we provide an overview of two publicly available benchmark retinal image datasets, which are suitable for supervised DL approaches for image level DR grading, i.e. have sufficient number images with image-level annotations for DR severity grades.

8.3.1.1 Kaggle EyePACS Dataset

Kaggle EyePACS dataset [1] consists of total 88,702 images with 5 DR stages labeled, with 35,126 images in the train set and 53,576 images in the test set. It provides

Table 8.1
DR class distribution in the Kaggle EYEPACs [1] and APTOS dataset [2]

Class	DR Stage	No. of Image	Accuracy	No. of Image	Accuracy
0	No DR	25810	73.48%	1805	49.29%
1	Mild DR	2443	6.96%	999	27.28%
2	Moderate DR	5292	15.07%	370	10.10%
3	Severe DR	873	2.48%	295	8.05%
4	Proliferative DR	708	2.01%	193	5.27%

high-resolution retinal images taken under different conditions and was provided by EyePACS clinics. Each patient has two images of right and left eyes. The image level annotation was provided by expert ophthalmologists, and each image has been assigned a DR grade on the scale of 0 to 4 that vary from no retinopathy to proliferative retinopathy. The distribution of classes in the dataset is depicted in Table 8.1, it is apparent that the dataset suffers from class imbalance.

8.3.1.2 APTOS Dataset

Asia Pacific Tele-Ophthalmology Society (APTOS) dataset [2] is the most recent dataset on Indian cases with five-class DR grading annotations. It provides with a large set of retinal images taken using fundus photography under a variety of imaging conditions and each image is graded by expert ophthalmologists for the severity of diabetic retinopathy on a scale of 0 to 4. It consists of 3662 training images and 1928 test images with varying resolutions from maximum resolution of 3216x2136 to minimum resolution of 640x480. The details of the dataset is depicted in Table 8.1, it also suffers from class imbalance.

Most the contributions in the field of DR-grading have been trained and evaluated on Kaggle EyePACS dataset. Therefore, we use the Kaggle EyePACS dataset to train our model and APTOS dataset to test the performance. This helps to validate the cross-dataset robustness of the preprocessing strategies.

8.3.2 DEEP CONVOLUTIONAL NEURAL NETWORK (DCNN)

The reviewed works indicate that, DCNN is the most popular choice among the researchers for DR detection tasks, as they are specially designed to efficiently learn and extract meaningful features from the images. CNNs are much computationally efficient compared to fully connected networks in handling images, as they require fewer learnable parameters with the same number of hidden layers, due to the weight sharing and spare connectivity in the convolution layers. Filters or kernels in the convolution layers employ convolution operations to encode local spatial information to detect significant patterns and objects within the image. The lower level convolutional layers learn to detect edges and structures by aligning the filters as edge and

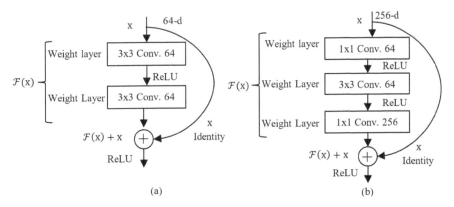

Figure 8.3 ResNet building blocks [9]. (a) Building block for ResNet 34; (b) A bottleneck building block learning a deeper residual function for ResNet 50/101/152.

blob detectors. On the other hand, deeper convolutional layers learn to detect more and more abstract structures and objects, which are scale, rotation and translation invariance, by aligning themselves as high-end feature extractors or image descriptors. It is observed from the reviewed literatures, that VGG-Net, ResNet [9] and their variants have been extensively adopted by the researchers and observed reasonably good performances in all three types of DR classification tasks. Other important models which are successfully used are GoogLeNet (Inception V1), Incetion-V3, Inception-ResNet (Inception V4) and Alex-Net. It is found that ResNet and its variants have outperformed other state-of-the-art CNNs, in both DR screening and DR-gradation tasks. The ResNet model apparently has the better ability to learn the most expressive and discriminative features from the retinal images, which probably contributed in better classification results. This is the rationale behind selecting ResNet as the baseline classification model for the experimental setup. The building blocks for learning of the residual function F in ResNet 34 and ResNet 50/101/152 are depicted in Figure 8.3(a) and 8.3(b), respectively. Taking into account the promising performance of the shallower models like VGG- Net variants and AlexNet in DR-grading and in DR-screening, we select a moderately deep ResNet-50 CNN as the baseline DCNN architecture for the classification tasks.

8.3.3 PREPROCEESSING PIPELINES AND STAGING

Preprocessing strategies for enhancing and standardizing the retinal images precede the feature extraction and DR classification steps in CNN [Figure 8.1].

8.3.3.1 Threshold, Smoothing, Cropping and Resizing

This work performs the following steps:

I k-Means clustering (k = 3 for background, retinal region and bright artifacts) is applied on the histogram equalized and median blurred input image.

II Then the threshold on the clustered image using the minimum non-zero cluster value and also smoothened the resultant binary mask using morphological opening and closing operations.

III Another binary mask is generated by threshold the input image by 10% of maximum intensity and the resultant binary mask is smoothened by re-moving holes and white island using morphological opening and closing operations.

IV The final retinal ROI mask is generated by minimal intersection of these two masks. The boundary of the circular mask is also eroded to remove illumination artifacts around the edges.

V Finally, the intersection output of the retinal images with their corresponding final retinal ROI masks are cropped around the inner retinal circle using the mask boundaries to obtain retinal ROI and to reject the unwanted back-ground and noisy artifacts.

VI The ROI extracted images are scale-normalized and resized to 256256 pixels using bi-linear interpolation, while retain fine-grain artifacts for better feature extraction.

8.3.3.2 Combining Enhancement and Normalization

To identify the most effective preprocessing strategy for DR classification, we select some commonly used preprocessing strategies, which have shown promising results in the reviewed DR works. In the preprocessing pipeline, we consider seven Contrast and Edge Enhancement Strategies (CEE) –

Five existing methods –

I LGI color space conversion (LGI)

II Contrast and Edge Enhancement Method proposed by Graham et al. [7] (GRAHAM)

III Non-Local Means Denoising (NLMD)

IV Median filtering followed by CLAHE (MDNCLAHE)

V Illumination Equalization based on Distance to Center (DBIE) introduced by of Zhou et al. [22] followed by enhancement based on Graham et al. [7] (DBIE_GRAHAM).

Two new methods –

I Illumination Equalization on median filtered CLAHE output (MDNCLAHE_IE)

II Contrast and edge enhancement based on Graham et al. [7] on median filtered CLAHE output (MDNCLAHE_GRAHAM)

Table 8.2
The {CEE, NORM} Pairs of the different Preprocessing Pipelines

SL. No.	{CEE, NORM}	SL. No.	{CEE, NORM}
1	{NONE, ZScr}	13	{MDNCLAHE, MnMx}
2	{LGI, ZScr}	14	{MDNCLAHE_IE, MnMx}
3	{GRAHAM, ZScr}	15	{DBIE_GRAHAM, MnMx}
4	{NLMD, ZScr}	16	{MDNCLAHE_GRAHAM, MnMx}
5	{MDNCLAHE, ZScr}	17	{NONE, Rscl}
6	{MDNCLAHE_IE, ZScr}	18	{LGI, Rscl}
7	{DBIE_GRAHAM, ZScr}	19	{GRAHAM, Rscl}
8	{MDNCLAHE_GRAHAM, ZScr}	20	{NLMD, Rscl}
9	{NONE, MnMx}	21	{MDNCLAHE, Rscl}
10	{LGI, MnMx}	22	{MDNCLAHE_IE, Rscl}
11	{GRAHAM, MnMx}	23	{DBIE_GRAHAM, Rscl}
12	{NLMD, MnMx}	24	{MDNCLAHE_GRAHAM, Rscl}

In addition, no-enhancement after ROI extraction (NONE) is also considered as an option.

We used three intensity normalization strategies (NORM) –

I Z-score normalization (ZScr)

II Min-Max normalization (MnMx)

III Re-scaling (Rscl)

The different preprocessing pipelines consisting of distinct combination of the enhancement and normalization pairs {CEE, NORM}, are listed in Table 8.2.

The pipeline goes as follows: ROI extraction and resizing are performed on the raw retinal images, then the output goes to the enhancement step (CEE) and then enhanced image goes to the normalization (NORM) step. Each of the distinct preprocessing pipeline is applied on the train, validation and test dataset, before feeding the result to the ResNet-50. The output of ROI extraction and different Contrast and Edge Enhancement Strategies are illustrated in Figure 8.4.

8.3.4 IMPLEMENTATION DETAILS

The baseline ResNet-50 (pretrained on ImageNet [5]) is first trained on the preprocessed retinal images from the Kaggle EyePACs dataset with 70%-30% split between train and validation data. For each of the 24 preprocessing pipelines, the model is separately trained for 100 epochs. Then, each of the Kaggle EyePACs pretrained ResNet-50 model of the 24 preprocessing pipelines are further fine-tuned on preprocessed retinal images from APTOS training dataset with 70%-10%-20% split between train, validation and test data, for another 100 epochs for each preprocessing pipeline.

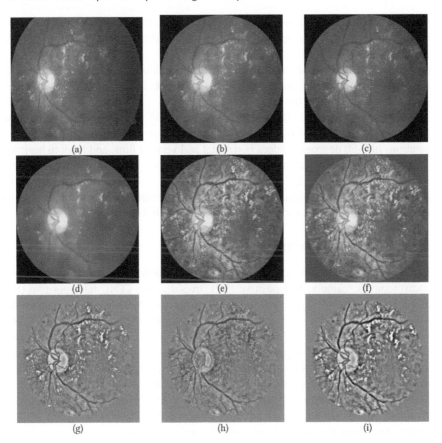

Figure 8.4 Output of ROI extraction and different Contrast and Edge Enhancement Strategies (a) original image (80e7cc0a0649.png); (b) ROI output; (c) ROI followed by LGI; (d) ROI followed by NLMD; (e) ROI followed by MDNCLAHE; (f) ROI followed by MD-NCLAHE_IE; (g) ROI followed by GRAHAM; (h) ROI followed by DBIE_GRAHAM; (i) ROI followed by MDNCLAHEGRAHAM.

The top layers [9] after the global average pooling layers of the pretrained model are dropped off and replaced by a dense layer with 1024 neurons followed by a batch-normalization layer, ReLU activation layer, and a dropout layer (dropout rate of 0.2). The final layer's weights are initialized according to He et al. [8]. Finally a 5-class softmax classifier is added for complete DR grading.

For binary classification of DR screening and referable DR, the predicted labels and probabilities from the softmax classifier are grouped accordingly to produce the predicted classes and their probabilities. The schematic overview of the preprocessing pipelines and DCNN framework for the classification task is illustrated in Figure 8.2. All the models are trained and tested on a single NVIDIA GeForce GTX 1650 GPU using Keras 2.3.1 on Tensorflow 1.14.0 backend. For each classification task and for each preprocessing pipeline, the DCNN is fine-tuned in end-to-end manner with SGD momentum optimizer with an initial learning rate of 0.001 and a fixed batch size of 8.

The learning rate is scheduled with a decrease rate of 0.1, when validation accuracy fails to drop for 10 consecutive epochs. L2 weight decay regularizer with factor of 0.001 is applied to all the layers.

We also increase the effective number of training images in order to increase generalization and reduce over-fitting. Random data augmentations such as random rotations of 0-90 degrees, random horizontal and vertical flips, and random horizontal and vertical shifts are employed to enforce rotation and translation invariances in the deep feature. It also helps to increases heterogeneity in the samples while preserving prognostic characteristics. Random oversampling of minority classes and augmentation together is used to address the class imbalance problem.

8.3.5   EVALUATION METRICS

In a binary classification settings, the evaluation metrics are based on four basic measurements, namely, true-positive (TP), true-negative (TN), false-positive (FP) and false-negative (FN). For measuring the performance of classification tasks like DR-grading, sensitivity (SN) or recall (RE), specificity (SP), accuracy (ACC), precision (PR), Area under the Receiver Operating Characteristic curve (AUC-ROC) and quadratic weighted kappa (κ) score are commonly used. Quadratic weighted kappa (κ) score is an effective weighted measure, especially in assessing classification accuracy in multiclass classification like DR-grading where datasets suffer from class imbalance problems. Equations (1.1), (1.2) and (1.3) depict the three metrics – Accuracy, Quadratic weighted kappa (κ) score and AUC-ROC used in this work to compare performances of different preprocessing approaches.

8.4   RESULTS AND DISCUSSION

In this section, we provide a comprehensive performance analysis of the different preprocessing pipelines explored in this study. Different preprocessing pipelines are evaluated on the APTOS dataset for all the three DR classification tasks and their performances are reported in Table 8.3. The results indicates their effectiveness in aiding the successful feature extraction by the DCNN models. Although all the preprocessing pipelines performed well in supporting the DCNN-based DR classification tasks, the preprocessing pipeline composed of ROI extraction, followed by edge and contrast enhancement using Graham's method (GRAHAM), and intensity normalization through z-score (ZScr) technique {ROI, GRAHAM, ZScr} has outperformed all other methods in majority of the classification tasks, by achieving highest accuracy of 98.5%, 96.51% and 90.59% in DR-screening, referable-DR and DR gradation tasks, respectively.

For multiclass classification tasks like five class DR severity grading, on a dataset with huge class imbalance, the effectiveness and robustness of a classifier model is best reflected by quadratic weighted kappa score (κ). The preprocessing pipeline

Table 8.3

Performance of different preprocessing pipelines in APTOS dataset [2]

SL. No.	ROI+{CEE, NORM}	DR Screening ACC	AUC	Referable-DR ACC	AUC	DR-Stage ACC	AUC	κ
1	ROI+{NONE, Rscl}	0.9741	0.9942	0.9523	0.9858	0.8772	0.9698	0.921
2	ROI+{NONE, MnMx}	0.9795	0.9938	0.9604	0.9867	0.8909	0.9705	0.911
3	ROI+{NONE, ZScr }	0.9659	0.9778	0.94	0.9801	0.8486	0.9615	0.916
4	ROI+{LGI, Rscl}	0.9536	0.9838	0.9632	0.987	0.8527	0.9646	0.889
5	ROI+{LGI, MnMx}	0.9659	0.9922	0.9604	0.9873	0.8813	0.9714	0.911
6	ROI+{LGI, ZScr}	0.9768	0.9949	0.9618	0.9908	0.9045	0.9795	0.932
7	ROI+{NLMD, Rscl}	0.9686	0.9834	0.9441	0.9854	0.8759	0.9697	0.937
8	ROI+{NLMD, MnMx}	0.9536	0.9741	0.9277	0.977	0.8199	0.9504	0.904
9	ROI+{NLMD, ZScr}	0.9741	0.986	0.9523	0.9731	0.8677	0.9631	0.91
10	ROI+{MDNCLAHE, Rscl}	0.9741	0.9856	0.9454	0.9804	0.8759	0.9676	0.932
11	ROI+{MDNCLAHE, MnMx}	0.9741	0.9865	0.9454	0.9807	0.8649	0.9655	0.922
12	ROI+{MDNCLAHE, ZScr}	0.9768	0.9917	0.9645	0.9843	0.9031	0.9734	0.942
13	ROI+{MDNCLAHE_IE, Rscl}	0.9768	0.9884	0.9345	0.976	0.884	0.9671	0.932
14	ROI+{MDNCLAHE_IE, MnMx}	0.9741	0.9847	0.9386	0.9775	0.8745	0.9648	0.923
15	ROI+{MDNCLAHE_IE, ZScr}	0.9686	0.9869	0.925	0.9805	0.8636	0.9686	0.917
16	ROI+{GRAHAM, Rscl}	0.9768	0.9904	0.9209	0.9813	0.8336	0.968	0.884
17	ROI+{GRAHAM, MnMx}	0.9836	0.9923	0.9454	0.9821	0.8799	0.9703	0.923
18	ROI+{GRAHAM, ZScr}	0.985	0.9981	0.9651	0.9882	0.9059	0.98	0.945
19	ROI+{DBIE_GRAHAM, Rscl}	0.9727	0.9965	0.9495	0.9839	0.8636	0.9682	0.905
20	ROI+{DBIE_GRAHAM, MnMx}	0.9754	0.9922	0.955	0.9784	0.869	0.9671	0.915
21	ROI+{DBIE_GRAHAM, ZScr}	0.9714	0.9961	0.9509	0.9817	0.8622	0.9726	0.892
22	ROI+{MDNCLAHE_GRAHAM, Rscl}	0.9495	0.9786	0.9345	0.9775	0.8131	0.9557	0.894
23	ROI+{MDNCLAHE_GRAHAM, MnMx}	0.97	0.9916	0.9563	0.9848	0.8868	0.9745	0.925
24	ROI+{MDNCLAHE_GRAHAM, ZScr}	0.9604	0.9848	0.9372	0.9758	0.8213	0.9617	0.897

{ROI, GRAHAM, ZScr} has achieved the best quadratic weighted kappa score (κ) of 0.945 in DR grading task, followed by the preprocessing method composed of median filtering, CLAHE and z-score intensity normalization {MDNCLAHE, ZScr}, which achieved quadratic weighted kappa score (κ) of 0.942. {MDNCLAHE, ZScr} also achieved comparable accuracy and AUC-ROC score of 90.31% and 0.9734, respectively, in DR grading. However, the pre-processing pipeline {ROI, GRAHAM, ZScr} has been less computationally expensive and has achieved better performance results in all three DR classification tasks than the later combination. {ROI, GRAHAM, ZScr} has achieved the highest AUC-ROC score of 0.98 and 0.9981, in both DR grading and DR screening tasks.

In DR Screening classification task, the pre-processing pipeline based on distance-based intensity equalization and GRAHAM's method of contrast and edge enhancement {ROI, DBIE_GRAHAM, Rscl} has also shown commendable performance by achieving an AUC-ROC score of 0.9965. Whereas in accuracy, the pipeline composed of only ROI extraction and min-max normalization without any contrast and edge enhancement has ended up being second highest, by achieving an accuracy of 97.95%.

In referable-DR classification task, although {ROI, GRAHAM, ZScr} has shown an overall better performance by achieving the best accuracy, and second best

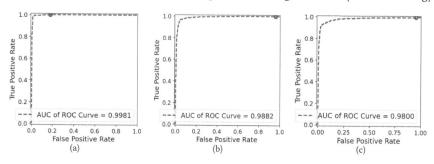

Figure 8.5 ROC curves for the preprocessing pipeline {ROI, GRAHAM, ZScr} for (a) DR Screening; (b) Referable-DR; (c) DR-grading tasks.

AUC-ROC score of 0.9882, the pre-processing pipeline based on the hybrid - LGI color space conversion{ROI, LGI, ZScr} has also proved to be very much effective by achieving the highest AUC-ROC score of 0.9908. In other tasks also, the LGI conversion followed by z-score intensity normalization has manifested promising and comparable results, proving it as an effective contrast and edge enhancement technique. We have tried to improve the performance of LGI conversion-based pipelines, by applying median filtering before the conversion, with and without the CLAHE enhancement of the LGI converted output, but none of such resulted any substantial improvement to its performance. We have also tried to enhance the contrast and edges in the LGI converted output by GRAHAM's methods, but this resulted degradation in the performance than their individual ones, therefore advocated for their incompatibility. Among the intensity normalization techniques, we have found that the z-score normalization has demonstrated overall better performance than the others.

Therefore, from the results summarized in Table 8.2, it is evident that the pre-processing pipeline composed of ROI extraction, followed by edge and contrast enhancement using Graham's method (GRAHAM), and intensity normalization through z-score (ZScr) technique {ROI, GRAHAM, ZScr} has been the most effective preprocessing strategy for all three DR classification tasks from retinal fundus images. The ROC curves of the best performing preprocessing method {ROI, GRAHAM, ZScr} for all three DR classification tasks are depicted in Figure 8.5.

8.5 CONCLUSION

DCNN model with correct preprocessing approaches is the key to rapid and accurate detection and classification of DR stages. This study contributes to the need of the missing comparative study on the impact of the different preprocessing strategies. In this study, we have evaluated performances of combining different contrast, edge enhancement and noise reduction techniques with different intensity normalization techniques. All combinations are tested with DCNN-based DR classification tasks. It is observed that for DR-grading, the pre-processing pipeline composed of ROI extraction, followed by edge and contrast enhancement using Graham's method [7]

(GRAHAM), and then z-score intensity normalization (ZScr) have outperformed all the other preprocessing strategies by achieving highest accuracy in all three classification tasks. It also achieved the best AUC-ROC in majority of the classification tasks and highest quadratic weighted kappa score in DR grading task.

Among the contrast enhancement methods, the LGI colour space conversion has also proved its potential of been effective in all three classification tasks after the GRAHAM's methods. Among the intensity normalization methods, the z-score normalization has manifested overall better performance than others. It is evident from the results that these pre-processing strategies play a significant role in the extraction of meaningful deep features, so that the baseline CNN model has managed to achieve at par performance as that of the other rich deep learning models, in all the three classification tasks. However, performances of the preprocessing pipelines could further increase if the DCNN is trained in an end-to-end manner with more number of epochs and larger batch sizes, which would be investigated in our future work. We would also per-form a comparative study of the performances of different benchmark DCNNs in DR detection, considering the finding of this work (ROI Extraction + GRAHAM + ZScr) as the baseline preprocessing method.

REFERENCES

1. Kaggle diabetic retinopathy detection competition. `https://www.kaggle.com/c/diabetic-retinopathy-detection/data`, 2016. Accessed: 2020-10-22.

2. Kaggle aptos blindness detection competition. `https://www.kaggle.com/c/aptos2019-blindness-detection/data`, 2019. Accessed: 2020-10-23.

3. K. L. Bhutia, N. Lomi, and S. C. Bhutia. Prevalence of diabetic retinopathy in type 2 diabetic patients attending tertiary care hospital in sikkim. *The Official Scientific Journal of Delhi Ophthalmological Society*, 28:19–21, 2017.

4. Y.W. Chen, T.Y. Wu, W. H. Wong, and C.Y. Lee. Diabetic retinopathy detection based on deep convolutional neural networks. *2018 IEEE International Conference on Acoustics, Speech and Signal Processing (ICASSP)*, pages 1030–1034, 2018.

5. J. Deng, W. Dong, R. Socher, L.J. Li, K. Li, and L. Fei-Fei. Imagenet: A large-scale hierarchical image database. *2009 IEEE Conference on Computer Vision and Pattern Recognition*, pages 248–255, 2009.

6. D. Doshi, A. Shenoy, D. Sidhpura, and P. Gharpure. Diabetic retinopathy detection using deep convolutional neural networks. *2016 International Conference on Computing, Analytics and Security Trends (CAST)*, pages 261–266, 2016.

7. B. Graham. Kaggle diabetic retinopathy detection competition report. `https://www.kaggle.com/c/diabetic-retinopathy-detection/discussion/15801`, 2016. Accessed: 2020-10-22.

8. K. He, X. Zhang, S. Ren, and J. Sun. Delving deep into rectifiers: Surpassing human-level performance on imagenet classification. *2015 IEEE International Conference on Computer Vision (ICCV)*, 1:1026–1034, 2015.

9. K. He, X. Zhang, S. Ren, and J. Sun. Deep residual learning for image recognition. *2016 IEEE Conference on Computer Vision and Pattern Recognition (CVPR)*, pages 770–778, 2016.

10. C. K. Lam, D. Y., M. Guo, and T. Lindsey. Automated detection of diabetic retinopathy using deep learning. *AMIA Summits on Translational Science Proceedings*, 2018:147 – 155, 2018.

11. X. Li, X. Hu, L. Yu, L. Zhu, C.W. Fu, and P.A. Heng. Canet: Cross-disease attention network for joint diabetic retinopathy and diabetic macular edema grading. *IEEE Transactions on Medical Imaging*, 39:1483–1493, 2020.

12. H. Looker, S. Nyangoma, D. Cromie, J Olson, G. Leese, M. Black, J. Doig, N. Lee, R. Lindsay, J. Mcknight, A. Morris, S. Philip, N. Sattar, S. Wild, and H. Colhoun. Diabetic retinopathy at diagnosis of type 2 diabetes in scotland. *Diabetologia*, 55:2335–42, 06 2012.

13. R. F. Mansour. Deep-learning-based automatic computer-aided diagnosis system for diabetic retinopathy. *Biomedical Engineering Letters*, 8:41–57, 2018.

14. J. I. Orlando, E. Prokofyeva, M. D. Fresno, and M. B. Blaschko. An ensemble deep learning based approach for red lesion detection in fundus images. *Computer Methods and Programs in Biomedicine*, 153:115–127, 2018.

15. G. Quellec, K. Charrière, Y. Boudi, B. Cochener, and M. Lamard. Deep image mining for diabetic retinopathy screening. *Medical Image Analysis*, 39:178–193, 2017.

16. R. Raman, P. K. Rani, S. R. Rachepalle, P. Gnanamoorthy, S. Uthra, G. Kumaramanickavel, and T. Sharma. Prevalence of diabetic retinopathy in india: Sankara nethralaya diabetic retinopathy epidemiology and molecular genetics study report 2. *Ophthalmology*, 116:311–318, 2009.

17. M. Rema and R. Pradeepa. Diabetic retinopathy: an indian perspective. *The Indian Journal of Medical Research*, 125:297–310, 2007.

18. E. J. Sussman, W. G. Tsiaras, and K. A. Soper. Diagnosis of diabetic eye disease. *JAMA*, 247 23:3231–4, 1982.

19. H. H. Vo and A. Verma. New deep neural nets for fine-grained diabetic retinopathy recognition on hybrid color space. *2016 IEEE International Symposium on Multimedia (ISM)*, pages 209–215, 2016.

20. S. Wan, Y. Liang, and Y. Zhang. Deep convolutional neural networks for diabetic retinopathy detection by image classification. *Computers & Electrical Engineering*, 72:274–282, 2018.

21. S. Zachariah, W. N. Wykes, and D. Yorston. Grading diabetic retinopathy (dr) using the scottish grading protocol. *Community Eye Health*, 28:72 – 73, 2015.

22. Y. Zhou, X. He, L. Huang, L. Liu, F. Zhu, S. Cui, and L. Shao. Collaborative learning of semi-supervised segmentation and classification for medical images. *2019 IEEE/CVF Conference on Computer Vision and Pattern Recognition (CVPR)*, pages 2074–2083, 2019.

9 Cancer Diagnosis from Histopathology Images Using Deep Learning: A Review

Vijaya Gajanan Buddhavarapu
College of Information and Computer Sciences,
University of Massachusetts Amherst, Amherst, MA, USA

J. Angel Arul Jothi
Department of Computer Science, Birla Institute of Technology
& Science, Pilani - Dubai Campus, Dubai, United Arab Emirate

CONTENTS

DOI: 10.1201/9781003246688-9

9.1 INTRODUCTION

Cancer is a group of diseases that involve abnormal growth of cells that could potentially spread to other parts of the body. It is the second major cause of death that led to 9.6 million deaths in 2018 [102]. Over 50% of cancers can be prevented by committing to lifestyle changes in diet, exercise and other habits. Significant drop in cancer-related deaths has been attributed to early screening, diagnosis and treatment [20].

Histopathology is the microscopic examination of tissue structures in order to study the characteristics of diseases. For most types of cancer, histopathology remains the 'gold standard' for the diagnosis of cancer. In this context, a pathologist is tasked with providing a comprehensive and accurate diagnosis of the tissue samples to affirm or deny the malignancy of the sample [14]. Computer-aided diagnosis (CADx) and computer-aided detection (CADe) refer to computer systems and software that assist medical professionals in the interpretation of medical images [19]. CADe/CADx systems are used in cancer diagnostic systems, involving histopathology images, in the following ways: (1) detection and segmentation of region of interest (ROI) or structures such as mitosis, nuclei and stroma, (2) classification of tissue samples as diseased/malignant or benign, and (3) classification of sub-types of a particular type of cancer [79, 1]. It should be noted that CAD systems do not aim to replace medical professionals. The objective of such systems is to assist medical professionals such that the final decision can be made with confidence [28].

9.1.1 RELATED SURVEYS

This article aims to survey the various studies in the literature pertaining to histopathology image analysis for cancer diagnosis (based on cancer type) using deep learning (DL). There are other recent surveys on the application of conventional machine learning and image processing methods for histopathology image analysis [3, 46, 93]. Madabhushi et al. survey different digital pathology tasks that have benefited from the use of machine learning [55]. Jimenez-del-Toro et al. [38] provide an overview of different machine learning methods used for histopathology image analysis. Jothi et al. [39] discuss image processing methods and algorithms for automated cancer diagnosis from histopathology images. There are also surveys [53, 78, 11]

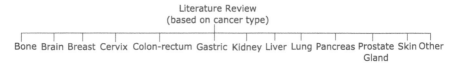

Figure 9.1 Cancer types reviewed.

that discuss the developments, challenges and opportunities of DL in medical image analysis. In line with this study, work has been done to survey histopathology image analysis using DL [16, 9, 6].

9.1.2 MOTIVATION AND OBJECTIVES

Even though research has been underway for developing CAD systems involving histopathology images for over a few decades, there has been a recent surge in interest in this area due to developments in DL research. Hence, the motivation for this review is to provide the reader a comprehensive overview of DL and review of the literature where DL is used for cancer diagnosis from histopathology images.

The objectives of this study are as follows:

- To provide the reader with an overview of DL that includes discussion on the various DL models, state-of-the-art Convolutional Neural Network (CNN) architectures, techniques such as data augmentation and transfer learning, evaluation metrics associated with classification and segmentation, and datasets commonly used to benchmark models for histopathology image analysis.
- To provide the reader a review of the literature on DL application using hematoxylin and eosin (H&E) stained histopathology images for cancer diagnosis.

The works on the following types of cancers are covered in this paper: the bone, brain, breast, cervix, colon-rectum, gastric, kidney, liver, lung, pancreas, prostate gland and skin as shown in Figure 9.1. We also provide a separate section where tasks that don't fall within the previously mentioned cancer types are explored. To the best of our knowledge, there is no previous work that summarizes DL applications in cancer diagnosis using H&E stained histopathology images and categorizes work done by type of cancer.

9.1.3 ORGANIZATION OF THE PAPER

The remainder of this paper is organized as follows: Section 2 presents an overview of DL; Section 3 explains some popular evaluation metrics used to quantify the performance of DL models; Section 4 introduces some of the publicly available histopathology datasets commonly used for benchmarking models; Section 5 describes the literature on DL applications for cancer diagnosis using H&E stained histopathology images, categorized by cancer type; Section 6 discusses recent trends

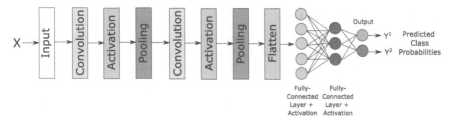

Figure 9.2 A CNN with all its layers. X is the input to the network. Y^1 and Y^2 are the output class labels.

and future scope of DL for histopathology image analysis; Section 7 concludes the paper with a summary.

9.2   DEEP LEARNING: AN OVERVIEW

DL architectures are the state of the art for several vision-related problems. Many studies have shown that these architectures are able to produce remarkable results in various domains. One of the primary reasons for the significant achievements of DL architectures is their ability to learn complex features automatically from the input data. It can adaptively learn features from the input data irrespective of the domain. The forerunner of DL architectures are the neural networks. The initial neural networks had three layers: the input layer, the hidden layer and the output layer. Each layer is made up of neurons. A 'deep' neural network refers to a neural network having several hidden layers. The deep networks or models are superior when compared to their shallow counterparts as they can learn inherent features from the data implicitly as it advances through several layers of processing.

This section introduces various DL neural network architectures used for cancer diagnosis from histopathology images. We will primarily focus on convolutional neural networks (CNNs) and popular named CNNs as they are primarily used for image analysis. Additionally, we will give a brief overview of Recurrent Neural Networks (RNNs), Generative Adversarial Networks (GANs) and Autoencoders (AE) as they have recently found use in histopathology image analysis.

9.2.1   CONVOLUTIONAL NEURAL NETWORK (CNN)

The most commonly used architecture for image-based tasks is convolutional neural network. There are different types of layers in a CNN: the input layer, the convolutional layer, the non-linearity (activation) layer, the pooling/down-sampling layer, the fully connected layer (FC) and the output layer. As opposed to the traditional neural networks, the neurons in the layers of a CNN are three dimensional. Thus, each layer of a CNN takes a 3D input volume, process and transforms the input volume to an output volume. Figure 9.2 shows the general structure of a CNN.

9.2.1.1   Input layer

The input layer will contain the input image.

9.2.1.2 Convolutional layer

The convolutional layer takes a 3D input volume as input. It contains of a set of filters also called as kernels. A filter has a width w, height h and depth d that extends across the depth of the input volume. Each kernel is placed on the top left most corner of the input volume, and the convolution operation is performed between the kernel and the region of the input volume over which the kernel is placed. This process is repeated until the entire input volume is convolved by shifting the kernel over the input volume. The output of a convolution layer is a set of feature maps corresponding to the number of filters in the convolution layer. The feature maps produced by all the filters of a convolution layer are stacked together along the depth dimension to produce the output volume which is then passed to the non-linearity layer. Thus, the convolution layer is tasked with extracting low-level features, and subsequently higher-level abstractions from the input images by applying a set of filters.

9.2.1.3 Activation layer

Generally, real-world problems are non-linear. Therefore, in order to remove linearity from the feature maps calculated during the convolution step, non-linear activation functions are used on the output of each neuron. This step ensures that the output of a neuron is not a linear combination of the input(s). The most commonly used activation functions in neural networks are Rectified Linear unit (ReLu) 9.1, Sigmoid $(\sigma(x))$, Tanh, and Softmax $(f_i(x))$. These activation functions are given by Eqs. 9.1 - 9.4 where x represents the input.

$$ReLu(x) = max(0,x) \tag{9.1}$$

$$\sigma(x) = \frac{1}{1+\exp(-x)} \tag{9.2}$$

$$tanh(x) = \frac{exp(x) - exp(-x)}{exp(x) + exp(-x)} \tag{9.3}$$

$$f_i(x) = \frac{\exp(x_i)}{\sum_j \exp(x_j)} \tag{9.4}$$

Figure 9.3 shows an example convolution and activation operations performed on an input.

9.2.1.4 Pooling layer

The pooling layer is tasked with reducing the spatial size of its input along the width and the height leaving the depth unaltered. This layer is usually placed between two convolutional layers. Since, the size of the input is reduced, this layer reduces the amount of computation performed by the network. Moreover, it helps to avoid over-fitting as the total number of parameters are reduced. Max pooling and average pooling are the most commonly used types of pooling in convolutional neural networks

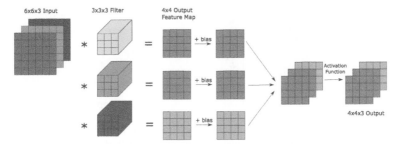

Figure 9.3 Example of convolution and activation operations performed by a CNN. Here, the input is a $6\times6\times3$ image that is convolved with three filters each of dimension $3\times3\times3$. Note that the depth of each filter is equal to the depth of the input image. Convolution of each filter with the input image outputs a corresponding feature map. These feature maps are then stacked together. The feature maps are then passed to an activation layer where an activation function is applied to them. The final output of the convolution layer and activation layer is a $4\times4\times3$ volume.

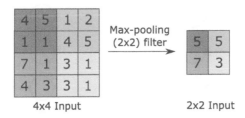

Figure 9.4 Example of the max pooling operation performed by a max pooling layer of a CNN. Assume the input is an activation map of size 4×4, and the max pool filter size is 2×2. The filter divides the input into four regions each of 2×2 size. The maximum value from each of these regions is taken and the output is produced. This results in an output of size 2×2. The size of the output is reduced by half due to this operation. This operation is done on all the activation maps, and the results are stacked together to get the output volume of a pooling layer.

where a filter of some width and height is used to reduce the size of the input volume. Figures 9.4 and 9.5 show the max pooling and average pooling operations, respectively.

9.2.1.5 Fully connected layer

A fully connected layer is a layer where all the neurons in that layer are connected to all the neurons of the previous layer. In a CNN, since the output of the layers (convolution and max pooling) is a 3D volume, the output 3D volume of the layer before the fully connected layer is flattened to obtain a linear set of neurons. The total number of neurons in the flattened output is the product of the width, height and depth of the output volume being flattened. Figure 9.6 shows the structure of a fully connected network.

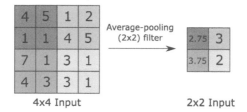

Figure 9.5 Example of the average pooling operation performed by a average pooling layer of a CNN. Assume the input is an activation map of size 4×4, and the average pool filter size is 2×2. The filter divides the input into four regions each of 2×2 size. The mean value from each of these regions is taken and the output is produced. This results in an output of size 2×2.

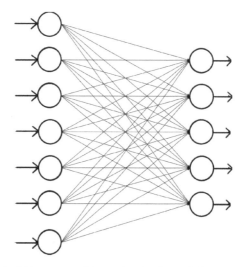

Figure 9.6 A sample fully connected layer showing how all the neurons from one layer are connected to all the neurons of the next layer.

9.2.1.6   Output layer

The output layer is the final layer of a CNN which has a total of C neurons where C is the total number of classes used for a classification task if the problem is a multi-class classification otherwise a single neuron if it is a binary classification. If the problem is binary classification problem a sigmoid activation is used otherwise a softmax activation is used.

The general architecture of a CNN can contain an input layer followed by one or more convolution layers followed by a max pooling layer. There can be several sets of convolutional layers followed by max pooling layers depending upon the depth of the network. Finally, the output volume is flattened which can be followed by few fully connected layers. This is followed by an output layer.

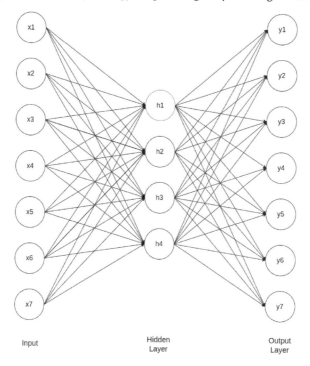

Figure 9.7   An autoencoder network.

9.2.2   AUTOENCODER (AE)

Autoencoders are a feed-forward neural networks that learn features from unlabeled input data in an unsupervised manner. An autoencoder learns an identity function f from the input x and output y. In order to learn useful features from the input, the network is restricted by limiting the number of neurons (bottleneck) in the hidden layers. An autoencoder primarily consists of two parts: encoder network and decoder network. The encoder network encodes its input to a lower dimension and then the decoder network decodes or reconstructs the output from the encoded representation. The output image is compared with the input and the reconstruction error is computed. The network is trained to minimize the reconstruction error. Autoencoders are used to extract features, reduce noise or ignore irrelevant information by the process of reducing the dimension of the input [25, 94]. Figure 9.7 shows an autoencoder network. Common tasks associated with autoencoders and histopathology image analysis are nuclei detection, segmentation of region of interests and classification of images which are further discussions in Section 9.5.

Some of the common types of autoencoders include spare autoencoders (SAE), denoising autoencoders (DAE) and variational autoencoders (VAE).

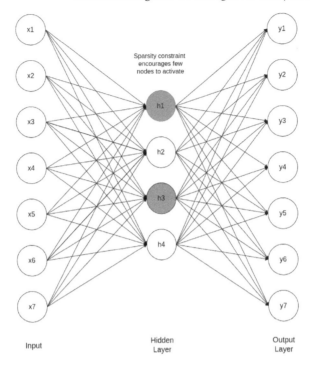

Figure 9.8   Sparse autoencoder (SAE).

9.2.2.1   Sparse Autoencoder (SAE)

A sparse autoencoder consists of an encoder, a decoder and a hidden layer that may have more nodes than the input layer. There is a sparsity constraint on the hidden layer nodes that allows for the learning of important features [62]. L1-regularisation and KL-divergence are common methods applied to enforce sparsity of the representation. Figure 9.8 shows the structure of a SAE.

9.2.2.2   Denoising Autoencoder (DAE)

A denoising autoencoder initially corrupts the input data with noise which is then denoised and reconstructed at the output end. This step inhibits the autoencoder from copying the input data without learning important features [95]. Figure 9.9 shows the structure of a DAE.

9.2.2.3   Variational Autoencoder (VAE)

A variational autoencoder is an autoencoder that follows a latent variable model. VAE is called a generative model because while training, the model learns the

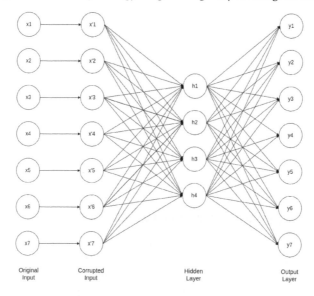

Figure 9.9   Denoising autoencoder (DAE).

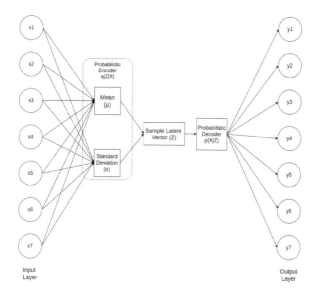

Figure 9.10   Variational autoencoder (VAE).

probability distribution of the dataset. The input is encoded into a latent distribution. A sample is then drawn from the distribution and is then decoded to calculate reconstruction error. Finally the error is backpropagated through the network [45, 73, 18]. Figure 9.10 shows the structure of a VAE.

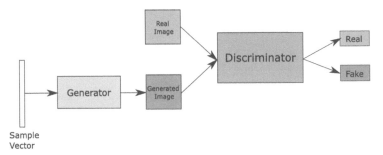

Figure 9.11   Generative adversarial network (GAN).

9.2.3   GENERATIVE ADVERSARIAL NETWORK (GAN)

Generative adversarial networks (GANs) are deep generative models that consist of two main structures: generator and discriminator [26]. The discriminator is a classifier network. The generator network transforms the input sample vector into a possibly meaningful output. During training, the following steps take place:

1. The generator takes in input a sample vector (usually random noise) and outputs transformed sample. When the model is image-based, this transformed sample is a fake image.

2. The transformed sample is then fed into the discriminator network that is tasked with distinguishing whether the sample is real or fake.

3. Using the discriminator output, the weights of discriminator and generator networks are updated.

4. The discriminator is penalized when fake data is predicted as real data and the generator is penalized when fake data is predicted as fake data.

When the generator produces fake data that is close to real data, the accuracy of the discriminator will begin to worsen as it is unable to distinguish between fake and real data. Figure 9.11 shows a GAN. GANs are used for data augmentation in histopathology image analysis which is further discussed in Section 9.5.

9.2.4   RECURRENT NEURAL NETWORK (RNN)

Recurrent neural network (RNN) is another class of artificial neural network that is used to analyze sequence data. The difference between a traditional feed-forward network and a RNN is that in a RNN, the output of a node acts as input to the same node. This gives the RNN two advantages: the ability to store information and the ability to learn sequential data.

RNNs are successfully used in natural language processing (NLP) applications such as image captioning, word level classification and language generation [111]. CNNs have difficultly in working on larger images but it has been shown that RNNs along with CNN can be used to improve image recognition of scale variable images [112]. Similarly, research has shown that combination of RNN and CNN can be used to generate natural sentences that describe an image [42]. RNNs, although not commonly used as compared to CNNs, have seen use in histopathology image analysis in attention-based models for classification [8].

9.2.5 WIDELY USED ARCHITECTURES

This section introduces a few CNN-based architectures that have been widely used and shown great success in various histopathology image processing applications.

9.2.5.1 Image Classification

9.2.5.1.1 LeNet-5

[49] proposed LeNet-5 for recognizing handwritten and machine-printed digits. It consisted of seven layers: three convolutional layers, two subsampling layers, one fully connected layer and one output layer. The input image was of 32×32 size.

9.2.5.1.2 AlexNet

[47] developed a much deeper CNN for classifying images from ImageNet[1]. ImageNet is a large-scale image database formed according to the WordNet[2] hierarchy to enable vision oriented research. AlexNet consisted of five convolutional layers, three max pooling layers and three fully connected layers. To reduce overfitting, data augmentation (translations, reflections, pixel intensity alterations of the RGB channels) and dropout [32] were performed. Dropout helps to eliminate complex co-adaptations between feature detectors by randomly omitting some neurons by setting their output to zero while training.

9.2.5.1.3 VGG-16 and VGG-19

[81] proposed two deep CNN architectures namely VGG-16 and VGG-19 for image classification. The main objective was to improve the accuracy of the CNNs. In this attempt, the authors built deep neural networks using very small convolutional filters (3×3) in all the convolutional layers. VGG-16 and VGG-19 had 16 and 19 weight layers (convolutional and fully connected layers), respectively. The input to the VGG CNNs is 224×224 size RGB image. The weight configuration of VGG-16 and VGG-19 is made publicly available in order to enhance research.

9.2.5.1.4 GoogLeNet

Researchers proposed a CNN architecture named GoogLeNet (Inception-v1) with a distinct basic unit called the Inception cell for the ImageNet Large-Scale Visual

---

[1] http://www.image-net.org/

[2] https://wordnet.princeton.edu/

Recognition Challenge (ILSVRC 2014) [89]. The purpose of the Inception cell was to perform convolutions at various scales which were then aggregated. The architecture had since been improved with initially batch normalization called Inception-v2 and later much more efficient Inception cells called Inception-v3 [90] and subsequently Inception-v4 [88]. GoogLeNet reduced the number of parameters in the network when compared to AlexNet and the variants of VGG.

9.2.5.1.5 ResNet

He et al. proposed a CNN architecture called residual neural network or ResNet that won the first place in ILSVRC 2015 image classification challenge. Increasing the depth of the neural network is often done to increase the accuracy and generalization of the model. However, it is noted that as the network grows deeper, it may reach a saturation point such that the gradient becomes infinitely small. This occurs when the gradient is back propagated through the network as the operations on the gradient cause it to shrink further and further. The implication of this is that it will make the training of earlier layers harder and therefore degrade the accuracy of the model. This would lead to worse test and training error rates. This problem is called as vanishing gradient problem.

ResNet avoids the vanishing gradient problem by employing residual blocks and skip connections or shortcut connections. ResNet hypothesized that letting the layers fit a residual mapping is better than making the layers fit an underlying mapping as seen in regular/simple networks, and it is much easier to optimize a residual mapping. ResNet has various variants like ResNet-18, ResNet-50 and ResNet-101 differing in the depth [29].

9.2.5.1.6 DenseNet

Huang et al. proposed a CNN architecture called dense convolutional neural network or DenseNet [34]. In this network, the feature maps of a layer are available as input to all subsequent layers and the feature maps of all preceding layers are also inputs to that layer. Since the network can use previously calculated feature maps, it has significant advantages like vastly reduces the number of parameters, alleviates the vanishing gradient problem, promotes feature reuse and favours feature propagation.

9.2.5.2 Object Recognition and Object Detection

9.2.5.2.1 Regions with CNN features (R-CNN)

Girshick et al. [24] proposed an object detection network called R-CNN. Firstly, selective search is run on the input image for the generation of 2000 region proposals. Secondly, each region proposal is fed into a CNN (after warping) to extract a 4096-dimensional feature vector. Finally, class-specific Support Vector Machine (SVM) is used to classify each region proposal. Improvements were made to R-CNN in Fast R-CNN [23]. In Fast R-CNN, the input image is processed by a CNN to produce a feature map. Next, region of interest pooling layer (RoIPool) uses this feature map along with object proposal to produce a feature vector. This feature vector is then used to output softmax probabilities and per-class bounding box offsets. Both R-CNN and Fast R-CNN use selective search for region proposal. Faster R-CNN [72], an improvement to the previous models, implements a region proposal network (RPN) for generating region proposals.

9.2.5.2.2   *You Only Look Once (YOLO)*

YOLO (You Only Look Once) is an object detection algorithm proposed by Redmon et al. [69]. YOLO requires only one forward pass through the deep neural network to predict bounding boxes and classes for the whole input image. The algorithm divides the input image into a $S \times S$ grid. Within each grid cell, the algorithm predicts B bounding boxes and confidence scores for those bounding boxes. Additionally, each grid cell also predicts C conditional class probabilities. Using the conditional class probabilities of each grid cell and the confidence scores of bounding boxes, bounding boxes that are weighted by their class-specific confidence scores are predicted. For the final prediction, non-max suppression and thresholding are utilized. Further improvements have been made to the YOLO algorithm [70, 71]. YOLOV2, the improved version of YOLO, was used to train YOLO9000 which can detect over 9000 object categories. YOLOV3 improves YOLOV2 with additions such as residual connections.

9.2.5.3   Image Segmentation

9.2.5.3.1   *U-Net*

U-Net is a fully convolutional deep neural network proposed by Ronneberger et al. [76] for semantic segmentation. As the name implies, the network is in the shape of a symmetric "U." It is divided into the contracting path on the left side and expansive path on the right side. The contracting path consists of multiple blocks of 3×3 convolutional layer and 2×2 max-pooling layer. The expansive path consists of multiple blocks of 2×2 transposed convolutional layer, concatenation of corresponding feature map from contracting path, and two 3×3 convolutional layers. Finally, a 1×1 convolution layer with desired number of classes is applied to predict the final segmentation map.

9.2.5.3.2   *Mask R-CNN*

He et al. [30] proposed Mask R-CNN, an object instance segmentation model that is based on Faster R-CNN. Mask R-CNN extends Faster R-CNN with the addition of a mask branch that runs in parallel with the existing class probabilities and bounding box regression branch. Instead of RoIPool, RoIAlign is used. The authors note that the quantizations in the RoIPool operation introduce misalignments between RoI and extracted features which can negatively impact mask prediction. Hence, RoIAlign, a quantization-free layer that preserves spatial locations is used.

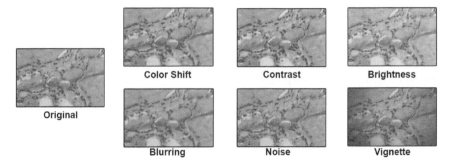

Figure 9.12   Examples of data augmentation.

9.2.6   DATA AUGMENTATION

Datasets of histopathology images are often limited in number of samples for training. As DL methods rely on rich source of data for better performance, one method for enhancing the already existing data samples is to perform data augmentation. Data augmentation consists of different transformations or methods applied to existing data to produce more data that can be used for training the model. Common transformations involve manipulation of data such as image rotations (through a range of angles), image crops, image zooms and image flips (horizontal and vertical flipping). Some of the other more advanced manipulations include vignetting, sharpening and blurring. Generation of new data can be manual or automatic using a generator such as GAN [80, 22]. Figure 9.12 shows a sample histopathology image along with the various transformations applied on it as a result of data augmentation.

9.2.7   TRANSFER LEARNING

Deeper networks are said to perform better than shallow networks. However, deeper the network, the more number of parameters need to be trained. Training a DL network from scratch requires a large amount of computational resources and large enough training set to prevent overfitting for better performance. To overcome the barrier of large training set or insufficient training resources, a technique called Transfer Learning (TL) is used. There are two types of TL: Feature Extractor (FE) and Fine Tuning (FT). Figure 9.13 gives an overview of TL. Both methods use pre-trained networks as the base. Pre-trained networks are DL networks that are already trained on a dataset for a particular task. These networks along with their weights can be used to solve a different task.

In FE, the layers of the pre-trained network except the fully connected layers and the classifier layer are retained. These layers are then frozen, i.e prevented from updating during back propagation. New fully connected and classifier layers are attached to the pre-trained layers. Thus, the pre-trained layers and the new attached layers form the new network. In this case, the pre-trained layers act as a feature

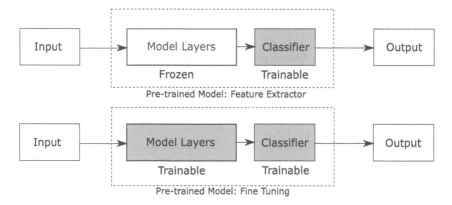

Figure 9.13   Overview of transfer learning.

extractor to the new fully connected and classifier layers. This method is effective when the new dataset is similar to the one used to pre-train the model.

In contrast to FE method of transfer learning, in FT, some of the layers of the pre-trained network are allowed to re-train on the new dataset. Many implementations of fine tuning freeze the first few blocks of the pre-trained network because they hold generic features. This method is effective when the domain of the new dataset is different from the dataset used to pre-train the network. However, this method requires sufficient data to re-train the pre-trained layers.

9.3   EVALUATION METRICS

This section gives a short summary of the common evaluation metrics used to measure the performance of DL-based classification and segmentation models.

9.3.1   EVALUATION METRICS: CLASSIFICATION

A confusion matrix is a table of size $n \times n$ where n is the total number of classes in a classification problem. The actual classes are presented along the row of a confusion matrix, whereas the predicted classes are presented along the column. Table 9.1 describes a confusion matrix for a binary classification problem where the two class labels are denoted as "Yes" and "No".

Let positive samples denote the samples that belong to the class of interest and negative samples denote the samples belonging to other class(es). True-positive (TP) are the positive samples that are correctly predicted as positive samples, false-positive (FP) are the negative samples that are incorrectly predicted as positive samples, true-negative (TN) are the negative samples that are correctly predicted as negative samples and false-negative (FN) are the positive samples that are incorrectly predicted as negative samples. From the confusion matrix, other evaluation metrics for classification are derived.

Table 9.1

Confusion matrix for a binary classification problem

		Predicted Class	
		Yes	No
Actual Class	Yes	True-postive (TP)	False-negative (FN)
	No	False-positive (FP)	True-negative (TN)

9.3.1.1 Accuracy (ACC)

It is the ratio of the total number of predictions that are correct to the total number of predictions made and is given by Eq. 9.5.

$$ACC = \frac{TP + TN}{TP + FP + TN + FN} \tag{9.5}$$

9.3.1.2 Recall

Recall also called sensitivity or True Positive Rate (TPR) is the ratio of true positives to actual positives and is given by Eq. 9.6. Out of the total number of positive samples, TPR gives the proportion of the samples that were correctly predicted. For better performance, it is expected to be high.

$$Recall = \frac{TP}{TP + FN} \tag{9.6}$$

9.3.1.3 Precision

Precision or Positive Predictive Value (PPV) is the ratio of TP to all predicted positives and is given by Eq. 9.7.

$$Precision = \frac{TP}{TP + FP} \tag{9.7}$$

9.3.1.4 F1 Score

F1 score is the weighted harmonic mean of recall and precision and is given by Eq. 9.8.

$$F1\,score = \frac{2 \times Precision \times Recall}{Precision + Recall} \tag{9.8}$$

9.3.1.5   False-positive Rate

False Positive Rate (FPR) is the ratio of false positives to actual negatives and is given by Eq. 9.9. Out of the total number of negative samples, FPR gives the proportion of the samples that do not belong to that class but are predicted as belonging to that class.

$$FPR = \frac{FP}{FP + TN} \tag{9.9}$$

9.3.1.6   Receiver Operator Curve (ROC) and Area Under Operator Curve (AUC)

Receiver operator curve (ROC) is a graph that is plotted between TPR and FPR at different classification thresholds in order to assess the performance of the classification model. The FPR is placed on the x-axis, and the TPR is placed on the y-axis of the graph. An ideal model will embrace the top left most side of the graph indicating a high TPR and low FPR. Area under operator curve (AUC) is the area underneath the ROC that calculates the models ability to discriminate samples from positive and negative classes. The value of AUC varies between the range from 0.5 to 1.0. A model which has perfect prediction has an AUC of 1.0.

9.3.2   EVALUATION METRICS: SEGMENTATION

For the segmentation tasks, a ground truth image is necessary which contains the manual annotation of the region of interest. Many of the previously mentioned metrics like accuracy, precision, recall and F1 score are used to evaluate the segmentation performance. In addition to these, metrics like Dice score co-efficient, Hausdorff distance, Jaccard similarity coefficient are also used to evaluate segmentation performance.

9.3.2.1   Dice Score Co-efficient (DSC)

Dice Score Co-efficient (DSC) or dice index is a measure of spatial overlap between two objects. Its value ranges from 0 to 1. DSC value of 0 indicates no spatial overlap and DSC value of 1 indicates complete overlap. Let A and B denote the set of pixels belonging to the object of interest in the ground truth and segmented image, respectively. DSC is defined as in Eq. 9.10.

$$DSC(A, B) = \frac{2|A \cap B|}{|A| + |B|} \tag{9.10}$$

9.3.2.2   Hausdorff Distance

The similarity between two shapes is given by Hausdorff distance. It is the maximum of the minimum distance between every point of one contour to every other point on the other contour. Hausdorff distance value of 0 indicates that the two shapes are similar, whereas the shapes are dissimilar as the value increases. Let A denote the automatically segmented contour and G denote the ground truth contour. Then the Hausdorff distance between the two contours A and G is given by Eq. 9.11.

$$d(A,G) = min_{a \in A}\{min_{g \in G}\{d(a,g)\}\}$$ (9.11)

9.3.2.3 Jaccard Similarity Cofficient

Jaccard similarity coefficent, also known as Intersection over Union (IoU), is another commonly used metric for segmentation tasks. The value of IoU ranges from 0 to 1. The measure of IoU emphasizes the similarity between the two finite sample sets. IoU of 0 indicates that the sets do not share any members and IoU of 1 indicates that the sets share all members. The Jaccard similar coefficient between A and B is given by Eq. 9.12.

$$J(A,B) = \frac{|A \cap B|}{|A \cup B|} = \frac{|A \cap B|}{|A| + |B| + |A \cap B|}$$ (9.12)

9.4 DATASETS

This section details some of the popular and publicly available datasets used by researchers in CAD systems for cancer diagnosis from histopathology images. Though, most of the studies obtain dataset privately from medical institutions and universities, the following datasets are publicly available either through research challenges or made accessible online by various research groups.

9.4.1 DATASETS AVAILABLE THROUGH RESEARCH GROUPS

One of the widely used publicly available source of histopathology images is The Cancer Genome Atlas (TCGA)[3]. It hosts over 10,000 H&E stained whole slide images (WSIs) from various cancer types. Another public source for histopathology images is Virtual Pathology at University of Leeds[4]. The Virtual Pathology Slide Library is an online source that hosts digital slide data. For research on tissue microarray data, The Standford Tissue Microarray Database (TMAD) hosts a variety of information that has been disseminated for use[5]. The UCSB Bio-Segmentation Benchmark dataset[6] provides about 50 H&E stained breast histopathology images used for breast cancer cell detection with associated ground truth images. The KIMIA Path960 dataset[7] is provided by a research group of the KIMIA Lab, University of Waterloo, Ontario, Canada. It consists of 960 image patches belonging to 20 different categories (tissue types).

---

[3]https://cancergenome.nih.gov/
[4]https://www.virtualpathology.leeds.ac.uk/
[5]https://tma.im/cgi-bin/home.pl
[6]https://bioimage.ucsb.edu/research/bio-segmentation
[7]http://kimia.uwaterloo.ca/kimia_lab_data_Path960

Table 9.2
BreakHis dataset breakdown

Magnification	Benign	Malignant	Total
40×	652	1,370	1,995
100×	644	1,437	2,081
200×	623	1,390	2,013
400×	588	1,232	1,820
Total number of images	2,480	5,429	7,909

Breast Cancer Histopathological Image Classification (BreakHis)[8] is one of the most popular datasets used for automated breast histopathology image analysis purposes [86]. It consists of over 9,000 images of breast tumor tissues from over 80 patients. All images are of 700×460 pixels, 3-channel RGB, 8-bit depth in each channel and PNG format. The images are divided into groups based on magnification (40×, 100×, 200×, 400×) and class (benign or malignant) as shown in Table 9.2. The tissue samples are H&E stained. The types of benign tumors considered are adenosis (A), fibroadenoma (F), phyllodes tumor (PT) and tubular adenoma (TA). The types of malignant tumors considered are carcinoma (DC), lobular carcinoma (LC), mucinous carcinoma (MC) and papillary carcinoma (PC).

Janowczyk et al. [37] presented a tutorial to aid researchers on applying DL to several digital pathology tasks like: nuclei segmentation, epithelium segmentation, tubule segmentation, Invasive Ductal Carcinoma (IDC) segmentation, lymphocyte detection, mitosis detection and lymphoma sub-type classification. Among these tasks, nuclei segmentation, epithelium segmentation, IDC segmentation, lymphocyte detection, mitosis detection and lymphoma sub-type classification used breast cancer histopathology images. Tubule segmentation task uses colorectal cancer histopathology images. The datasets (images) used in this study along with the necessary ground truth is made available publicly[9].

9.4.2 DATASETS AVAILABLE THROUGH RESEARCH CHALLENGES

Research challenges pose certain objectives to be achieved and are a source of publicly available datasets. Some of the popular histopathology challenges in the past are summarized as follows:

Mitosis detection in breast cancer histological images MITOS-12[10] was a contest to detect mitosis from 100 H&E stained breast cancer histological images. MITOS-ATYPIA-14 research challenge[11] had two objectives: 1) to detect mitosis 2) to score

---

[8] https://web.inf.ufpr.br/vri/databases/breast-cancer-histopathological-database-breakhis/
[9] http://www.andrewjanowczyk.com/deep-learning/
[10] http://ludo17.free.fr/mitos_2012/index.html
[11] https://mitos-atypia-14.grand-challenge.org/Home/

nuclear atypia from H&E stained breast cancer histological images. The dataset contained 284 image frames at 20× magnification for nuclear atypia scoring and 1136 image frames at 40× for mitosis detection. All images were RGB images and were in Tag Image File Format (TIFF) format.

The Bioimaging 2015 challenge[12] aimed at classifying H&E stained breast histology images into four classes: normal, benign, in situ carcinoma and invasive carcinoma. This dataset has been extended by increasing the number of images in the dataset[13].

The Gland Segmentation (GlaS) challenge[14] was hosted in 2015 to segment gland structures from H&E stained colorectal cancer histopathology images. The dataset consisted of 165 images from different cancer grades along with the ground truth in bitmap image file (BMP) format.

CAMELYON16[15] and CAMELYON17[16] are a series of challenges whose objective was to evaluate algorithms to detect and classify cancer metastasis in lymph node histopathology images. The data in these challenges contained a total of 400 and 1000 H&E stained whole-slide images of lymph node sections, respectively.

The objectives of ICIAR2018 BACH challenge[17] were twofold as follows: 1) To automatically classify H&E stained breast histology images into four classes: normal, benign, in situ carcinoma and invasive carcinoma and 2) to perform pixel-wise labelling of whole-slide breast histopathology images into the above four classes. The challenge provided around 400 labeled microscopy images (extension of Bioimaging 2015 challenge dataset), and 10 pixel-wise labeled and 20 non-labeled WSIs.

Multiorgan nuclei segmentation (MoNuSeg) 2018[18] was a challenge to segment nuclei from H&E stained tissue images of various organ cancers. The dataset was built by sourcing tissue images at 40× magnification from the TCGA archive. The dataset consists of 30 training images with around 22,000 nuclear boundary annotations and 14 test images with around 7,000 nuclear boundary annotations.

9.5   REVIEW OF THE LITERATURE

This section details the work done on each cancer type. Under each cancer type, the works are further divided into three sections: Classification tasks, segmentation tasks, and combined and other tasks. In the combined and other tasks section, work that involves both classification and segmentation or work that does not fall under the mentioned categories are detailed.

---

[12] http://www.bioimaging2015.ineb.up.pt/challenge_overview.html
[13] https://rdm.inesctec.pt/dataset/nis-2017-003
[14] https://warwick.ac.uk/fac/sci/dcs/research/tia/glascontest/
[15] https://camelyon16.grand-challenge.org/
[16] https://camelyon17.grand-challenge.org/
[17] https://iciar2018-challenge.grand-challenge.org/
[18] https://monuseg.grand-challenge.org/Home/

9.5.1 BONE

This section covers the works done on histopathology images of bone and bone marrow tissue. Research has been done to detect osteosarcoma cells and different bone marrow cells.

9.5.1.1 Classification Tasks

Mishra et al. proposed a convolutional neural network for the classification of tumor and non-tumor from osteosarcoma histopathology slides [57]. The authors experimented with different architectures of CNN for performance and selected the five-layer model for osteosarcoma classification. The model utilized the simplicity of LeNet and data augmentation methods of AlexNet while being faster than both in running time. The proposed method achieved significantly better results than LeNet and AlexNet, obtaining accuracy, precision, recall and F1 score of 0.84, 0.89, 0.84 and 0.86, respectively.

9.5.1.2 Combined and Other Tasks

Song at al. proposed a parallel network called a synchronized asymmetric hybrid deep autoencoder network (Syn-AHDA) that simultaneously detected and classified erythroid and myeloid cells from bone marrow trephine histology images [84]. This model was novel as it combined both detection and classification of cells using a single DL network thereby reducing the training time. The architecture was based on the fact that both the detection and the classification networks rely on the same high level features. The Syn-AHDA network was built using a stacked autoencoder (AE) neural network. The network consisted of the following four subnetworks: an input network, a connection network, a classification network and a detection network. The input network extracted high level features from the input image (I). These features were then fed to the classification and the detection network via the connection network. The F1 scores of the proposed model for the detection and the classification tasks were 0.9466 and 0.8795, respectively. The model was accurate when compared with other DL architectures for detection and classification tasks and also exhibited reduced training time. Moreover, the model performed better in identifying irregularly shaped cells.

The research works on bone histopathology images are summarized in Table 9.3.

9.5.2 BRAIN

This section covers the works done on brain histopathology images. Studies included in this section have proposed methods to detect glioma, a type of malignant brain tissue, and classify the glioma into disease stages.

9.5.2.1 Classification Tasks

Xu et al. proposed a CNN and support vector machine (SVM) model for the tasks of segmentation and classification of brain tumor histopathology images [104]. The

Table 9.3
Summary of work done on bone cancer histopathology images

Reference	Objectives	Datasets	Architectures	Approaches Used	Results	Contributions	Potential Improvements
Mishra et al. (2017)	Classification of tumors in osteosarcoma histopatholoy slides.	40x Magnification	Proposed CNN LeNet AlexNet	Data Augmentation	**Proposed CNN** Accuracy = 0.84 Precision = 0.89 Recall = 0.84 F1 Score = 0.86	Accurate and efficient method for osteosarcoma classification.	The method currently uses tiles obtained from WSI. It could be extended to classification of WSIs.
Song et al. (2017)	Simultaneous detection of and classification of cells from bone marrow trephine histology images.	52 images University Hospitals Coventry and Warwickshire(UHCW) 40x Magnification	Syn-AHDA		**Detection Task** Precision = 0.9266 Recall = 0.9674 F1 Score = 0.9466 **Classification Task** Precision = 0.8712 Recall = 0.8879 F1 Score = 0.8795	Obtained better detection and classification than conventional and well-known deep learning networks. Reduced number of layers.	The high-level features of detection or classification may become inaccurate because of the interaction between classification and detection mechanisms. Improvement could be explored.

classification challenge was to classify the given brain histopathology image into two classes, namely, Glioblastoma Multiforme (GBM) or Low Grade Glioma (LGG). The segmentation challenge was to segment regions of tissue as necrotic and non-necrotic. The proposed method for classification extracted features from the input image patches using a CNN. Relevant features were then identified and selected from the extracted features. Finally a SVM classifier was used to classify the images as GBM or LGG. For the segmentation challenge, the features of the input patches from the CNN were given to a linear SVM classifier which classified the patches as necrotic or non-necrotic. This is based on the reasoning that if the necrotic area of the patch is greater than 50% it is considered as a necrotic patch. Classification confidence maps were then generated for each pixel within a patch. The final segmentation was obtained by the threshold generated by cross-validation. The accuracy of the classification and segmentation tasks was 97.5% and 84%, respectively. Since this study had small training dataset, the authors transferred the deep convolutional activation features learned from ImageNet to classify and segment hitopathology images.

Yonekura et al. proposed a CNN architecture that classified two stages of brain tumor using histopathology images [108]. LGG is a type of lower stage malignant brain tumor that has better prognosis than high-grade glioma like GBM. GBM is a type of High Grade Glioma (HGG) with poor prognosis. The proposed CNN model had two convolutional layers, two pooling layers, one fully connected layer and finally a softmax layer. The proposed model exhibited accuracies of 99.8% and 99.3% for training and testing, respectively.

In a follow-up paper, Yonekura et al. proposed a modified CNN that performed classification of glioma histopathology images into LGG and HGG [109]. The objective of this study was to improve the generalization of the proposed CNN by considering more number of input images as opposed to the previous study. The proposed CNN had three convolution layers, three pooling layers, a fully connected layer and a final softmax layer. The filter size was 5×5 for the convolutional layers and 2×2 for the pooling layers. The accuracies for training and testing for this study were 98.2% and 87.2%, respectively. The authors of the study noted that the results were less accurate compared to their previous CNN. Also, the classification accuracy of the proposed CNN was lower than the training accuracy which indicated that generalization is insufficient. Hence, the authors realized the need for a better model.

In another followup study, Yonekura et al. built a much more deeper CNN to classify the images between the two stages of GBM in brain tumor histopathology images [110]. The proposed CNN architecture was made up of seven convolution layers, eight ReLU and six pooling layers. The filter size was 5×5 for the convolutional layers and 3×3 for the pooling layers. The number of filters in the seven convolutional layers were 32, 32, 48, 48, 64, 64 and 128, respectively. Similarly, the number of filters in the 6 pooling layers were 32, 32, 48, 48, 64 and 64 respectively. The authors compared the proposed CNN architecture with that of LeNet, VGGNet and ZFNet in classifying the images into LGG and GBM. The results showed that the proposed CNN architecture yielded an accuracy of 96.5%. Among the other architectures used for comparison, the VGGNet achieved the highest accuracy of 97.0%.

The research works on brain histopathology images are summarized in Table 9.4.

9.5.3 BREAST

This section covers works done on breast cancer histopathology images. The studies in this section focus on classification of malignant and benign breast tissue images, and segmentation of important histological structures from breast histopathology images.

9.5.3.1 Classification Tasks

Xu et al. classified patches of breast cancer histopathology images into nuclei and non-nuclei using stacked sparse autoencoder (SSAE) [103]. The advantage of using an autoencoder is its ability to learn representation of images from low-level features. The method used two layers of stacked autoencoder followed by a softmax classification layer. Initially the input was presented to the first sparse autoencoder to obtain the primary features. The primary features were then fed to the second sparse autoencoder to obtain the secondary features. The feature learning ability was then tested by sending the secondary features to a softmax classifier. The system showed positive results in learning high-level features without supervision with a classification accuracy of 83.7%.

Sun et al. compared different deep architectures like: Caffenet, GoogLeNet and ResNet-50 for classifying breast cancer histopathology images from the BreakHis dataset, with and without transfer learning while taking into account crop-size [87]. The results showed that transfer learning has an edge over training the network from the scratch. It was found that transfer learning coupled with proper amount of contextual information improved performance. Additionally, it was noted that it is important to match between the scale of the kernel and the scale of discriminatory structures in the histopathology images.

Tumor associated stroma is a diagnostic biomarker that was previously neglected in the diagnosis of breast cancer. Bejnordi et al. utilized stromal morphological features to classify WSI breast tissue images into cancer or benign using CNN [4]. The authors exploited two CNN architectures: The first CNN named as CNN I was an altered VGGNet architecture, and the second CNN named as CNN II was the standard VGG architecture. CNN I classified the WSI into three regions: epithelium, fat and stroma, whereas CNN II operated on the stromal regions and classified it into tumor associated stroma or normal stroma. Features representing each regions distinguished by both CNNs were then extracted. A random forest classifier was used to classify the features. The overall classification accuracy was 0.921. CNN I and CNN II achieved pixel level classification accuracies of 95.5% and 92.0%, respectively. Therefore, the results are in favour of the fact that stromal features can be used to distinguish normal and cancerous breast tissues.

Araújo et al. proposed a method for both binary classification and multi-class classification of breast histopathology images [2]. The images were divided into patches and then augmented. Then a CNN and a CNN + SVM architectures were

Table 9.4

Summary of work done on brain cancer histopathology images

Reference	Objective	Dataset	Architecture	Approach Used	Result	Contribution)	Potential Improvement
Xu et al. (2015)	Classification of histopathology images into GBM and LGG. Segmentation of tissue into necrotic and non-necrotic.	TCGA	AlexNet derived CNN	Transfer Learning	Classification Accuracy = 97.5% Segmentation Accuracy = 84 %	Proposed effective classification method for GBM and LGG. Proposed segmentation method for necrotic and non-necrotic tissue.	The model could be extended to classify sub-grades of glioma.
Yonekura et al. (2017a)	Classification of histopathology images into LGG and HGG.	TCGA	Self-proposed Architecture CNN		Training Accuracy = 99.8% Testing Accuracy =99.3%	Proposed a highly accurate model for the classification of LGG and HGG.	The model could be extended to classify sub-grades of glioma. The effects of pre-processing could be studied further.
Yonekura et al. (2017b)	Classification of histopathology images into LGG and HGG.	TCGA	Self-proposed Architecture CNN		Training Accuracy = 98.2% Testing Accuracy =87.2%	Follow up study that utilized more patient data and deep model for classification of LGG and HGG.	Generalization of model could be improved further.
Yonekura et al. (2018)	Classification of histopathology images into LGG and HGG. Compared results with LeNet, ZFNet and VGGNet.	TCGA	Self-proposed Architecture CNN, LeNet, ZFNet & VGGNet		Classification Accuracy (CNN) = 96.5% Best performing architecture (VGGNet) = 97%	Follow up study that studied various architectures for classification of LGG and HGG.	The proposed model could be extended to classify subtypes.

trained on the augmented patches. The patch-wise results were used whole slide classification using either: majority voting, maximum class probability or sum of probabilities. The results indicated that the performance of both networks were comparable. The performance of the networks was comparable to state-of-the-art and the networks showed a much higher sensitivity that is of interest to pathologists.

Wahab et al. proposed a DL method for detection of mitotic cells using a two-phase model [96]. The first phase used a CNN to identify easy and hard mitotic and non-mitotic cells. The easy non-mitotic images were undersampled and all mitotic images and hard non-mitosis images were augmented for balancing. The images together form the modified training set which is passed into another CNN that is trained to identify mitotic and non-mitotic cells. The results were compared to a CNN with imbalanced dataset and a CNN with randomly sampled dataset. The results showed that the balanced dataset performed better than methods using handcrafted features and CNN derived features.

For effective classification or detection using DL models, a sizable amount of annotated training data must be available. But, annotating regions of WSIs is a long and expensive process. In response, a deep manifold autoencoder (DMAE) was implemented to extract hierarchical features from unlabeled breast histopathology image patches [21]. The encoder layers were then connected to a softmax classifier. The model was then trained on labeled images using fine-tuning. The trained network was used to predict the class label of test patches. This method was compared to SSAE + softmax model and six other machine learning models. The proposed method showed better classification accuracy for all magnification levels at both image and patient level than SSAE + softmax model. The proposed model also outperformed the other machine learning models.

Usually, WSI classification involves extraction of multiple image patches and annotating each patch for decision making. This can be overcome by multiple instance learning (MIL) which does not require the time consuming patch level labelling. Das et al. [17] proposed a MIL-CNN model to classify breast WSI images into malignant or benign. The MIL-CNN used the VGGNet architecture along with a pooling layer called multiple instance pooling (MIP). The network was trained using transfer learning on bags along with the class labels of the bags. A bag contained set of image patches. The MIP layer aggregated the patch level predictions to predict the class outcome for a bag. The method was experimented with images of different magnifications from the BreakHis dataset. It obtained 89.52%, 89.06%, 88.84% and 87.67% classification accuracies for $40\times$, $100\times$, $200\times$ and $400\times$ magnifications respectively.

Naylor et al. proposed and compared different methods for prediction of Triple Negative Breast Cancer (TNBC) treatment efficiency using histopathology WSIs [60]. Residual Cancer Burden (RCB) is a metric used to quantify residual disease after neoadjuvant chemotherapy. In the order of increasing amount of residual disease, it is divided into four classes: RCB-0 (pCR), RCB-I, RCB-II and RCB-III. RCB-0 (pCR) denotes complete pathological response and RCB-III denotes extensive amount of residual disease. The first task was to classify between pCR and RCB-I/RCB-II/RCB-III. The second task was to classify between pCR/RCB-I and

RCB-II/RCB-III. There were three approaches used: (1) Manual feature extraction and random forest models, (2) encoded features extracted from ResNet-152 with uniform sampling + K-means algorithm and random forests and (3) encoded features extracted from ResNet-152 and neural network with bottleneck. The results indicate that the supervised algorithm involving ResNet was able to capture important features without manual feature extraction.

Romero et al. proposed a method for the classification of Invasive Ductal Carcinoma (IDC) slides by modifying the InceptionNet architecture [75]. The modified inception block incorporated batch normalization after each convolution. The proposed method obtained a balanced accuracy of 0.890 while the original InceptionNet architecture obtained a balanced accuracy of 0.868.

Romano et al. proposed an architecture for classification of IDC histopathology images [74]. The architecture consisited of a CNN with a Accept-Reject pooling layer. The balanced accuracy obtained was 0.8541.

Murtuza et al. proposed Biopsy Microscopic Image Cancer Network (BMIC_Net) model for eight-class classification of BreakHis dataset [59]. The model consisted of hierarchical classification approach that consisted of three models. The first level model predicted whether the image was benign or malignant and the second level models predicted the four subtypes associated with benign and malignant. The study compared the performance of extracted features of BMIC_Net with traditional machine learning algorithms. The results showed that the proposed hierarchical approach performed better than one-level classification models.

9.5.3.2   Segmentation Tasks

Naylor et al. trained three different architectures of CNNs, namely, PangNet [65], Fully Convolutional Net (FCN) and DeconvNet [63] to automate the segmentation of nuclei from H&E stained tissue samples of breast cancer histopathology images [61]. Additionally, the authors performed post-processing using watershed algorithm that split touching cells, preventing them from being recognized as a single object. Among the experimented architectures, DeconvNet had the best performance with a classification accuracy of 0.954. An ensemble network, consisting of DeconvNet and FCN was also trained and tested, which yielded a classification accuracy of 0.944. The authors also noticed that the simpler PangNet failed when the nuclei were not homogeneous. On the other hand deeper networks like FCN and DeconvNet were able to capture the nuclei accurately even in the presence of variability. This is because the deeper networks were able to learn better features.

9.5.3.3   Combined and Other Tasks

Janowczyk et al. applied DL for several digital pathology tasks like nuclei segmentation, epithelium segmentation, IDC segmentation, lymphocyte detection and mitosis detection from breast cancer histopathology images [37]. The authors utilized the AlexNet architecture, with task specific modifications, for the tasks mentioned. **(a) Nuclei Segmentation Task:** Nuclear morphology plays a huge role for the detection of cancer and nuclei detection is challenging due to overlapping nuclei border. The

Table 9.5

Summary of work done on cervix cancer histopathology images

Reference	Objective	Dataset	Architecture	Approach Used	Result	Contribution	Potential Improvement
Tian et al. (2019)	Classification of WSIs from patch tiles	800 WSIs (300 patients) Xijing hospital, Fourth Military Medical University 4x and 40x Magnification	Inception-v3 InceptionResNet-v2 ResNet-50	Data Augmentation Image Processing	**AUC** Inception-v3 = 0.9476 InceptionResNet-v2 = 0.9621 ResNet-50 = 0.9593 Ensemble = 0.9784 **FROC Score** Inception-v3 = 0.5279 InceptionResNet-v2 = 0.5417 ResNet-50 = 0.5338 Ensemble = 0.5609	Proposed DL method for locating tumor regions and classifying WSIs	The localization of tumor regions could be improved.
Song et al. (2014)	Segmentation of nuclei from cervical histopathology images	53 Slides (200 patients) Sixth People's Hospital of Shenzhen 40x Magnification	Self-proposed Architecture CNN	Image Processing	Precision = 0.9143 ± 0.0202 Recall = 0.8726 ± 0.0008	Proposed state-of-the-art method for segmentation of nuclei that outperformed other methods.	Experiments on different scales of magnification could be performed.
Wang et al. (2017)	Segmentation of basal membrane for diagnosis of MIC	200 Images International Peace Maternity & Child Health Hospital of China Welfare Institute, Shanghai	Self-proposed GAN	Data Augmentation Transfer Learning	AUC = 0.64	Proposed DL method for segmentation of basal membrane for detection of MIC.	Task could be extended to WSIs and segmentation of blurry basal membrane.
Yang et al. (2018)	Segmentation of WSI into nuclei from segmented lesion region	138 WSIs International Peace Maternity & Child Health Hospital of China Welfare Institute, Shanghai 4x and 20x Magnification	U-Net MTC-Net SDAE ED	Data Augmentation	D1 = 0.8426 D2 = 0.7338 Score = 0.7792	Proposed DL workflow that segments both lesion region and nuclei.	The accuracy of the model could be improved.

results of this task showed a F1 score of 0.83 across 12,000 nuclei. (**b**) **Epithelium Segmentation Task:** The separation of epithelium and stroma is important because regions of cancer usually appear in the epithelium. The results of this task showed a F1 score of 0.84 across 1735 regions. (**c**) **IDC segmentation:** IDC is a common subtype of cancer and in this task, the slide is given an aggressiveness score. The aggressiveness grade is based on regions containing IDC and therefore need to be segmented. The results of this task showed a F1 score of 0.7648 on 50,000 testing patches. (**d**) **Mitosis Detection:** The aggressiveness of cancer correlates to the number of mitoses occurring per area. Accurately identifying mitoses is challenging as pathologists need to make sure that they eliminate false positives. The results of this task showed a F1 score of 0.53 across 550 mitotic events. (**e**) **Lymphocyte Detection:** The number of lymphocytes in an area increase in the presence of disease. Therefore, the identification, location and density of lymphocytes in an area is important in recommending correct treatment to patients. The results of this task showed a F1 score of 0.90 across 3064 lymphocytes.

CNN models that process WSIs are very hard to train and test. Consequently, researchers opt for CNN models that operate on image patches. However, with patch-based models, the inclusion of non-relevant areas is frequent and this is computationally inefficient. To address this, [8] proposed a system, with a recurrent visual attention model, that classified tissues as well as learned to discriminate between areas in a limited number of observations or glimpses. The proposed system was used to predict macro- and micro-metastases in sentinel lymph nodes of breast cancer patients on a slide-level. The proposed method was effective at discrimination and classification with a WSI-level classification accuracy of 0.95 ± 0.2.

The research works on breast histopathology images are summarized in Tables 9.6 and 9.7.

9.5.4   CERVIX

This section covers the work done on cervical histopathology images. Studies have been done to segment the basal membrane from histopathology images of the cervix, nuclei from cytoplasm in cervix cells, nuclei from lesion regions and to classify WSI based on patches.

9.5.4.1   Classification Tasks

Tian et al. proposed a method for locating squamous cell carcinoma tumor regions and classifying tumor WSIs from normal WSIs [92]. WSIs were pre-processed by extracting the tissue regions (ROI) using Otsu's thresholding and morphological operations. Later 256 x 256 patches were obtained from the ROIs. Three popular CNNs (Inception-v3, InceptionResNet-v2, ResNet-50) and their ensemble were experimented for classifying between tumor and normal patches. After this, every WSI was transformed to a probability heatmap. Twenty-eight morphological and geometrical features were extracted from the heatmap. A random forest classifier was used to discern the tumor WSIs from the normal WSIs using the extracted features.

Table 9.6
Summary of work done on breast cancer histopathology images (1)

Reference	Objective	Dataset	Architecture	Approach Used	Result	Contribution	Potential Improvement
Xu et al. (2014)	Classification of nuclei using autoencoder from breast histopathology images	37 WSIs (from 17 patients) 20x Magnification	Self-proposed Architecture Autoencoder SSAE		**SSAE+SM** Precision = 76% Recall = 89% F1 Score = 82% AUC-0.8992 Accuracy = 83.7%	Proposed unsupervised method for feature extraction and classification.	The task could be extended to WSIs for nuclei classification
Sun et al. (2017)	Compared the effect of fine tuning and cropping on various CNN architectures	BreakHis	ResNet-50 Caffenet GoogLeNet	Transfer Learning	Full image average patient-rate over multi-resolution GoogLeNet = 95.00±3.64 Caffenet = 94.29±4.84 ResNet-50 = 95.00±3.64	Compared performance of different architectures.	The task could be extended to include effect of batch size, data augmentation and other factors.
Bejnordi et al. (2017)	Proposed method for classification of breast cancer images using tumor-associated stroma	646 sections (from 444 women) 20x Magnification	CNN1 (based on VGG) CNN2 (VGG-16)	Data Augmentation	**CNN1** Pixel-level Accuracy = 95.5% AUC = 0.882 **CNN2** Pixel-level Accuracy = 92.0% AUC = 0.904 System AUC = 0.921	Proposed system that utilized tumor-associated stroma as a discriminator for Breast Cancer images.	Models could be explored for improving the accuracy.

(Continued on next page)

Reference	Proposal	Dataset	Architecture	Preprocessing	Results	Remarks	Future Work
Araújo et al. (2017)	Proposed CNN for classification of breast cancer histopathology images	Bioimaging 2015 breast histology classification challenge[a]	Self-proposed CNN and CNN + SVM architectures	Data Augmentation	**Accuracy:** Binary Classification = 83.3% Four class Classification = 77.8%	Proposed effective method for classification of breast histopathology images using CNN and SVM classification methods.	This method could be extended to other datasets.
Wahab et al. (2017)	Proposed two-phase CNN for classification of mitotic and non-mitotic cell in breast cancer histopathology images.	MITOS12[b] TUPAC16[c]	Self-proposed two-phase CNN architecture	Image Processing Data Augmentation	F1 score = 0.79	Proposed method for classification of mitotic and non-mitotic cells and class balancing using CNN.	This method could be extended to other datasets.
Feng et al. (2018)	Proposed multiple instance learning framework for CNN for classification of breast cancer histopathology images	BreakHis	Deep manifold persevering autoencoder (DMAE)		Image-level classification accuracy $40\times$ Malignant = 94.42 ±1.02 Benign = 83.71 ±1.09 $100\times$ Malignant = 93.15 ±0.89 Benign = 82.98 ±0.98 $200\times$ Malignant = 99.36 ±0.20 Benign = 83.57 ±1.01 $400\times$ Malignant = 94.86 ±1.03 Benign = 83.45 ±1.15	Proposed effective method for classification of breast histopathology images using DMAE.	Hyperparameter optimization could be conducted.

(Continued on next page)

Table 9.6 (Continued)

				Accuracy:		
Das et al. (2018)	Proposed multiple instance learning framework for CNN for classification of breast cancer histopathology images	BreakHis	Proposed CNN model (based on VGG)	Data Augmentation Multiple instance pooling	40× Magnification = 89.52 100× Magnification = 89.06 200× Magnification = 88.84 400× Magnification = 87.67	Achieved state-of-the-art results using multiple instance learning approach	Experiments on other datasets could be conducted for verification.
Naylor et al. (2019)	Prediction of TNBC treatment efficiency using histopathology WSIs.	122 WSIs Curie Institute 40× Magnification	Self-proposed architecture (based on ResNet)	Data Augmentation	pCR vs RCB K-means + mean augmentation Accuracy = 59.9% pCR-RCB-I vs RCB-I/RCB-II/RCB-III Neural network + mean augmentation Accuracy = 60.6%	Proposed a method to predict RCB score of TNBC patients from biopsy images.	Methods could be explored to improve the accuracy of prediction.

[a]http://www.bioimaging2015.ineb.up.pt/rules.html
[b]Self-proposed CNN architecture
[c]http://tupac.tue-image.nl/node

Table 9.7
Summary of work done on breast cancer histopathology images (2)

Reference	Objective	Dataset	Architecture	Approach Used	Result	Contribution	Potential Improvement
Romero et al. (2019)	Proposed multi-level batch normalization in Inception model for prediction of IDC	Cruz-Roa et al. [15]	Proposed CNN Model (derived from Inception)		F1 score = 0.897 Balanced Accuracy = 0.890 AUC = 0.956	Multi-level batch normalization surpassed original model. Reduced internal covariances for stability.	Metrics could be further improved.
Romano et al. (2019)	Proposed CNN with accept-reject pooling layer for prediction of IDC.	Cruz-Roa et al. [15]	Proposed CNN Architecture	Data Augmentation	F1 score = 0.8528 Balance Accuracy = 0.8541	Proposed Accept-Reject pooling layer in CNN for prediction of IDC.	Model could be improved for metrics.
Murtuza et al. (2019)	Classification of breast cancer histopathology images using proposed architecture.	BreakHis	BMIC_Net	Transfer Learning Data Augmentation	First-level classifier Accuracy = 95.48% Second-level classifiers Accuracy of first classifier = 94.62% Accuracy of second classifier = 92.45%	Proposed accurate hierarchical classification model that classifies breast cancer histopathology images into eight classes.	A generic classification model could be explored.
Naylor et al. (2017)	Segmentation of nuclei from breast cancer images	H&E stained histopathology images[a]	PangNet [65] DeConvNet [63] FCN Ensemble of DeConvNet and FCN	Data Augmentation	**Accuracy** PangNet = 0.924 DeConvNet = 0.954 FCN = 0.944 Ensemble = 0.944	Proposed segmentation workflow for nuclei from breast cancer images.	This method could be extended to other datasets.

(Continued on next page)

Table 9.7 (Continued)

			F-Score	
Janowczyk et al. (2016)	Applied AlexNet for numerous tasks on breast histopathology images: **(a) Nuclei Segmentation (b) Epithelium Segmentation (c) IDC Segmentation (d) Lymphocyte Detection (e) Mitosis Detection**	**(a) Nuclei Segmentation** 141 images 2000x2000 ROIs of ER+ Breast Cancer 40x Magnification **(b) Epithelium Segmentation** 42 images 1000x1000 1735 regions from ROIs of ER+ Breast Cancer 20x Magnification **(c) IDC Segmentation** 162 WSIs (from Breast Cancer) 40x Magnification **(d) Lymphocyte Detection** 100 images 100x100 3064 lymphocyte centers from ROIs of Breast Cancer 40x Magnification **(e) Mitosis Detection** 311 images 2000x2000 550 mitosis centers from ROIs of Breast Cancer 40x Magnification	AlexNet	**F-Score** **(a)** = 0.83 **(b)** = 0.84 **(c)** = 0.7648 **(d)** = 0.90 **(e)** = 0.53	Applied a single architecture for a variety of tasks.
BenTaieb et al. (2018)	Proposed attention-based model for classification of breast cancer histopathology images	Camelyon16	Attention-based CNN Model	Proposed 3 Glimpses WSI Accuracy = 0.95±0.2	Achieved state-of-the-art results using an Attention-based CNN	Improvements to aggregation of micro-metastases could follow.

a http://cbio.mines-paristech.fr/pnaylor/BNS.zip

9.5.4.2   Segmentation Tasks

Song et al. applied superpixel and CNN for segmentation of nuclei from cervical histopathology images. The input images were pre-processed using trimmed mean filter [85]. Initially, a coarse segmentation of the cytoplasm was obtained using Otsu's thresholding. After coarse segmentation, the superpixel algorithm was used to get a fine-grained segmentation of the nucleus along with the cytoplasm. In the next step, a CNN was used to fine tune the segmentation by detecting the nuclei regions. This method achieved a precision of 0.9143 ± 0.0202 and a recall of 0.8726 ± 0.0008 for nucleus segmentation.

Segmentation of basal membrane is an essential prerequisite in the diagnosis of microinvasive carcinoma of the cervix (MIC). Wang et al. modeled an adversarial training network with the generator replaced by a CNN to implement basal membrane segmentation [97]. The proposed architecture had two parts: the segmentation network and the discriminator network. The segmentation network in turn was composed of an encoder-decoder architecture. The pre-trained weights of the VGG16 model was transferred to the encoder part of the segmentation network. The output of the segmentation network was a probability map for each pixel. The probability map denoted if a pixel belonged to background or basal membrane. The discriminator network compared the ground truth images with the generated probability maps and the loss is propagated back into the network. The method yielded an F1 score of 0.624. The authors noted that in future, this method can be extended to WSIs and for segmenting ultra-blurry basal membrane by improving the model.

It's important to find an efficient method of lesion detection, based on morphology of structures in WSIs, because they are subject to observer variability. Yang et al. proposed a DL workflow for the segmentation of lesion region and the segmentation of nuclei from lesion region [106]. The workflow consisted of using U-Net to segment the lesion from the WSI. The lesion area, after zooming in the WSI, was then sent to multi-task-cascade network (MTC) consisting of four stages: U-Net1, Stack desnoiding autoencoder (SDAE), U-Net2 and encoder-decoder (ED) for nuclei segmentation. The network obtained an overall score (average of dice coefficients D1 and D2) of 0.7792.

The research works on cervix histopathology images are summarized in Table 9.5.

9.5.5   COLON-RECTUM

This section details the work done on colon and rectum histopathology images. CAD systems have been developed to detect and classify colon adenocarcinoma, cell nuclei and colo-rectal liver metastases. Also segmentation of gland structures and tubules has also been addressed. Experiments have also been done to verify the importance of stain normalization of images in increasing the accuracy of the classifiers. A method has also been proposed to predict the patient survival rate from tissue images.

9.5.5.1 Classification Tasks

Ponzio et al. experimented on CNN architectures to classify tissue images of colorectal cancer (CRC) into three classes, namely, adenoma carcinoma, healthy tissue and benign tumors [67]. This study compared the results of the following: 1) a VGG-16 CNN fully trained with annotated colorectal cancer (CRC) samples only, 2) a pre-trained VGG-16 CNN (trained using samples from ImageNet) used to generate features from the CRC dataset and the generated features are classified using a SVM classifier and 3) a pre-trained VGG16 CNN (trained using samples from ImageNet) trained with CRC samples using transfer learning and fine tuning. It was observed that the fully trained network obtained a classification accuracy of 90%, whereas the pre-trained network obtained a classification accuracy of 96%, outperforming the former. The results showed that despite the network being trained on a dataset from a different domain (ImageNet), it was able to extract low level features that would generalize to CRC image classification, without much of the computational and overfitting problems that a fully trained CNN would encounter.

Bychkov et al. proposed a combination of convolutional and recurrent deep neural networks, based on pre-trained VGG-16, for predicting the five-year survival rate of patients with colorectal cancer [10]. The dataset consisted of H&E stained colorectal cancer Tissue-Micro-Array (TMA) samples. Initially the input images were fed to the CNN which is a VGG-16 architecture pre-trained on ImageNet. The output of the CNN was the feature vectors that act as inputs to the long short-term memory (LSTM) neural network otherwise known as the RNN. The recurrent neural network predicts the five-year disease-specific survival. The proposed model was compared with traditional supervised learning methods such as SVM, Logistic Regression and Naive Bayes. The results of the study, in terms of AUC for the various models were 0.69 (LSTM), 0.64 (SVM), 0.65 (Logistic Regression), 0.61 (Naive Bayes).

9.5.5.2 Segmentation Tasks

The morphology of tubules indicates the aggressiveness of cancer. The aggressiveness can be observed as disorganized tubules in the sample. Janowczyk et al. used AlexNet (after pre-processing) for the task of tubule segmentation from colorectal tissue samples [37]. Pre-processing was done in order to predict potential false positives and negatives by means of feature extraction and naive-Bayes classifier. The results of this task showed a mean F-score of 0.836 ± 0.05 from 795 tubules.

9.5.5.3 Combined and Other Tasks

The ultimate aim of segmentation is classification but the idea that classification can also benefit from the results of segmentation was explored. BenTaieb et al. [7] proposed a method that involved a joint framework for segmentation and classification of colon adenocarcinoma using a deep CNN architecture. They introduced a multiloss objective function to accomplish the joint classification and the segmentation tasks. The framework classified between benign and malignant tissue images and also provided the relevant gland segmentation. The classification part of the CNN performed convolution and subsampling operations, whereas the segmentation part of the CNN performed deconvolution and upsampling. The proposed model obtained an accuracy of 89% for classification and 92% for segmentation.

Sirinukunwattana et al. proposed a spatially constrained convolutional neural network (SC-CNN) for nucleus center detection from colon cancer histology images [83]. Furthermore, the authors proposed a neighbouring ensemble predictor (NEP) coupled with CNN to predict the class label of detected nuclei that can belong to any of the four classes namely epithelial, inflammatory, fibroblast and miscellaneous. The approach was novel as it does not segment the nuclei. The SC-CNN consisted of a standard CNN with two new layers at the last, namely, the parameter estimation layer (S1) and the spatially constrained layer (S2). These layers generated a probability map that denoted if a pixel belonged to the center of a nuclei. The probability maps were constrained such that pixels close to the center of a nucleus had higher probability value than those farther away. The NEP used the prediction output of all the spatially neighbouring patches to predict the class of a patch. This joint detection and classification produced highest F1 score of 0.802.

The process of staining samples varies from laboratory to laboratory. This variability in the staining might not limit the ability of a histopathologist but it may affect the ability of an automated image analysis system to accurately determine inferences. In order to reduce the variability in staining, normalization is essential which aims to match the color of input images to that of a template. Ciompi et al. investigated if a CNN trained on a dataset (Rectal Cancer data) obtained from a source could be used by another dataset (Colorectal Cancer data) acquired from another source with and without stain normalization [13]. A CNN was trained to classify WSI images of rectal cancer (RC) dataset into six classes. The trained CNN was then used to classify colorectal cancer (CRC) dataset. The CNN used in this study had six convolutional layers, four max pooling layers and a final softmax layer. Authors showed that applying the CNN directly on CRC data gave poor results (accuracy = 50.96%). However, when they applied two stain normalization algorithms SN1 [5] and SN2 [54] on the CRC dataset (using an image from the RC dataset as a template), the classification accuracies improved to 75.55% and 73.99%, respectively. Also, when stain normalization algorithms were applied to both training and testing data, the classification accuracy became 79.66%. This study showed that stain normalization is a necessary step in histopathology image analysis.

Xu et al. proposed two multi-scale context aware convolutional neural networks MCCNet1 and MCCNet2 for classifying and segmenting regions from colorectal liver metastases (CRLM) images [105]. The input images to these networks were of same size but were taken at $10\times$, $20\times$ and $40\times$ magnifications. The high magnification level allowed the networks to learn texture features and the low magnification level enabled them to learn contextual information. Each network had three subnetworks that were trained separately on the images obtained at different magnification levels. The difference between the two MCCNets is when the feature maps are merged (early fusion for MCCNet1 and late fusion for MCCNet2). The two networks MCCNet1 and MCCNet2 reached tissue classification accuracies of 96.43%

and 96.51%, respectively, which outperformed the other recent DL models for image classification like Alexnet and VGG-16.

Gland formation is an important feature for grading of cancer in colon. Therefore, automated algorithms for segmenting gland structures from histopathology images are essential. Manivannan et al. combined hand-crafted features with deep convolutional features to segment gland structures from colon histopathology images. A fully convolutional neural network was used to extract features from input patches. Similarly, zoom out features from concatenated window descriptors were also extracted. These two features were concatenated. The proposed method finds a mapping from the features to a set of labeled exemplars using a linear stucture classifier like linear SVM. While testing, if the labels of an image location is given, the proposed method directly predicts the local structure of the labels [56].

The research works on colon-rectum histopathology images are summarized in Table 9.8.

9.5.6 GASTRIC

This section covers works done on the gastric histopathology images. Experiments have been performed to classify gastric WSIs into cancerous or non-cancerous, multi-class classification of gastric WSIs, and detection of cancerous and necrotic tissue.

9.5.6.1 Classification Tasks

Sharma et al. designed a deep convolutional neural network model for two tasks namely cancer classification and necrosis detection from H&E stained histopathology whole-slide images of gastric carcinoma [77]. The CNN used for the tasks had three convolution layers, three max pooling layers and three fully connected layers. The three convolutional layers had filter sizes 7×7, 5×5 and 3×3 with 24, 16 and 16 feature maps, respectively. The kernel size of the max pooling layers was 2×2. The three fully connected layers had 256, 128 and 3 neurons, respectively. The proposed architecture had a classification accuracy of 0.6990 and a necrosis detection accuracy of 0.8144. The experimental results showed that the proposed DL model performed on par with conventional hand-engineered feature extraction and classification techniques for the classification task, whereas the former outperformed the latter for the necrosis detection task.

Li et al. presented a DL framework based on CNN called GastricNet to classify between cancerous and non-cancerous histopathology gastric slices [50]. The GastricNet consisted of convolution layer, max pooling layer, multi-scale module (MSM), network in network (NIN) module, fully connected layer and slice-based classification unit (SCU). The MSM used dilated convolutions which increased the receptive field so that the network could simultaneously extract multiscale features. The NIN module was made of convolution and max pooling layers and fused the concatenated features from the MSMs. The slice-based classification unit then predicts the label for the entire slice. The GastricNet achieved a 100% classification

Table 9.8
Summary of work done on colon-rectum cancer histopathology images

Reference	Objective	Dataset	Architecture	Approach Used	Result	Contribution	Potential Improvement
Ponzio et al. (2018)	Classification of colo-rectal histopathology images.	27 TMA WSIs (from 27 patients) Virtual Pathology University of Leeds	VGG-16	Transfer Learning	**Patch Scores** Fully-trained CNN = 0.9037 Fine-tuned CNN = 0.9682 CNN+SVM = 0.9646 **Patient Scores** Fully-trained CNN = 0.9022(±0.0155) Fine-tuned CNN = 0.9667(±0.0082) CNN+SVM = 0.9678(±0.00092)	Demonstrated the effectiveness of transfer learning for colo-rectal image classification at both patch and patient levels.	The work could be extended to WSIs and classification of tissue subtypes.
Bychkov et al. (2018)	Prognosis prediction from colo-rectal cancer TMAs using DL approach.	420 TMA Slides (from 420 patients) Helsinki University Central Hospital	Recurrent Network (LSTM)	Transfer Learning	AUC LSTM = 0.69 SVM = 0.69 Logistic Regression = 0.65 Naïve Bayes = 0.61	Proposed a DL method prognosis prediction without intermediate classification.	Prognostic accuracy could be improved.
Janowczyk et al. (2016)	Segmentation of tubules from colo-rectal ROIs	85 ROIs 40x Magnification	AlexNet	Data Augmentation Transfer Learning	F-score = 0.836±0.05	Segmented tubules with good generalization.	The experiment could be extended to WSIs.
BenTaieb et al. (2016)	Classification and segmentation of colon adeno-carcinoma and glands respectively.	Warwick-QU dataset 20x Magnification	Self-proposed FCN architecture	Data Augmentation	**Accuracy (ACC)** Class ACC = 89.0% Pixel ACC = 92.0% **Dice coefficient(DC)** Benign DC = 0.90 Malignant DC = 0.76	Proposed a DL architecture that utilizes multi -loss function for simultaneous classification and segmentation.	The experiment could try utilizing weakly annotated data.

(Continued on next page)

Table 9.8 (Continued)

Sirinukunwattana et al. (2016)	Detection and classification of nuclei in colon histology images	100 WSIs of colon adenocarcinoma	Proposed Spatially-Constrained CNN (SC-CNN)	Data Augmentation	Combined nucleus detection and classification F1 score = 0.692	Proposed spatially constrained CNN and neighboring ensemble predictor for detection and classification of cell nuclei.	The improvement in the time efficiency of the proposed model using GPUs could be analyzed. Use of spacial smoothing to correct mis-classification of nucleus can be explored.
Ciompi et al. (2017)	Investigates the importance of stain normalization for classification of colorectal cancer	**Rectal cancer data:** 74 slides (from 74 patients) 200X magnification **Colorectal cancer data:** [43]	Self-proposed CNN architecture	Stain normalization	**Accuracy** Test and train with stain normalization = 79.66% Test and train without stain normalization = 50.96%	Experimentally demonstrated the importance of stain normalization in colorectal cancer classification.	Experiments could be extended to other datasets.
Xu et al. (2018)	Classification of colorectal liver metastases using multi-scale context-aware networks	18 WSIs 100,000x100,000 40X Magnification	Multi-scale context-aware CNN networks: MCCNet1 & MCCNet2	Data Augmentation	**Tissue Classification Accuracy:** MCCNet1 = 96.43% MCCNet2 = 96.51%	Introduced context-aware networks that are able to integrate information from different magnifications for better accuracy.	This method could be extended to other datasets.
Manivannan et al. (2018)	Segmentation of glandular structures from colon histopathology images	GlaS challenge dataset [82]	Proposed FCN architecture	Data Augmentation Transfer Learning	**F1 Score** Test A = 0.892 Test B = 0.801 **Objective Dice** Test A = 0.887 Test B = 0.853 **Hausdorff Distance** Test A = 51.175 Test B = 86.987	Proposed state-of-the-art method for gland segmentation.	The improvement in the time efficiency of the proposed model using GPUs could be analyzed.

accuracy on slice level and 97.93% on patch level using the publicly available BOT gastric slice dataset.

Wang et al. proposed a two-stage framework for gastric WSI classification of normal, dysplasia and cancer [99]. The first stage consisted of a localization network that selected discriminative instances. The selected instances were passed into the second stage that consisted of a recalibrated multi-instance DL network (RMDL). The proposed method was able to explore the relationship between different patches and how each patch contributed to classification at image-level. The proposed framework obtained an average classification score of 0.923 and accuracy of 86.5%.

9.5.6.2   Segmentation Tasks

Liang et al. proposed a reiterative learning approach with FCN for the segmentation of gastric tumor images [51]. The framework uses Overlapped Region Forecast (ORF) to remove boundary error between patches and to increase accuracy of prediction. The framework obtained a mean IoU of 88.3 and mean accuracy of 91.09%

9.5.6.3   Combined and Other Tasks

Wang et al. proposed a deep learning method for predicting cancer outcome using Tumor Area to Metastatic Lymph Node area (T/MLN) ratio calculated from gastric cancer histopathology images [101]. The method used a segmentation network (U-Net) to extract lymph node regions and then passed those regions to a classification network (ResNet-50) that detected tumor regions. The T/MLN ratio was calculated based on the output heatmaps. The results validated the use of T/MLN ratio as a prognostic marker for gastric cancer. The framework obtained a sensitivity of 98.5% and specificity of 96.1%.

The research works on gastric histopathology images are summarized in Table 9.9.

9.5.7   KIDNEY

This section covers works done on kidney histopathology images. Current research has shown work on classification of renal cell carcinoma, its subtypes and risk-assessment of carcinoma.

9.5.7.1   Classification Tasks

Tabibu et al. proposed a method that used pre-trained networks for the classification of renal cell carcinoma (RCC) using kidney histopathology images [91]. The first deep model used pre-trained ResNet for the binary classification of cancer vs normal image patches extracted from WSIs. The second deep model used pre-trained ResNet for extraction of features that are sent to a direct acyclic graph support vector machine (DAG-SVM) for cancer subtype classification of clear cell RCC (KIRC), papillary RCC (KIRP) and chromophobe RCC (KICH). The first model was also uses to generate probability heatmaps for assessing the risk index of the WSI.

Table 9.9

Summary of work done on gastric cancer histopathology images

Reference	Objective	Dataset	Architecture	Approach Used	Result	Contribution	Potential Improvement
Sharma et al. (2017)	Classification and necrosis detection in WSIs of gastric histopathology images	11 annotated WSIs for cancer classification 4 annotated WSIs for necrosis detection	Self-proposed Architecture CNN	Data Augmentation	**Classification Accuracy** Cancer Classification = 0.6990 Necrosis Detection = 0.8144	Proposed CNN that performed on-par with traditional ML methods.	Classification accuracy could be improved.
Li et al. (2018)	Classification of cancerous and non-cancerous gastric slices	BOT gastric slice dataset 20x Magnification	GatricNet	Data Augmentation	**Average Classification Accuracy** Patch-based = 97.93% Slice-based = 100%	Proposed highly accurate CNN for gastric slice classification.	Model could be extended to other datasets.
Wang et al. (2019)	Proposed two-stage framework for the classification of gastric WSIs	608 WSIs The Sixth Affiliated Hospital of Sun Yat-sen University 40x Magnification	Proposed (1) Localization network (2)RMDL network	Image Processing Data Augmentation	Average classification score = 0.923 Accuracy = 86.5%	Built gastric WSI dataset and proposed accurate RMDL network for gastric WSI classification.	Parallel model technique could be explored for training RMDL network.
Liang et al. (2018)	Segmentation of gastric cancer images using reiterative learning approach	1400 gastric tumor images 2017 China Big Data and Artificial Intelligence Innovation and Entrepreneurship Competition.	FCN	Data Augmentation Image Processing	IOU = 0.883 Mean Accuracy = 91.09%	Proposed reiterative learning method for gastric cancer segmentation.	Recognition of small tumors could be improved.
Wang et al. (2021)	Proposed deep learning framework for analyzing gastric cancer lymph node WSIs and calculating T/MLN.	WSIs from Changhai Hospital and Jiangxi Provincial Cancer Hospital	ResNet-50 (Feature Extraction) Neural Conditional Random Field (Classification Network) U-Net (Segmentation Network)	Data Augmentation	Sensitivity = 98.5% Specificity = 96.1%	Proposed the use of T/MLN as a prognostic factor.	Study could be extended to data obtained from other regions.

The research works on kidney histopathology images are summarized in Table 9.10.

9.5.8 LIVER

This section covers work done on liver histopathology images.

Liao et al. proposed a method for classification and mutation prediction of hepatocellular carcinoma (HCC) using H&E-stained WSIs and TMAs [52]. A residual CNN was used to classify WSI data into HCC and normal and another residual CNN was used to predict mutation. The results using CNN showed improvement over feature-based methods.

Kiani et al. developed a deep learning assistant that assisted pathologists in distinguishing two subtypes of primary liver cancer and evaluated the impact of the assistant on pathologist performance [44]. A DenseNet model classified image patches into hepatocellular carcinoma (HCC) and cholangiocarcinoma (CC). The results showed that the performance of a subset of pathologists improved with the assistant. However, it was noted that assistance significantly decreased if the model was inaccurate and correspondingly, assistance significantly increased if the model was accurate. These effects were seen across all pathologist experience levels and case difficulty levels. The DenseNet model used achieved testing accuracy of 0.842.

The research works on liver histopathology images are summarized in Table 9.11.

9.5.9 LUNG

This section covers work done on lung histopathology images. Work has been done for segmentation of microvessels and for segmentation and classification of nuclei.

9.5.9.1 Classification Tasks

Yi et al. proposed a DL network that is fully convolutional to segment microvessels from H&E stained lung adenocarcinoma histopathology images [107]. The outcomes of the experiment indicated that the model performed well in associating microvessel density with patient survivability.

9.5.9.2 Combined and Other Tasks

Qu et al. proposed a framework that segmented and classified nuclei in lung histopathology images [68]. The framework consisted of a prediction network and a perpetual loss network. The prediction network segmented and classified nuclei. The perpetual loss network was used to improve segmentation accuracy by utilizing both cross-entropy and perpetual loss. The results showed that with transfer learning, the model was able to simulataneously obtain good segmentation and classification results of 0.886 (Dice coefficient) and 84.75% (Accuracy). The research works on lung histopathology images are summarized in Table 9.12.

Table 9.10

Summary of work done on kidney cancer histopathology images

Reference	Objectives	Datasets	Architectures	Approaches Used	Results	Contributions	Potential Improvements
Tabibu et al. (2019)	Classification of kidney histopathology into cancerous vs normal and subtypes.	TCGA **KIRC** Cancer = 1027 Normal = 379 **KIRP** Cancer = 303 Normal = 47 **KICH** Cancer = 254 Normal = 83 20x and 40x Magnification	ResNet-18 ResNet-34	Data Augmentation Transfer Learning	**Accuracy** ResNet-18 KIRC = 93.39% ResNet-34 KICH = 87.34% ResNet-18 + DAG-SVM = 92.62%	Proposed a DL method for both prediction and prognosis estimation.	The experiment could be extended to WSIs.

Table 9.11

Summary of work done on liver cancer histopathology images

Reference	Objectives	Datasets	Architectures	Approaches Used	Results	Contributions	Potential Improvements
Liao et al. (2020)	Proposed a method for classification and mutation prediction of HCC using WSIs and TMAs.	**WSIs:** TCGA **TMAs:** The Biobank of West China Hospital	Self-proposed architecture InceptionV3		Highest per slide AUC after aggregation = 1.00 (Method 1 on TCGA dataset)	Proposed a method for HCC diagnosis using CNN.	Experiments on larger datasets could be conducted for verification.
Kiani et al. (2020)	Proposed a deep learning assistant that helps pathologists differentiate two subtypes of liver cancer.	**Training/Validation/ Tuning:** TCGA **Testing:** Department of Pathology Stanford University Medical Center	DenseNet-121	Transfer Learning	Testing accuracy = 0.842	Demonstrated the effectiveness of tool-assisted pathologists in diagnosing two subtypes of liver cancer.	Model could be extended to run on WSI.

Table 9.12
Summary of work done on lung cancer histopathology images

Reference	Objectives	Datasets	Architectures	Approaches Used	Results	Contributions	Potential Improvements
Yi et al. (2019)	Segmentation of microvessels and prediction of patient survivability from lung adenocarcinoma images	WSIs of lung adenocarcinoma National Cancer Center /Cancer Hospital /Chinese Academy of Medical Sciences Peking Union Medical College, China 20x Magnification	Proposed Fully Convolutional Network (FCN)		Pixel Accuracy = 0.952 Mean Accuracy = 0.833 Mean Intersection over Union = 0.755		The network could be extended to WSIs.
Qu et al. (2019)	Segmentation and classification of nuclei from lung histopathology WSIs.	40 WSIs of lung adenocarcinoma or lung squamous cell carcinoma	Proposed network based on U-Net, ResNet-34 and VGG-16	Data Augmentation Transfer Learning	**Segmentation:** F1 = 0.0886 Dice = 0.876 Hausdorff distance = 4.14 **Classification:** Accuracy 84.75%	Proposed framework that segmented and classified nuclei. Combined cross entropy loss and perpetual loss for better segmentation details.	The network could be extended to WSIs.

9.5.10 PANCREAS

This section covers work done on pancreas histopathology images. Work has been done for the classification of cells in pancreas histological images.

9.5.10.1 Classification Tasks

Chang et al. proposed a convolutional neural network called DeepNC to classify cancerous and normal cells from pancreas histological images [12]. As DL methods demand large datasets, immunofluoresence (IF) images were used in addition to pathologists' annotated dataset for creating large dataset of cells. Initially, the authors experimented with several convolutional neural network architectures with the pathologists' annotated dataset to decide on an optimal network for training on the large dataset. The optimal network exhibited an accuracy of 99.5%. The optimal network consisted of three convolutional layers containing 32, 48 and 64 filters with convolution kernels of dimensions 7×7, 5×5, and 3×3, respectively, followed by 2×2 max-pooling layers, four fully connected layers with 31 neurons, and an output layer with a single neuron. The optimal network was then trained and tested using the large dataset which exhibited a classification accuracy of 91.3%.

The research work on pancreas histopathology images is summarized in Table 9.14.

9.5.11 PROSTATE GLAND

This section covers works done on the prostate gland histopathology images. Work has been done to predict the metastatic capacity, scale and gleason grades of prostate histopathology images.

9.5.11.1 Classification Tasks

There is an important need to identify high-risk patients with high-grade prostate cancer from prostate needle biopsy images. Ing et al. hypothesized that the morphological features of prostate tumor cells could be used to predict the aggressiveness of tumor and possible metastases. The authors applied a multi-instance learning approach which used a CNN, whose architecture was similar to AlexNet, to encode images. The encodings were passed on to a two layer feed-forward neural network classifier which classified the metastatic state of the image. The results support the hypothesis that information related to metastatic capacity of prostate cancer cells can be obtained through analysis of nuclei. The method yielded an average AUC of 0.71 ± 0.08. [35].

The scale of a histopathology image is important because certain cellular structures are visible and clear at certain magnifications. In addition to this fact, many researchers who depend on open-source repositories for DL data face the challenge of working with images with variable staining methods, image formats and unknown scale of images. In order to reduce the variability of images in a dataset with respect to scale, two CNN architectures were tested in classifying (detecting) the scale of various patches extracted from prostate WSI [64]. The first architecture was a shallow 4-layer CNN and the second architecture was the DenseNet. The shallow CNN had 32, 3×3 convolutional kernels and max pooling of a 2×2 neighborhood. The DenseNet used was called as DenseNet-BC 121 which was a 121-layer variation of DenseNet. The various patch sizes tested were 5×, 10× and 20×. The fine-tuned DenseNet architecture with ImageNet pre-trained weights achieved a Cohen's kappa coefficient of 0.9897, whereas the shallow CNN achieved a Cohen's kappa coefficient of 0.9617.

A Gleason score assists in evaluating prostate cancer prognosis. The score ranges from Grade 1 to Grade 5. In Grade 1, the glands are well differentiated and uniform in structure and in Grade 5, the glands are very poorly differentiated. The higher the grade, the higher the risk of mortality. Karimi et al. proposed a DL pipeline for Gleason grading of prostate cancer [41]. The pipeline consists of training three different CNNs that correspond to different patch sizes for two different tasks: (1) Classification of benign (Gleason grades 1 and 2) vs cancerous (Gleason grades 3, 4 and 5) patches and (2) classification of l ow-grade (Gleason grade 3) vs high-grade (Gleason grades 4 and 5) patches. The decisions of all three CNNs are aggregated using logistic regression. Data augmentation methods such as elastic deformation, generation of patches using GANs, rotations, flips, Synthetic Minority Over-sampling TEchnique (SMOTE) and intensity jitter were applied. The data augmented result was significantly better than the result without it. The classification accuracies for task (1) and task (2) were 92% and 86% with data augmentation, respectively.

The research works are summarized in Table 9.13.

9.5.12   SKIN

This section reviews work done on skin cancer such as melanoma. So far work has been done for the classification of histopathological melanoma images.

9.5.12.1   Classification Tasks

Hekler et al. compared the performance of CNN with that of 11 board certified pathologists for the classification of skin histopathology images into nevi and melanoma [31]. The authors utilized pre-trained ResNet-50 and used cropped images (10× magnification) for training and testing. The CNN achieved comparable results with the pathologists.

The research work is summarized in Table 9.15.

9.5.13   OTHER

This section covers work done on mixed cancer types and other tasks.

Using an end-to-end convolutional regression network, a time efficient and accurate model was proposed that would detect cells in histopathology images [98]. The model was trained and tested on bone marrow and breast images. The F1 score of the method proposed on the bone marrow dataset was 88.10 ± 2.71. The F1 score of the method proposed on the breast cancer tissue dataset was 90.60.

Table 9.13
Summary of work done on prostate gland cancer histopathology images

Reference	Objectives	Datasets	Architectures	Approaches Used	Results	Contributions	Potential Improvements
Ing et al. (2018)	Tested prediction of metastatic capacity of prostate cell from morphology using CNN	Data from 171 patients 40x Magnification Los Angeles Veterans Affairs Healthcare System	AlexNet		AUC = 0.71	Hypothesis supported. Morphology of prostate tumor cell nuclei can be used to predict the aggressiveness of cancer.	Accuracy of the model could be improved.
Otálora et al. ()	Classified patches based on their scale.	50 WSIs 5x, 10x and 20x Magnification	DenseNet-BC-121 ShallowNet	Transfer Learning	ShallowNet Kappa = 0.9617 DenseNet-BC-121 Kappa =0.9477 Fine-tuned DenseNet-BC -121 Kappa = 0.9897	The scale of patches can be effectively determined using deep learning methods	Testing based on real-world factors such as different staining and multiple organ images can be done.
Karimi et al. (2019)	Classification of TMAs: (1) Benign vs Cancerous (2) Low-grade vs High-grade	333 TMAs (231 Patients) 40x Magnification Vancouver Prostate Centre	CNN(Large) CNN(Medium) CNN(Small)	Data Augmentation	With Data Augmentation: (1) Accuracy = 92% (2) Accuracy = 86% Without Data Augmentation: (1) Accuracy = 73% (2) Accuracy = 67%	Proposed two data augmentation methods and DL method for automatic Gleason grading.	Experiments on similar datasets could be conducted for comparison.

Table 9.14

Summary of work done on pancreas cancer histopathology images

Reference	Objectives	Datasets	Architectures	Approaches Used	Results	Contributions	Potential Improvements
Chang et al. (2017)	Classification of nuclei extracted from pancreas histology WSIs.	Pancreas Histology ImageNet (extracted from WSIs)	DeepNC	Data Augmentation	Accuracy = 91.23% Sensitivity = 89.9% Specificity = 92.8% Precision = 92.6%	Proposed a method for extracting nuclei using IF and pathologist-annotated images. Proposed a convolutional neural network for classification of nuclei.	Different strategies could be applied for training large networks for better accuracy.

Table 9.15

Summary of work done in skin section

Reference	Objectives	Datasets	Architectures	Approaches Used	Results	Contributions	Potential Improvements
Hekler et al. (2019)	Classification of skin histopathology images using CNN and compared results with those of 11 pathologists.	595 images (from 695 slides) 10x Magnification	ResNet-50	Transfer Learning	**Mean of 11 test runs** Sensitivity = 76% Specificity = 60% Accuracy = 68% **Mean of 11 pathologists** Sensitivity = 51.8% Specificity = 66.5% Accuracy = 59.2%	Proposed a DL method (CNN) for classification of skin histopathology images. CNN was shown to be more accurate than board-certified pathologists.	The experiment could be extended to WSIs.

Höfener et al. proposed a FCN for the segmentation of nuclei [33]. The authors trained and tested the network using colorectal adenocarcinoma and breast tumor tissue dataset. The intermediate map generated by the segmentation network is called PMap. PMap was post-processed and the impact factors related to post-processing are then compared. The results indicated with the right parameters, it is possible to achieve state-of-the-art results. The best-performing configuration obtained F1 scores of 0.816 for colorectal adenocarcinoma dataset and 0.819 for breast tumor tissue dataset.

Lal et al. proposed a deep learning architecture called NucleiSegNet which consisted of a robust residual block, a bottleneck block and an attention decoder block for segmentation of nuclei from liver cancer histopathology images and multi-organ histopathology images [48]. The residual block extracted high-level feature maps, the bottleneck block reduced the number of features and the attention decoder block reconstructed the features. The proposed method was able to handle touching nuclei and nuclei with different shapes. It also outperformed other methods with fewer parameters. On 10-fold cross validation of the multiorgan Kumar dataset, the model achieved average F-1 score of 81.12 ± 0.18 and IoU of 68.50 ± 0.24.

The research work is summarized in Table 9.16.

9.6 DISCUSSION: RECENT TRENDS AND FUTURE SCOPE

This section discusses the current trends of DL-based CADe/CADx systems in histopathology image analysis and possible paths this domain can take in the near future.

One of the biggest advantages that DL algorithms possess that other machine learning algorithms do not is the ability to perform feature extraction during the training of the algorithm. This has reduced the dependence on building hand-crafted feature extractors. Although neural networks and the science of building complex networks have been around for more than a decade, the widespread use of DL techniques for solving various image and computer vision problems has begun recently.

This is mainly due to the following reasons:

1. Widespread availability and affordability of GPUs and multicore processors in recent years that enable faster complex computations and parallelism.

2. Widespread availability of cloud-based storage and computing resources such as Microsoft Azure, Amazon Web Services (AWS), and Google Cloud Platform (GCP) that has helped in overcoming the issue of having on-site resources.

3. Widespread availability and affordability of storage capacity that has helped DL to progress and foster interest in its applicability to various fields including CADe/CADx.

Studies have shown that deeper the network, better the learning ability of the network. However, networks that contain several number of hidden layers increase the

Table 9.16
Summary of work done in other section

Reference	Objectives	Datasets	Architectures	Approaches Used	Results	Contributions	Potential Improvements
Wang et al. (2017)	Proposed end-to-end cell detection pipeline using convolutional regression neural networks.	Bone Marrow dataset [40] Breast Cancer tissue dataset [27]	Self-proposed convolutional regression neural network	Data Augmentation	**Bone Marrow dataset** F1 score = 88.10±2.71 **Breast cancer tissue dataset** F1 score = 90.60	End-to-end method that is time efficient and applicable to WSIs.	Experiments on other large datasets could be conducted for verification.
Höfener et al. (2018)	Detection of nuclei using fully convolutional network.	Colorectal adenocarcinoma [83] Breast Tumor Tissue [58]	Self-proposed FCN	Data Augmentation PMap Post-Processing	F1 score of Best Performing Configuration Colorectal Adenocarcinoma = 0.816 Breast Tumor Tissue = 0.819	Experiments showed that PMap Post-processing had an impact on detection quality.	Experiments on other datasets could be conducted for verification.
Lal et al. (2021)	Proposed deep learning architecture for nuclei segmentation.	**KMC Dataset:** 80 liver histopathology images 40x Magnification **Kumar dataset:** 44 multi-organ histopathology images	Self-proposed architecture		**KMC Dataset:** F1 score = 83.59 IoU = 72.06 **Kumar Dataset:** F1 score = 81.363 IoU = 68.883	Proposed model achieved state-of-the-art performance for nuclei segmentation.	Experiments on large datasets could be conducted for verification.

complexity of operations performed and require more computational power and large volume of data. DL models require large number of training samples to generalize. Otherwise, the models may overfit and do not perform well on unseen data. Additionally, there is another reason why DL models overfit: Due to the layers of abstraction, DL models may learn to model outliers and rare dependencies. A few methods that are used to combat overfitting include regularization such as L1 regularization, L2 regularization and dropout regularization. Depending upon the architecture, models may be prone to vanishing and exploding gradients. Therefore, DL does come with its own caveat: It requires ample amounts of computational resources and data to unlock its full potential.

When the original dataset is very small, researchers rely on data augmentation, a process in which the number of images in the dataset is artificially increased by performing transformations on the original dataset. Data augmentation reduces overfitting and invariability by increasing the sample size of the training and validation sets [66].

In histopathology image-based CADe/CADx systems, the collection of data is done using digital slide scanners. However, due to the number of different scanners used to scan and upload slides and different stain preparation and application protocols, there could be color variability in the digitized slides that could affect the performance of the model. Another important step that could improve the generalization and accuracy of the model is stain normalization as previously seen in some works [13, 36].

DL models are usually trained from scratch with large datasets that require a lot of time and computation resources. On the contrary, histopathology datasets are often small in number, large in slide size and involve high annotation. As discussed in Section 9.2.6, TL can be implemented in order to overcome the high costs of training a large network from scratch. For more specialized tasks, fine-tuning can be implemented to re-train layers of the pre-trained model. This method of training has shown great success [87].

With respect to processing whole-slide histopathology images using DL models, the huge size of images makes applying DL to WSI processing a costlier task. Therefore, researchers usually prefer patch-based processing where small patches at higher magnification levels are extracted from the WSIs. This also supports the cause that the visual features used for diagnosis are highly visible only under higher magnification scales. Later, the patch-based results are aggregated together to arrive at the slide level results. Many studies have suggested that patch-based processing is not an optimal solution as it ignores context. In addition, processing power and time is spent on processing areas of non-interest because ROIs usually compose a small area of the entire slide. Even if a patch-based framework is used, there needs to be a sufficient number of images to train the network. Despite their remarkable performance on various tasks in histopathology image analysis, the applicability of DL on processing the WSIs without breaking them down into patches is highly challenging as of now. In order to overcome the drawbacks of patch-based processing, some studies have proposed WSI-level processing using multiple instance learning and recurrent visual attention model [100, 17, 15, 92, 77]. But, the suitability of such methods to clinical workflow is yet to be evaluated and verified.

Finding the best network for very specific objectives, especially in histopathology, is also another challenge. Researchers empirically test multiple networks in order to find out the best results. Finding theoretical justification for choosing an optimal network architecture is still an ongoing field of research.

Recent trends in DL-based medical image analysis include the generation and use of synthetic data by using models such as GANs, real-time inference and visualization of medical images, and analysis of 3D image data. In future, developments in hardware technology and the availability of large labeled datasets may pave the way for DL-based end-to-end digital pathology workflow.

9.7 CONCLUSION

In this review, we have presented a detailed overview of deep learning and we have compiled and segregated papers based on type of cancer. This will benefit researchers who are interested in developing applications using DL in histopathology image analysis in their specific area of interest. Additionally, we have discussed the various trends in DL for histopathology image analysis and possible future areas of work.

REFERENCES

1. Abbas K. Alzubaidi, Fahad B. Sideseq, Ahmed Faeq, and Mena Basil. Computer aided diagnosis in digital pathology application: Review and perspective approach in lung cancer classification. In *2017 Annual Conference on New Trends in Information & Communications Technology Applications (NTICT)*. IEEE, 2017.

2. Teresa Araújo, Guilherme Aresta, Eduardo Castro, José Rouco, Paulo Aguiar, Catarina Eloy, António Polónia, and Aurélio Campilho. Classification of breast cancer histology images using convolutional neural networks. *PloS one*, 12(6):e0177544, 2017.

3. MA Aswathy and M Jagannath. Detection of breast cancer on digital histopathology images: present status and future possibilities. *Informatics in Medicine Unlocked*, 8:74–79, 2017.

4. Babak Ehteshami Bejnordi, Jimmy Lin, Ben Glass, Maeve Mullooly, Gretchen L Gierach, Mark E Sherman, Nico Karssemeijer, Jeroen Van Der Laak, and Andrew H Beck. Deep learning-based assessment of tumor-associated stroma for diagnosing breast cancer in histopathology images. In *Biomedical Imaging (ISBI 2017), 2017 IEEE 14th International Symposium on*, pages 929–932. IEEE, 2017.

5. Babak Ehteshami Bejnordi, Geert Litjens, Nadya Timofeeva, Irene Otte-Höller, André Homeyer, Nico Karssemeijer, and Jeroen AWM van der Laak. Stain specific standardization of whole-slide histopathological images. *IEEE Transactions on Medical Imaging*, 35(2):404–415, 2016.

6. Babak Ehteshami Bejnordi, Mitko Veta, Paul Johannes Van Diest, Bram Van Ginneken, Nico Karssemeijer, Geert Litjens, Jeroen AWM Van Der Laak, Meyke Hermsen,

Quirine F Manson, Maschenka Balkenhol, et al. Diagnostic assessment of deep learning algorithms for detection of lymph node metastases in women with breast cancer. *Jama*, 318(22):2199–2210, 2017.

7. A. BenTaieb, J. Kawahara, and G. Hamarneh. Multi-loss convolutional networks for gland analysis in microscopy. In *2016 IEEE 13th International Symposium on Biomedical Imaging (ISBI)*, pages 642–645, April 2016.

8. Aicha BenTaieb and Ghassan Hamarneh. Predicting cancer with a recurrent visual attention model for histopathology images. In *International Conference on Medical Image Computing and Computer-Assisted Intervention*, pages 129–137. Springer, 2018.

9. Dragan Bosnacki, Natal van Riel, and Mitko Veta. Deep learning with convolutional neural networks for histopathology image analysis. In *Automated Reasoning for Systems Biology and Medicine*, pages 453–469. Springer, 2019.

10. Dmitrii Bychkov, Nina Linder, Riku Turkki, Stig Nordling, Panu E Kovanen, Clare Verrill, Margarita Walliander, Mikael Lundin, Caj Haglund, and Johan Lundin. Deep learning based tissue analysis predicts outcome in colorectal cancer. *Scientific Reports*, 8(1):3395, 2018.

11. Gustavo Carneiro, Yefeng Zheng, Fuyong Xing, and Lin Yang. Review of deep learning methods in mammography, cardiovascular, and microscopy image analysis. In *Deep Learning and Convolutional Neural Networks for Medical Image Computing*, pages 11–32. Springer, 2017.

12. Young Hwan Chang, Guillaume Thibault, Owen Madin, Vahid Azimi, Cole Meyers, Brett Johnson, Jason Link, Adam Margolin, and Joe W Gray. Deep learning based nucleus classification in pancreas histological images. In *Engineering in Medicine and Biology Society (EMBC), 2017 39th Annual International Conference of the IEEE*. IEEE, 2017.

13. Francesco Ciompi, Oscar Geessink, Babak Ehteshami Bejnordi, Gabriel Silva De Souza, Alexi Baidoshvili, Geert Litjens, Bram Van Ginneken, Iris Nagtegaal, and Jeroen Van Der Laak. The importance of stain normalization in colorectal tissue classification with convolutional networks. *2017 IEEE 14th International Symposium on Biomedical Imaging (ISBI 2017)*, 2017.

14. James L. Connolly, Stuart J. Schnitt, Helen H. Wang, Janina A. Longtine, Ann Dvorak, and Harold F. Dvorak. Role of the surgical pathologist in the diagnosis and management of the cancer patient. In Donald W Kufe, Raphael E Pollock, Ralph R Weichselbaum, Robert C Bast, Ted S Gansler, James F Holland, and Emil Frei, editors, *Holland-Frei Cancer Medicine*. Hamilton (ON):BC Decker, 2003.

15. Angel Cruz-Roa, Ajay Basavanhally, Fabio González, Hannah Gilmore, Michael Feldman, Shridar Ganesan, Natalie Shih, John Tomaszewski, and Anant Madabhushi. Automatic detection of invasive ductal carcinoma in whole slide images with convolutional neural networks. In *Medical Imaging 2014: Digital Pathology*, volume 9041, page 904103. International Society for Optics and Photonics, 2014.

16. M. Dabass, R. Vig, and S. Vashisth. Review of histopathological image segmentation via current deep learning approaches. In *2018 4th International Conference on Computing Communication and Automation (ICCCA)*, pages 1–6, Dec 2018.

17. Kausik Das, Sailesh Conjeti, Abhijit Guha Roy, Jyotirmoy Chatterjee, and Debdoot Sheet. Multiple instance learning of deep convolutional neural networks for breast histopathology whole slide classification. In *Biomedical Imaging (ISBI 2018), 2018 IEEE 15th International Symposium on*, pages 578–581. IEEE, 2018.

18. Carl Doersch. Tutorial on variational autoencoders. *arXiv preprint arXiv:1606.05908*, 2016.

19. Kunio Doi. Computer-aided diagnosis in medical imaging: historical review, current status and future potential. *Computerized Medical Imaging and Graphics*, 31(4–5):198–211, June 2007.

20. K M Emmons and G A Colditz. Realizing the potential of cancer prevention - the role of implementation science. *New England Journal of Medicine*, 376(10):986 – 990, March 2017.

21. Y. Feng, L. Zhang, and J. Mo. Deep manifold preserving autoencoder for classifying breast cancer histopathological images. *IEEE/ACM Transactions on Computational Biology and Bioinformatics*, 17(1):91–101, Jan 2020.

22. Maayan Frid-Adar, Eyal Klang, Michal Amitai, Jacob Goldberger, and Hayit Greenspan. Synthetic data augmentation using gan for improved liver lesion classification. In *2018 IEEE 15th International Symposium on Biomedical Imaging (ISBI 2018)*, pages 289–293. IEEE, 2018.

23. Ross Girshick. Fast r-cnn. In *Proceedings of the IEEE International Conference on Computer Vision*, pages 1440–1448, 2015.

24. Ross Girshick, Jeff Donahue, Trevor Darrell, and Jitendra Malik. Rich feature hierarchies for accurate object detection and semantic segmentation. In *Proceedings of the IEEE Conference on Computer Vision and Pattern Recognition*, pages 580–587, 2014.

25. Ian Goodfellow, Yoshua Bengio, and Aaron Courville. *Deep Learning*. MIT Press, 2016. http://www.deeplearningbook.org.

26. Ian Goodfellow, Jean Pouget-Abadie, Mehdi Mirza, Bing Xu, David Warde-Farley, Sherjil Ozair, Aaron Courville, and Yoshua Bengio. Generative adversarial nets. In *Advances in Neural Information Processing Systems*, pages 2672–2680, 2014.

27. Metin N Gurcan, Anant Madabhushi, and Nasir Rajpoot. Pattern recognition in histopathological images: An icpr 2010 contest. In *International Conference on Pattern Recognition*, pages 226–234. Springer, 2010.

28. M.N. Gurcan, L.E. Boucheron, A. Can, A. Madabhushi, N.M. Rajpoot, and B. Yener. Histopathological image analysis: A review. *IEEE Reviews in Biomedical Engineering*, 2:147–171, Oct 2009.

29. K. He, X. Zhang, S. Ren, and J. Sun. Deep residual learning for image recognition. *2016 IEEE Conference on Computer Vision and Pattern Recognition (CVPR)*, pages 770–778, 2016.

30. Kaiming He, Georgia Gkioxari, Piotr Dollár, and Ross Girshick. Mask r-cnn. In *Proceedings of the IEEE International Conference on Computer Vision*, pages 2961–2969, 2017.

31. Achim Hekler, Jochen Sven Utikal, Alexander H Enk, Carola Berking, Joachim Klode, Dirk Schadendorf, Philipp Jansen, Cindy Franklin, Tim Holland-Letz, Dieter Krahl, et al. Pathologist-level classification of histopathological melanoma images with deep neural networks. *European Journal of Cancer*, 115:79–83, 2019.

32. Geoffrey E. Hinton, Nitish Srivastava, Alex Krizhevsky, Ilya Sutskever, and Ruslan R. Salakhutdinov. Improving neural networks by preventing co-adaptation of feature detectors. *CoRR*, abs/1207.0580, 2012.

33. Henning Höfener, André Homeyer, Nick Weiss, Jesper Molin, Claes F Lundström, and Horst K Hahn. Deep learning nuclei detection: A simple approach can deliver state-of-the-art results. *Computerized Medical Imaging and Graphics*, 70:43–52, 2018.

34. Gao Huang, Zhuang Liu, Laurens Van Der Maaten, and Kilian Q Weinberger. Densely connected convolutional networks. In *2017 IEEE Conference on Computer Vision and Pattern Recognition (CVPR)*. IEEE, 2017.

35. Nathan Ing, Jakub M Tomczak, Eric Miller, Isla P Garraway, Max Welling, Beatrice S Knudsen, and Arkadiusz Gertych. A deep multiple instance model to predict prostate cancer metastasis from nuclear morphology. In *Conference on Medical Imaging with Deep Learning. Amsterdam*, 2018.

36. Andrew Janowczyk, Ajay Basavanhally, and Anant Madabhushi. Stain normalization using sparse autoencoders (stanosa): Application to digital pathology. *Computerized Medical Imaging and Graphics*, 57:50–61, 2017.

37. Andrew Janowczyk and Anant Madabhushi. Deep learning for digital pathology image analysis: A comprehensive tutorial with selected use cases. *Journal of Pathology Informatics*, 7(1):29, July 2016.

38. Oscar Jimenez-Del-Toro, Sebastian Otálora, Mats Andersson, Kristian Eurén, Martin Hedlund, Mikael Rousson, Henning Müller, and Manfredo Atzori. Chapter 10 - analysis of histopathology images: From traditional machine learning to deep learning. In Adrien Depeursinge, Omar S. Al-Kadi, and J.Ross Mitchell, editors, *Biomedical Texture Analysis*, pages 281–314. Academic Press, 2017.

39. J Angel Arul Jothi and V Mary Anita Rajam. A survey on automated cancer diagnosis from histopathology images. *Artificial Intelligence Review*, 48(1):31–81, June 2017.

40. Philipp Kainz, Martin Urschler, Samuel Schulter, Paul Wohlhart, and Vincent Lepetit. You should use regression to detect cells. In *International Conference on Medical Image Computing and Computer-Assisted Intervention*, pages 276–283. Springer, 2015.

41. D. Karimi, G. Nir, L. Fazli, P. C. Black, L. Goldenberg, and S. E. Salcudean. Deep learning-based gleason grading of prostate cancer from histopathology images - role of multiscale decision aggregation and data augmentation. *IEEE Journal of Biomedical and Health Informatics*, pages 1–1, 2019.

42. Andrej Karpathy and Li Fei-Fei. Deep visual-semantic alignments for generating image descriptions. In *Proceedings of the IEEE conference on computer vision and pattern recognition*. IEEE, 2015.

43. Jakob Nikolas Kather, Cleo-Aron Weis, Francesco Bianconi, Susanne M Melchers, Lothar R Schad, Timo Gaiser, Alexander Marx, and Frank Gerrit Zöllner. Multi-class texture analysis in colorectal cancer histology. *Scientific Reports*, 6:27988, 2016.

44. Amirhossein Kiani, Bora Uyumazturk, Pranav Rajpurkar, Alex Wang, Rebecca Gao, Erik Jones, Yifan Yu, Curtis P Langlotz, Robyn L Ball, Thomas J Montine, et al. Impact of a deep learning assistant on the histopathologic classification of liver cancer. *NPJ Digital Medicine*, 3(1):1–8, 2020.

45. Diederik P Kingma and Max Welling. Auto-encoding variational bayes. *arXiv preprint arXiv:1312.6114*, 2013.

46. Daisuke Komura and Shumpei Ishikawa. Machine learning methods for histopathological image analysis. *Computational and Structural Biotechnology Journal*, 16:34–42, 2018.

47. Alex Krizhevsky, Ilya Sutskever, and Geoffrey E Hinton. Imagenet classification with deep convolutional neural networks. *Advances in Neural Information Processing Systems*, 25:1097–1105, January 2012.

48. Shyam Lal, Devikalyan Das, Kumar Alabhya, Anirudh Kanfade, Aman Kumar, and Jyoti Kini. Nucleisegnet: Robust deep learning architecture for the nuclei segmentation of liver cancer histopathology images. *Computers in Biology and Medicine*, 128:104075, 2021.

49. Yann LeCun, Léon Bottou, Yoshua Bengio, and Patrick Haffner. Gradient-based learning applied to document recognition. *Proceedings of the IEEE*, 86(11):2278–2324, 1998.

50. Y. Li, X. Li, X. Xie, and L. Shen. Deep learning based gastric cancer identification. In *2018 IEEE 15th International Symposium on Biomedical Imaging (ISBI 2018)*, pages 182–185, April 2018.

51. Qiaokang Liang, Yang Nan, Gianmarc Coppola, Kunglin Zou, Wei Sun, Dan Zhang, Yaonan Wang, and Guanzhen Yu. Weakly supervised biomedical image segmentation by reiterative learning. *IEEE Journal of Biomedical and Health Informatics*, 23(3):1205–1214, 2018.

52. Haotian Liao, Yuxi Long, Ruijiang Han, Wei Wang, Lin Xu, Mingheng Liao, Zhen Zhang, Zhenru Wu, Xuequn Shang, Xuefeng Li, et al. Deep learning-based classification and mutation prediction from histopathological images of hepatocellular carcinoma. *Clinical and Translational Medicine*, 10(2), 2020.

53. Geert Litjens, Thijs Kooi, Babak Ehteshami Bejnordi, Arnaud Arindra Adiyoso Setio, Francesco Ciompi, Mohsen Ghafoorian, Jeroen Awm Van Der Laak, Bram Van Ginneken, and Clara I Sánchez. A survey on deep learning in medical image analysis. *Medical Image Analysis*, 42:60–88, February 2017.

54. Marc Macenko, Marc Niethammer, James S Marron, David Borland, John T Woosley, Xiaojun Guan, Charles Schmitt, and Nancy E Thomas. A method for normalizing histology slides for quantitative analysis. In *Biomedical Imaging: From Nano to Macro, 2009. ISBI'09. IEEE International Symposium on*, pages 1107–1110. IEEE, 2009.

55. Anant Madabhushi and George Lee. Image analysis and machine learning in digital pathology: Challenges and opportunities. *Medical Image Analysis*, 33:170 – 175, 2016.

56. Siyamalan Manivannan, Wenqi Li, Jianguo Zhang, Emanuele Trucco, and Stephen J. Mckenna. Structure prediction for gland segmentation with hand-crafted and deep convolutional features. *IEEE Transactions on Medical Imaging*, 37:210–221, 2018.

57. Rashika Mishra, Ovidiu Daescu, Patrick Leavey, Dinesh Rakheja, and Anita Sengupta. Histopathological diagnosis for viable and non-viable tumor prediction for osteosarcoma using convolutional neural network. In Zhipeng Cai, Ovidiu Daescu, and Min Li, editors, *Bioinformatics Research and Applications*, pages 12–23, Cham, 2017. Springer International Publishing.

58. Jesper Molin, Anna Bodén, Darren Treanor, Morten Fjeld, and Claes Lundström. Scale stain: Multi-resolution feature enhancement in pathology visualization. *arXiv preprint arXiv:1610.04141*, 2016.

59. Ghulam Murtaza, Liyana Shuib, Ghulam Mujtaba, and Ghulam Raza. Breast cancer multi-classification through deep neural network and hierarchical classification approach. *Multimedia Tools and Applications*, pages 1–31, 2019.

60. Peter Naylor, Joseph Boyd, Marick Laé, Fabien Reyal, and Thomas Walter. Predicting residual cancer burden in a triple negative breast cancer cohort. In *2019 IEEE 16th International Symposium on Biomedical Imaging (ISBI 2019)*, pages 933–937. IEEE, 2019.

61. Peter Naylor, Marick Laé, Fabien Reyal, and Thomas Walter. Nuclei segmentation in histopathology images using deep neural networks. In *Biomedical Imaging (ISBI 2017), 2017 IEEE 14th International Symposium on*. IEEE, 2017.

62. Andrew Ng et al. Sparse autoencoder. *CS294A Lecture Notes*, 72(2011):1–19, 2011.

63. Hyeonwoo Noh, Seunghoon Hong, and Bohyung Han. Learning deconvolution network for semantic segmentation. In *Proceedings of the IEEE International Conference on Computer Vision*, pages 1520–1528, 2015.

64. Sebastian Otálora, Oscar Perdomo, Manfredo Atzori, Mats Andersson, Ludwig Jacobsson, Martin Hedlund, and Henning Müller. Determining the scale of image patches using a deep learning approach. In *Biomedical Imaging (ISBI 2018), 2018 IEEE 15th International Symposium on*. IEEE, 2018.

65. Baochuan Pang, Yi Zhang, Qianqing Chen, Zhifan Gao, Qinmu Peng, and Xinge You. Cell nucleus segmentation in color histopathological imagery using convolutional networks. In *2010 Chinese Conference on Pattern Recognition (CCPR)*, pages 1–5. IEEE, 2010.

66. Luis Perez and Jason Wang. The effectiveness of data augmentation in image classification using deep learning. *arXiv preprint arXiv:1712.04621*, 2017.

67. Francesco Ponzio., Enrico Macii., Elisa Ficarra., and Santa Di Cataldo. Colorectal cancer classification using deep convolutional networks - an experimental study. In *Proceedings of the 11th International Joint Conference on Biomedical Engineering Systems and Technologies - Volume 2 BIOIMAGING: BIOIMAGING*, pages 58–66. INSTICC, SciTePress, 2018.

68. Hui Qu, Gregory Riedlinger, Pengxiang Wu, Qiaoying Huang, Jingru Yi, Subhajyoti De, and Dimitris Metaxas. Joint segmentation and fine-grained classification of nuclei in histopathology images. In *2019 IEEE 16th International Symposium on Biomedical Imaging (ISBI 2019)*, pages 900–904. IEEE, 2019.

69. Joseph Redmon, Santosh Divvala, Ross Girshick, and Ali Farhadi. You only look once: Unified, real-time object detection. In *Proceedings of the IEEE Conference on Computer Vision and Pattern Recognition*, pages 779–788, 2016.

70. Joseph Redmon and Ali Farhadi. Yolo9000: better, faster, stronger. In *Proceedings of the IEEE Conference on Computer Vision and Pattern Recognition*, pages 7263–7271, 2017.

71. Joseph Redmon and Ali Farhadi. Yolov3: An incremental improvement. *arXiv preprint arXiv:1804.02767*, 2018.

72. Shaoqing Ren, Kaiming He, Ross Girshick, and Jian Sun. Faster r-cnn: Towards real-time object detection with region proposal networks. In *Advances in Neural Information Processing Systems*, pages 91–99, 2015.

73. Danilo Jimenez Rezende, Shakir Mohamed, and Daan Wierstra. Stochastic back-propagation and approximate inference in deep generative models. *arXiv preprint arXiv:1401.4082*, 2014.

74. A. M. Romano and A. A. Hernandez. Enhanced deep learning approach for predicting invasive ductal carcinoma from histopathology images. In *2019 2nd International Conference on Artificial Intelligence and Big Data (ICAIBD)*, pages 142–148, May 2019.

75. F. P. Romero, A. Tang, and S. Kadoury. Multi-level batch normalization in deep networks for invasive ductal carcinoma cell discrimination in histopathology images. In *2019 IEEE 16th International Symposium on Biomedical Imaging (ISBI 2019)*, pages 1092–1095, April 2019.

76. Olaf Ronneberger, Philipp Fischer, and Thomas Brox. U-net: Convolutional networks for biomedical image segmentation. In *International Conference on Medical Image Computing and Computer-Assisted Intervention*, pages 234–241. Springer, 2015.

77. Harshita Sharma, Norman Zerbe, Iris Klempert, Olaf Hellwich, and Peter Hufnagl. Deep convolutional neural networks for automatic classification of gastric carcinoma using whole slide images in digital histopathology. *Computerized Medical Imaging and Graphics*, 61:2–13, November 2017.

78. Dinggang Shen, Guorong Wu, and Heung-Il Suk. Deep learning in medical image analysis. *Annual Review of Biomedical Engineering*, 19:221–248, June 2017.

79. J Shiraishi, Q Li, D Appelbaum, and K Doi. Computer-aided diagnosis and artificial intelligence in clinical imaging. *Seminars in Nuclear Medicine*, 41(6), November 2011.

80. Connor Shorten and Taghi M Khoshgoftaar. A survey on image data augmentation for deep learning. *Journal of Big Data*, 6(1):60, 2019.

81. Karen Simonyan and Andrew Zisserman. Very deep convolutional networks for large-scale image recognition. *arXiv preprint arXiv:1409.1556*, 2014.

82. Korsuk Sirinukunwattana, Josien PW Pluim, Hao Chen, Xiaojuan Qi, Pheng-Ann Heng, Yun Bo Guo, Li Yang Wang, Bogdan J Matuszewski, Elia Bruni, Urko Sanchez, et al. Gland segmentation in colon histology images: The glas challenge contest. *Medical Image Analysis*, 35:489–502, 2017.

83. Korsuk Sirinukunwattana, Shan E Ahmed Raza, Yee-Wah Tsang, David RJ Snead, Ian A Cree, and Nasir M Rajpoot. Locality sensitive deep learning for detection and classification of nuclei in routine colon cancer histology images. *IEEE Transactions on Medical Imaging*, 35(5):1196–1206, 2016.

84. Tzu-Hsi Song, Victor Sanchez, Hesham EIDaly, and Nasir Rajpoot. Simultaneous cell detection and classification with an asymmetric deep in bone marrow histology images. In Maria Valdes Hernandez and Victor Gonzalez-Castro, editors, *Medical Image Understanding and Analysis*, pages 829–838, Cham, 2017. Springer International Publishing.

85. Youyi Song, Ling Zhang, Siping Chen, Dong Ni, Baopu Li, Yongjing Zhou, Baiying Lei, and Tianfu Wang. A deep learning based framework for accurate segmentation of cervical cytoplasm and nuclei. *2014 36th Annual International Conference of the IEEE Engineering in Medicine and Biology Society*, 2014.

86. Fabio A Spanhol, Luiz S Oliveira, Caroline Petitjean, and Laurent Heutte. A dataset for breast cancer histopathological image classification. *IEEE Transactions on Biomedical Engineering*, 63(7):1455–1462, October 2015.

87. Jiamei Sun and Alexander Binder. Comparison of deep learning architectures for h&e histopathology images. In *Big Data and Analytics (ICBDA), 2017 IEEE Conference on*. IEEE, 2017.

88. Christian Szegedy, Sergey Ioffe, Vincent Vanhoucke, and Alexander A Alemi. Inception-v4, inception-resnet and the impact of residual connections on learning. In *AAAI*, volume 4, page 12, 2017.

89. Christian Szegedy, Wei Liu, Yangqing Jia, Pierre Sermanet, Scott Reed, Dragomir Anguelov, Dumitru Erhan, Vincent Vanhoucke, and Andrew Rabinovich. Going deeper with convolutions. In *Proceedings of the IEEE Conference on Computer Vision and Pattern Recognition*. IEEE, 2015.

90. Christian Szegedy, Vincent Vanhoucke, Sergey Ioffe, Jon Shlens, and Zbigniew Wojna. Rethinking the inception architecture for computer vision. In *Proceedings of the IEEE Conference on Computer Vision and Pattern Recognition*. IEEE, 2016.

91. Sairam Tabibu, PK Vinod, and CV Jawahar. Pan-renal cell carcinoma classification and survival prediction from histopathology images using deep learning. *Scientific Reports*, 9(1):1–9, 2019.

92. Ye Tian, Li Yang, Wei Wang, Jing Zhang, Qing Tang, Mili Ji, Yang Yu, Yu Li, Hong Yang, and Airong Qian. Computer-aided detection of squamous carcinoma of the cervix in whole slide images. *arXiv preprint arXiv:1905.10959*, 2019.

93. Thaina A. Azevedo Tosta, Leandro A. Neves, and Marcelo Z. do Nascimento. Segmentation methods of h&e-stained histological images of lymphoma: A review. *Informatics in Medicine Unlocked*, 9:35 – 43, 2017.

94. Pascal Vincent, Hugo Larochelle, Yoshua Bengio, and Pierre-Antoine Manzagol. Extracting and composing robust features with denoising autoencoders. In *Proceedings of the 25th International Conference on Machine Learning*, ICML '08, pages 1096–1103, New York, NY, USA, 2008.

95. Pascal Vincent, Hugo Larochelle, Isabelle Lajoie, Yoshua Bengio, and Pierre-Antoine Manzagol. Stacked denoising autoencoders: Learning useful representations in a deep network with a local denoising criterion. *Journal of Machine Learning Research*, 11(Dec):3371–3408, 2010.

96. Noorul Wahab, Asifullah Khan, and Yeon Soo Lee. Two-phase deep convolutional neural network for reducing class skewness in histopathological images based breast cancer detection. *Computers in Biology and Medicine*, 85:86–97, 2017.

97. Du Wang, Chaochen Gu, Kaijie Wu, and Xinping Guan. Adversarial neural networks for basal membrane segmentation of microinvasive cervix carcinoma in histopathology images. *2017 International Conference on Machine Learning and Cybernetics (ICMLC)*, 2017.

98. Du Wang, Kaijie Wu, Chaochen Gu, and Xinping Guan. Time efficient cell detection in histopathology images using convolutional regression networks. In *Control Conference (CCC), 2017 36th Chinese*. IEEE, 2017.

99. Shujun Wang, Yaxi Zhu, Lequan Yu, Hao Chen, Huangjing Lin, Xiangbo Wan, Xinjuan Fan, and Pheng-Ann Heng. Rmdl: Recalibrated multi-instance deep learning for whole slide gastric image classification. *Medical Image Analysis*, 58:101549, 2019.

100. X. Wang, H. Chen, C. Gan, H. Lin, Q. Dou, E. Tsougenis, Q. Huang, M. Cai, and P. Heng. Weakly supervised deep learning for whole slide lung cancer image analysis. *IEEE Transactions on Cybernetics*, pages 1–13, 2019.

101. Xiaodong Wang, Ying Chen, Yunshu Gao, Huiqing Zhang, Zehui Guan, Zhou Dong, Yuxuan Zheng, Jiarui Jiang, Haoqing Yang, Liming Wang, et al. Predicting gastric cancer outcome from resected lymph node histopathology images using deep learning. *Nature Communications*, 12(1):1–13, 2021.

102. World Health Organization. Cancer. `http://www.who.int/en/news-room/fact-sheets/detail/cancer`, 2018.

103. Jun Xu, Lei Xiang, Renlong Hang, and Jianzhong Wu. Stacked sparse autoencoder (ssae) based framework for nuclei patch classification on breast cancer histopathology. In *2014 IEEE 11th International Symposium on Biomedical Imaging (ISBI)*. IEEE, 2014.

104. Yan Xu, Zhipeng Jia, Yuqing Ai, Fang Zhang, Maode Lai, and Eric L-Chao Chang. Deep convolutional activation features for large scale brain tumor histopathology image classification and segmentation. In *Acoustics, Speech and Signal Processing (ICASSP), 2015 IEEE International Conference on*. IEEE, 2015.

105. Zhaoyang Xu and Qianni Zhang. Multi-scale context-aware networks for quantitative assessment of colorectal liver metastases. *2018 IEEE EMBS International Conference on Biomedical & Health Informatics (BHI)*, 2018.

106. Qiuju Yang, Kaijie Wu, Hao Cheng, Chaochen Gu, Yuan Liu, Shawn Patrick Casey, and Xinping Guan. Cervical nuclei segmentation in whole slide histopathology images using convolution neural network. In *International Conference on Soft Computing in Data Science*, pages 99–109. Springer, 2018.

107. Faliu Yi, Lin Yang, Shidan Wang, Lei Guo, Chenglong Huang, Yang Xie, and Guanghua Xiao. Microvessel prediction in h&e stained pathology images using fully convolutional neural networks. *BMC Bioinformatics*, 19(1):64, 2018.

108. Asami Yonekura, Hiroharu Kawanaka, VB Surya Prasath, Bruce J Aronow, and Haruhiko Takase. Glioblastoma multiforme tissue histopathology images based disease stage classification with deep cnn. In *Informatics, Electronics and Vision & 2017 7th International Symposium in Computational Medical and Health Technology (ICIEV-ISCMHT), 2017 6th International Conference on*. IEEE, 2017.

109. Asami Yonekura, Hiroharu Kawanaka, VB Surya Prasath, Bruce J Aronow, and Haruhiko Takase. Improving the generalization of disease stage classification with deep cnn for glioma histopathological images. In *Bioinformatics and Biomedicine (BIBM), 2017 IEEE International Conference on*. IEEE, 2017.

110. Asami Yonekura, Hiroharu Kawanaka, VB Surya Prasath, Bruce J Aronow, and Haruhiko Takase. Automatic disease stage classification of glioblastoma multiforme histopathological images using deep convolutional neural network. *Biomedical Engineering Letters*, 8(3):321–327, June 2018.

111. Tom Young, Devamanyu Hazarika, Soujanya Poria, and Erik Cambria. Recent trends in deep learning based natural language processing. *IEEE Computational Intelligence Magazine*, 13(3):55–75, July 2018.

112. Dong-Qing Zhang. Image recognition using scale recurrent neural networks. *CoRR*, 2018.

10 Skin Lesion Classification by Using Deep Tree-CNN

Prakash Choudhary and Sameer Mansuri
Computer Science and Engineering, National Institute of Technology
Hamirpur, Himachal Pradesh, India

CONTENTS

10.1 INTRODUCTION

Skin cancer constitutes one out of three of the cancers diagnosed worldwide as estimated by the World Health Organization [5]. Melanoma is the most dangerous forms of skin cancer which usually begins developing in skin melanocytes. About Five million people are diagnosed with skin cancer in the US alone every year [18]. The skin cancer is broadly classified into basal cell and carcinomas of skin cancer. The first one (BCC) shows delayed growth to form and thus can be dealt easily. It grows in top skin layer and particularly at the bottom of it. Skin that has not been covered from sunlight for long periods of time shows it's development. It occurs as a thin, flat, glossy, white or waxy lump with hard, dried or scaly patches that can be red or brown in colour [14]. At early stage diagnosis, both of these forms are extremely

DOI: 10.1201/9781003246688-10

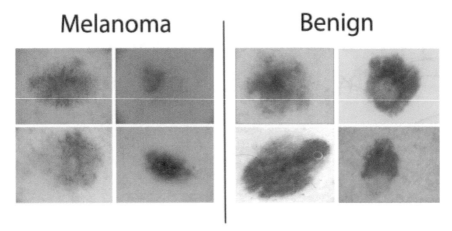

Figure 10.1 Dermoscopy of skin lesion image showing similarity between melanomous and benign skin lesions.

curable. In the US, 99 out of 100 people survive whose melanoma has been detected at an early stage [14].

Generally, visually examining the skin cancer is a very complex and time taking process. Furthermore, since there is a high degree of resemblance between various types of skin lesions as shown in Figure 10.1, visual inspection is difficult and can result in incorrect diagnosis of lesions (melanoma and non-melanoma) [13]. During the process of medical examination of skin lesions, the optical examination is needed for the discrimination between lesion and normal areas, which may influence the exact recognition process. Dermoscopy is examination of skin using surface microscopy, it is an non-evasive imaging technique to get color accurate image of concerned skin lesion. Dermoscopy also utilized some form gel medium to reduce skin reflection, artifacts, and shadow to get clear image of skin lesion. So the accuracy of diagnosis is improved by dermoscopy compared to manual inspection such as screening procedures to obtain and RGB image suitable for use in an tool for computer-aided diagnosis. Nowadays, we have widely available skin lesion datasets which have been curated and diagnosed by experts in the diagnostics field. Also each year newer data is recorded and is being added to already available data. Furthermore, digital cameras are also becoming very capable each year,

So implementing a computer-aided diagnostics methods for diagnosis of melanoma skin lesions will aid the healthcare professionals to assess the situation in reduced time. This type of method can also be used by people to get an estimate on personal level which can provide and early estimate the severity of any skin lesion. Thus developing this kind of CNN for detecting melanoma can increase the efficiency and accuracy of early recognition of melanoma.

A computer-based automated approach consists of these major steps : preprocessing, segmentation, extraction features and classification. Between these steps, segmentation lesion and correct extraction features are difficult steps due to high degree of similarity between lesions, lesion irregularities and variation in skin conditions. An incorrect method of lesion segmentation with irrelevant characteristics decreases the reliability of classification. The efficiency of melanoma detection has been optimized due to the usage of dermoscopic methods. Dermoscopy is a skin imaging procedure, in which the body is not cut open like in surgeries and is capable of collecting illuminated and magnified skin lesion images in order to increase spot visibility. The optical appearance of skin lesions can be improved if the reflection of the surface of the skin is eliminated [3].

Automated malignant melanoma observation and diagnosis is a very tedious task in the medical imaging sector due to several known problems which includes similarity of different classes, poor contrast of lesions, color similarities, coarse hair and texture between lesion and healthy areas [18]. In addition, extraction and choice of features is another obstacle for discovering new information through machine learning. For a number of applications nowadays, such as diagnostic imaging and cultivation deep CNN is gaining the focus of researchers with improved categorization and identification efficiency [10]. An important issue to tackle is the correct identification and diagnosis of skin lesions.

Here, we are considering a new multimodal approach for classification of skin lesions. Taking inspiration from hierarchical classifiers, we implement a Deep-Tree CNN for classification of dermoscopy skin images. Deep-Tree CNN compose of multiple CNN nodes joint in a tree like structure. Each node aims to classify the input image into one of its child node. We aim to focus more on relative embedding of feature among different classes by implementing feature learning in a hierarchical way. Another advantage of using a tree like structure is that after training on dataset, we can expand the tree horizontally as more data become available for newer classes.

10.2 LITERATURE SURVEY

Many methods for the segmentation, identification and classification of melanoma in the skin lesions were used in the sector of machine learning aided diagnostics. Segmentation methods are generally classified as threshold-based segmentation, region-based segmentation, shape and texture feature-based segmentation. Moreover, in computer vision machine learning-based implementations for lesion segmentation and lesion categorization have always been an goto approach [18]. Specifically convolutional neural networks (CNNs) are deep learning algorithm that can take an image as input, assign importance to various features in the input image, by implementing learnable weights and biases which can distinguish one image feature from the another [19]. Relative to other traditional classification algorithms, the amount of preprocessing needed in case of a Convolutional Neural Network is significantly less as compared to a primitive methods where image filters are hand-engineered, Convolutional Neural Networks can learn these filters/characteristics in an automated manner with enough experience. In computer vision tasks deep learning (CNN) has shown superior results compared to tradition machine learning solutions [18]. Because the architecture of a CNN is inspired by the organization of the Visual Cortex in human brain, also their communication pattern is similar to that of Neurons in the Human Brain. Specific neurons will only respond to stimulus in a particular area of the visual field called the Receptive Field. A set of such fields can be stacked on top of each other to fill the entire visual field. Similarly a number of Convolutional Network layers analogous to receptive fields are stacked on top of each other to form a CNN. Given dermoscopic images of skin lesion, Lequan et. al. [4] used a very deep residual neural network (VDRN) to classify skin lesions. Originally, the segmentation is done by implementing and using a fully convolutionary residual network (FCRN), which differentiates between the foreground lesion and background region of the dermoscopy image. Then the segmented image is further enhanced by multiscale contextual integration schemes. FCRN is then integrated with VDRN output for the classification of input lesions images. The proposed methodology is tested upon the ISBI 2016 dataset and has a peak identification accuracy of 85.5%. Compared to handcrafted features or CNN with shallow designs, the proposed approach worked effectively and provides improved accuracy. In the paper by SM Jaisakti and P. Mirunalini [14], they tackle the task of segmentation of melanoma of images. They achieve this task by utilising GrabCut and k- means algorithms. They implemented a system having an automatic method to sliver the affected areas by implementing a semi-supervised learning method. This method comprises of a couple of steps, which includes preprocessing with segmentation. In the meantime of first step, artefacts which include, coarse hair and marks on the skin, blemishes are eliminated from the image. The image visibility is initially improved and after that coarse hair are taken out using various algorithms, e.g. Frangi vesselness [16]. Then from the resulting image, malignant melanoma areas are segmented using clustering algorithms and the Grabcut method. This method slivers the forepart image (affected skin region) and then the other algorithms that include k-means as well as flood-fill algorihm are implemented to extract the required area having ameliorate boundaries. For both datasets ISIC 2017 and PH2, they achieved good accuracy of 0.91935, 0.96047, respectively. In the approaching years to come, other methods of deep learning can be implemented to improve the parameter of exactness and coefficient of dice values.

Wei et al. [6] suggested a new approach for identifying three different forms of skin diseases, including dermatitis, herpes and psoriasis. Dermoscopic images were initially pre-processed with filtering and transformation to remove noise and irrelevant background. After that, the segmentation of skin lesion from the background was achieved using the Grey-Level Co-occurrence Matrix (GLCM) approach. As a result, the texture and colour characteristics of diverse skin disease photos were carefully determined. Finally, an SVM was utilized to classify three different types of skin disorders.

In an another paper of this field, scrutinizing of skin has been done to recognize the presence of a lesion using Deep Learning Networks [5] in which two frameworks are used. In one framework, lesion indexing network simultaneous segmentation and coarse classification takes place using the fully convoluted residual network and further the lesion index calculation unit for refined classification [8]. Lesion Feature Network, which is a framework of CNN is applied to implement the another chore of drawing out dermoscopic features.

This deep learning frameworks have been implemented on the International Skin Imaging Collaboration dataset 2017 testing set. They used Jaccard Index as well as

Area Under Curve for detrminig the effectiveness of this approach. The performance metrics achieved by this method for melanoma slivering and categorization were 0.718, 0.823, respectively, which were comparable to competition winners. The used framework, Lesion Feature Network, achieves the best parametric score for precision and sensitivity, i.e. 0.409, 0.665, respectively, for drawing out the dermoscopic features, which explains that it can be a handy method of implementation for the given problem [5].

Lesion segmentation has been an challenging task, as different variety of shape, size,texture, skin color, skin type makes designing a robust segmentation algorithm very difficult. Hosny KM and Kaseem [5] implemented a transfer learning approach along with data augmentation for skin lesion classification. Lesion segmentation and feature extraction both are done using a Deep Convolution Neural Network (DCNN). They used AlexNet [10] which was pretrained on Imagenet dataset, so the weights in the internal layers of this model were well suited for extracting features such as boundry, color, contrast, etc. from any given image. As the model is alreay suited for feature extraction, they only needed to fine tune the weights to better fit the task operating on dermoscopy images.

After that dropping, the last few classification layers from AlexNet the model were integrated with a softmax classifier. The weight for softmax classifier was randomly assigned and then the whole network was trained for the classification task. They used various dataset augmentation methods such as image rotation, horizontal flip, vertical flip, image shift and shear on the dataset to increase the amount of effective images used for training. Then they trained AlexNet model while tweaking the parameters to fine tune the learned weights to better suite the task of lesion classification. The obtained features from AlexNet are then used to train the softmax classifier. This method achieved classification rate of 96.7% which outperformed the competing implementation significantly. Barata, Catarina Celebi [18] performed a survey of all the existing research done on lesion segmentation and classification. They came to the conclusion that using handcrafted features to train classification algorithm does not lead to desired performance as many important features are missed if an handcrafted filter, while using deep learning features automated feature extraction can efficiently gather all feature but there is no way to provide some clinically inspired features weightage depending upon their degree of importance in an clinical diagnosis. As it is very difficult to interpret the output of various layer of CNN [4], which makes it very difficult for dermatologists to work with deep learning-based feature extraction. Therefore, it is important but tedious tasks to efficiently segment the affected region to highlight important features and choosing the better characteristics so that the computer-aided diagnostics has increased sensitivity, specificity and accuracy.

10.3 METHODOLOGY

The explanation of Hierarchical Deep Convolutional Neural network is given in this section in detail. The method that we have proposed include image preprocessing and categorization. The first task of the above method increments the local lesion contrast in the given images of dermoscopic concern, initially. Hierarchical DCNN is then used to classify the images.

10.3.1   IMAGE CONTRAST STRETCHING

This step is an important step in the field of computer vision such as dermoscopic images, mammograms and MRI. Several types of irregularities exist in dermoscopic images that include artifacts, low contrast, bubbles and alikeness between lesions and non-affected regions. For tackling these differences, 3D box Filtering, decorrelation, weighted Gaussian technique and a contrast stretching solution is implemented. Detailed explanation includes

- **3D box filter** – Box filtering is effectively an image filtering type in which surrounding pixels are averaged to obtain the value of the pixel. It is basically a convolution filter that is a commonly used information processing method and is used for image filtering. In an image, it can sharpen, emboss, detect edge, smooth, movement-blur, etc. provided appropriate filter kernel is used [14] .
- **Decorrelation** – Decorrelation methods can be used in image processing to intensify or extend colour variations in each pixel of an image. It is usually called as 'decorrelation stretching' [14]. In many other fields, the concept of decorrelation can be applied. Decorrelation is used in neuroscience in the analysis of the human visual system in neural networks.
- **Gaussian filter** – We talk of a linear filter when we talk about a Gaussian filter. Typically it is used for blurring the image or reducing noise. We can use them for unsharp masking, if we use two of them and then take their difference (edge detection). Solely, the Gaussian filter will reduce the contrast and blur the edges[11].

10.3.2   FEATURE EXTRACTION

Extraction of features is one of the challenging and crucial step in computer-aided pattern recognition to represent an object in several applications such as engineering and medical imaging. Optical learning has become very important nowadays and plays an important role in the society. As a result, multiple extraction methods for classifier learning features like point features, shape features, texture features, color features geometric features and deep learning were introduced recently. In many computer vision questions, CNN have become the technique of choice [17]. This approach has also been adopted by the medical imaging community, with growing number of applications on using this method to identify or partition organs and structures in medical images. Deep learning methods are based on the ANN. This type of algorithm for learning consists of nodes (known as neurons) each having a specific activation function f and other parameters w, b where w is the weights set and b is biases set. Deep learning is now a days latest research areas that display evolutionary advances in applications for computer vision and machine learning. Deep learning is essential in medical imaging to accurately recognize tumors such as malignant or be-nign [1]. Therefore, for feature extraction, a new Deep CNN model is implemented. Implementation of model DCNN was inspired by design of ResNet CNN.

10.3.3   DEEP CONVOLUTIONAL NEURAL NETWORK (DCNN)

Deep Convolutional Neural Network is created by using various neural networks layers. This neural networks consists of a series of convolutional layers which can identify and differentiate between characteristics called features from an input im-age. A simple architecture of a DCNN is shown in Figure 10.3. The feature extract-ing components in a DCNN are a combination of *convolutional + pooling* layer [15]. Convolutional layers are borrowed from signal processing where matrix multiplica-tion is performed between the input image and a specified filter or kernel. Then this pooling layers are used to reduce the spacial size of the image representation which also reduces the amount of parameters and computation performed in the CNN. This components are repeated to capture different features from the input image at differ-ent levels of DCNN [10]. In early layers of DCNN basic features are learned from the images such as lines, boundary from the input image than mid level features such as shapes, as we move further they obtain high level features from the given image. Finally this features are flattened and connected to an output layer that then classifies this features into specified no of classes.

The word "deep" in DCNN refers to the number of layers in the network. Nor-mally a CNN will will have 5–10 such layers for feature learning but in case of a DCNN the number of layers can be more than 50–100 deep. The working of a CNN is fairly similar to the overly-simplified working of human visual cortex. So hav-ing greater number of layers will be able to extract features much more effectively [19]. Therefore, DCNNs can be implemented for the task of classifying skin lesion images. The reason behind using a CNN is that there is still lot of noise, artifacts and aberrations that make hand-crafted feature extraction very difficult. Also, there can be large variation of feature in same class plus the visual similarities of different classes of lesions. So, a dataset having large amount of images must be used to train a CNN.

In many computer vision tasks, CNN has been used to enhance performance. Various successfull DCNN architectures such as AlexNet, LeNet, GoogLeNet, VGGNet and ZFNet are available for use in a variety of computer vision tasks. Here Xception [2] has been used to implement the Tree-CNN in this study.

10.3.4   DEEP TREE CNN

Tree-CNN is inspired by hierarchical classifiers and structured in a tree-like way having multiple connected nodes. Every node has a DCNN (except leaf nodes) that is trained to classify the node's input into one of its child nodes. Here we use Xception CNN [2] for classification at tree nodes. The root node is the starting node at the highest level, where the initial classification occurs. The image is then moved, as per the classification, to its child node. This node further classifies the picture, the last

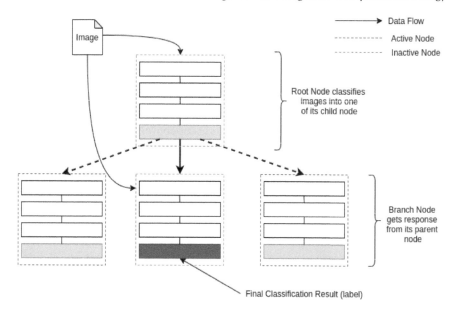

Figure 10.2   Shows the basic structure of deep tree CNN.

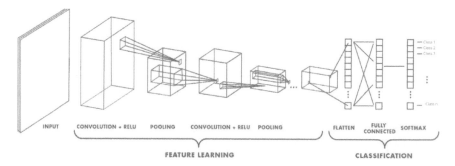

Figure 10.3   Shows a Convolutional Neural Network using convolution + pooling layer for feature learning and flatten + fully connected layer for classification of learned features.

stage of classification, before we reach a leaf node. Each internal nodes has a parent node and one or more than one child node. The final stage of the classification is when the images reaches one of the leaf node. Each class is uniquely assigned to one of the leaf nodes. Figure10.2 shows a Tree-CNN with two levels. Each CNN at the second level is a leaf node, also called the output node of the corresponding branch of tree structure.

At every CNN node, we get a 3D matrix of the format $O^{C \times N \times I}$. Where **C** is the count of corresponding children nodes, **N** is the count of associated classes and **I** is the count of input images. So, $O(c,n,i)$ represents the output corresponding to the ith image of nth class among the available classes. For every input image, softmax likelihood is calculated, and according to the softmax values, one of the child nodes is selected with the current image as input, explained in algorithm 2. By repeating the above algorithm recursively for every level obtain the predicted class for an input image at leaf nodes.

Algorithm 2 Evaluate for input

1: I = Input Image, node = root Node;
2: **procedure** PREDICT($I, node$)
3:    count = #child nodes;
4:    **if** $count = 0$ **then**
5:        label = current node's label
6:        **return** label
7:    **else**
8:        nextNode = EvaluateNext(I, node)
9:        **return** predict(I, nextNode)
10:   **end if**
11: **end procedure**

When adding new classes in the Deep-Tree CNN, we obtain the softmax matrix for the current image with respect to all the leaf nodes, indicating similarity with the existing classes already in the network. Here we define two threshold values α and β that are used to identify whether a child nodes can incorporate the new class otherwise we create a new child for new classes. For any new class, we define $v1, v2, v3$ where are in descending order such that $v1 > v2 > v3$. Depending on the values of $v1, v2, v3$ one of the following actions will be taken.

- **Add class to existing child node:** If $v1$ is greater by a threshold, α than the next value $v2$, the class implies a clear similarity with one of the child nodes. The corresponding child node is assigned the new class.
- **Add class after merging two child nodes:** If the new node associates with more than one child node such that $v1 - v2 < \alpha$ and $v2 - v3 > \beta$, then we combine those two children nodes into a new node and assign the new class to it.
- **Add new class to a new child node:** If no children nodes share any similarities with the new class such that $v1 - v2 < \alpha, v2 - v3 < \beta$, then expand current level horizontally by adding a new child node to current node and assign the new class to it.

As we add new classes, to keep the distribution of classes balanced we limit the maximum number of child nodes to two. To achieve this, we can set the softmax value of child nodes which are full to 0. Thus only child nodes having room to incorporate new classes are considered while expanding the Deep-Tree CNN.

Once the new child nodes are added in the tree. CNN training is done for only the newer nodes. This saves us from changing the entire network, and retraining is needed for only affected or modified parts of the network. As the root node is always

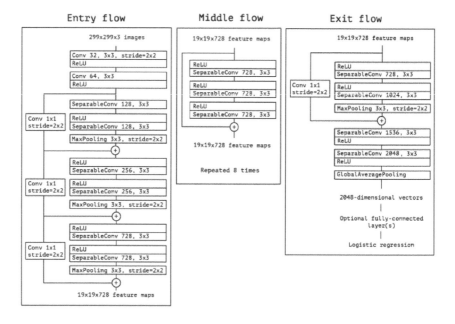

Figure 10.4   Xception [2] node structure for feature extraction.

affected by the change in available classes, we start training from root node with all
the classes and move down level wise to the modified internal nodes. During training
if an incorrect classification happens, we will not send it to the child nodes as its child
nodes will also miss classifying the image. Thus only assigned classes are considered
for every node in the neural network.

10.3.5   TREE-CNN NODE

Starting from the root node of the Tree-CNN, we implement each node as a DCNN.
The node structure contains a neural network such as Xception which is used for
feature extraction. Xception is an iteration of Inception family of DCNN [2], it
stands for Extreme version of Incepton. By having modified depthwise separable
convolutios, it improves upon its predecessor Inception-v3. Xception is an image
recognition DCNN which is widely used for computer-aided classification of images. This CNN has been shown to achieve greater than 79.0% accuracy on the ImageNet dataset [2]. Xception architecture comprises of symmetric and asymmetric
building blocks. These building block consists of convolutions, max pooling, average pooling, dropout, concatenate and fully connected layers. Depthwise separable
convolutions are implemented by using a point-wise convolution (1x1) followed by
depthwise convolutions (nxn spacial) with residual connections. A sample feature
extraction node structure is shown in Figure 10.4. These features are then flattened
and sent to a pair of fully connected layers which then classifies this features into the

Algorithm 3 Logic to add new node/class.

```
 1: procedure ADDNODE(C)
 2:     top3 = getTopClasses(C)              ▷ Holds top 3 Similar classes v1, v2, v3
 3:     if v1 − v2 ≥ α then
 4:         top3[0].add(C)                                        ▷ Add to First Node
 5:     else if (v1 − v2 ≤ α) && (v2 − v3 ≥ β) then
 6:         new = merge(top3[0], top3[1])
 7:         new.add(C)                                    ▷ merge and add new class
 8:     else
 9:         new = root.createChild()
10:         new.add(C)                            ▷ Create a separate node for class
11:     end if
12: end procedure
```

classes assigned to one of its children. The root CNN classifies the skin lesion images into unique labels available in the dataset. But as the data moves down to the child nodes, which only operate on a subset of the available classes which are uniquely assigned at a given level. This type of organization of class label among various CNN nodes make it difficult to use a global labeling scheme for the dataset. So, in order to ensure consistency of label among nodes at any level $l > 0$, each node maintains its own lookup table called *LabelTransform* that defines label mapping from global classes present in the dataset to the local classes assigned to the current node and its child nodes.

This lookup tables are utilized extensively when forwarding the input images, and when newer classes are seen by the Tree-CNN. Algorithm for adding newer classes to preexisting tree is shown in algorithm 3, we need update and modify *LabelTransform* table to maintain label consistency so that no images are incorrectly classified which is taken care by *add* and *merge* functions. For example, in case a node is to be merged or split, we use softmax value to find similarity between the child nodes and modify the *LabelTranform* table according to changes in the structure. In case of merging the *LableTransform*, table is also merged and the mapping is updated. This way of adding newer classes helps up accommodate newer classes into over network in a hierarchical manner while also retaining the knowledge learned from the previously seen classes by the Tree-CNN nodes.

10.4   RESULTS

10.4.1   DATASET

We used ISIC 2018 for evaluation for proposed methods. Datasets is provided by the International Skin Imaging Collaboration (ISIC) for 2018 [12] [17]. The ISIC 2018 dataset comprises of total 10015 RGB dermoscopic images. The dataset is vetted by recognised melanoma experts. This ISIC images are provided with a gold standard diagnostics information by these experts in medical field. The dataset has total of seven classes each of different skin cancer. Along with images metadata of patients is also provided such as age, sex, patient_id, anatomical_site. All the images provided in the ISIC 2018 dataset has been resized to a uniform 1024x1024. These classes in addition to the ISIC training and testing dataset groups were augmented by deploying various image augmentation strategies such as image rotation ranging from 0^0 to 90^0, image flip and image shear.

10.4.2 METRICES

The metrices used for evaluation and comparison of the implemented neural network are *f1-score, accuracy, precision and recall*. These metrices are evaluated by using the following formulas [18].

- **f1-score:** f1-score is the harmonic mean between precision and recall. It tells you how precise (correct classification) the classifier is as well as how robust (miss significantly less number of instances) its is. So, a high value of F1 score indicates that you have a low number of false-positives and false-negatives, indicating that you are correctly detecting actual threats and are not bothered by false alarms.
- **Specificity or Precision:** Precision tells us that how often the models correctly predicts the positive outcome of the problem statement, i.e. it is determined as the ratio of the number of correctly classified positive samples to the total number of samples classified as positive. This helps us identify the performance of the model in case the cost of false-positive is high such as skin lesion classification.

$$precision = \frac{t_p}{t_p + f_p}$$

$$f1 - score = 2 \times \frac{precision \times recall}{precision + recall}$$

- **Accuracy:** Accuracy refers to the amount of times the model correctly predicts an outcome. This generally shows that how our model performs among all the available classes. But this may lead to misleading interpretation when the dataset has severe class imbalance.

$$accuracy = \frac{t_p + t_n}{t_p + f_p + f_n + f_p}$$

where t_p, f_p, t_n, f_n are abbreviations for true-positive, false-positive, true-negative and false-negative, respectively. furthermore, a confusion matrix is also plotted in 10.6 that maps the lesion image's predicted label with its actual diagnostics.

10.4.3 TRAINING ANALYSIS

In the proposed methodology, we implement the model in such a way that incremental learning can be implemented. So the newer input images are continuously used

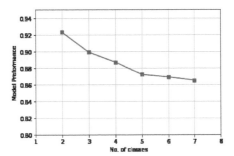

Figure 10.5   Performance variation of the hierarchical classifier as newer knowledge is added to CNN.

to expand upon the knowledge already learned by the network. In this section, we talk about how the performance of the network is affected as it is subjected to newer data/classes that are unknown to the network.

We start the construction and training of the network from the root CNN which acts as the entry point for all the data. This CNN at the root of hierarchy is initially trained with two classes, namely, Melanoma (*MEL*) and Bening Keratosis (*BKL*). Here the CNN achieves and accuracy of 0.923 during the training.

After this initial training, the CNN is presented with one newer classes in each iteration. The classes that were introduced after the initial step were Melanocytic Nevus *NV*, Besal Cell Carcinoma (*BCC*), Actinic Keratosis (*AKIEC*), Dermatofibrosis (*DF*) and Vascular Lesion (*VASC*). As the newer classes are added to the network in an hierarchical manner, an impact on the performance on the network is observed. Comparison of the performance during training as the number of classes are increased is shown in Figure 10.5 where a decrease in the performance can be seen as the number of classes are gradually increased from two initial classes to all seven classes available in the dataset.

10.5   CLASSIFICATION RESULTS

All experiments start by reading the colour images form the dataset, then by passing it to the augmentation step. According to the implemented Xception neural network, the input size of the image was kept $224 \times 224 \times 3$, where 3 is the depth for RGB color channel of the image. Therefore, after loading the images a resizing process is implemented to align the input images to specified dimensions for the acceptable input the neural network, i.e. $224 \times 224 \times 3$ for width, height and depth, respectively.

For the evaluation of the performance of proposed classification architecture on ISIC 2018 dataset, we opt a split of 9:1 for training and testing set of images. Initially, 90% samples are utilized for training the proposed Tree-CNN . The CNN is trained level by level. Initially we train the network on a subset of our dataset. Then after partially training is completed, whole dataset with remaining images is used for training to implement incremental learning. Later we observe, performance on

Table 10.1

Performance comparison of the proposed methodology

Method	f1_score	Specificity	Accuracy
[3] Deep ensemble	0.634	0.649	0.760
[1] Aggregate Deep CNN	-	-	**0.868**
[9] Deep Residual Network	0.658	0.637	0.855
[7] Neural Style Transfer	0.776	0.757	0.777
[11] Deep CNN fusion	0.844	0.840	0.851
Proposed Approach	**0.859**	**0.861**	0.865

test images set, validate and report the performance of our approach. For comparison purpose, we utilize different metrices such as *f1-score*, *precision* and *accuracy*.

The Deep Ensemble approach used by N. Codella [3] achieves an accuracy 76.0 by utilizing an ensemble of deep CNNs, whereas Aggregate CNN [1], Deep Resifdual Network [9] and Deep CNN fusion [11] achieves an accuracy score of 86.6, 85.5, 85.1 respectively. The proposed model at its final stage achieved an accuracy score of 86.3 which is competitive in comparison to all the different available methods, and only lagged behind Aggregate Deep CNN [1] by a difference value of 0.3 .

Furthermore in case of specificity comparison, the proposed methodology outperforms existing solutions by an average of 0.14 higher specificity.

The implemented CNN obtain maximum Accuracy of 0.865, specificity of 0.861 and f1_score of 0.859. To conclude the result analysis, a tabular performance comparison of proposed approach with existing methods is given in Table 10.1. Z. Yu. et. al. are using mean average precision, AUC and accuracy to measure the performance of Aggregate Deep CNN [1] approach, whereas we are emphasizing more on f1-score. Also the the confusion matrix for the classification on the test image dataset is shown in Figure 10.6. As we can conclude from the comparison that the

Figure 10.6　Classification confusion matrix.

proposed methodology achieves competitive performance in different metrices while also outperforms existing solutions in some scenarios.

10.6   CONCLUSION AND FUTURE SCOPE

Computer-aided classification is an ever developing area in the domain of computer vision. Automated computer-based systems are much important in helping early diagnostics of melanoma. In this paper, we propose a newer approach for solving the classification task of skin images. Here we created an hierarchical tree like CNN, that classifies the input images at every level of its hierarchy where the each node in the hierarchy is assigned to a particular class. Every node then classifies their respective input image into one of its child nodes. Here we also discuss how newer data that become available every year can be incorporated in the already trained network. Furthermore by changing and optimizing the node architecture and routing function for sending the images to its children, the performance of the Tree CNN can be further improved.

Every year newer methods are discovered to optimize upon the state of the art. As described in the dissertation, the type of approach used in the design of an CNN is very significant. Therefore, use of improved or efficient CNN can be done to further improve the learning ability of the whole network.

Also currently the training is done in an level by level order, in future rather than training the network in this manner it can be done in an end-to-end fashion. Where only the corresponding branches are trained as a whole,with a global loss function that gives different weightage to loss from individual nodes depending upon their position and level on the hierarchy.

REFERENCES

1. Said Bahassine, Abdellah Madani, Mohammed Al-Sarem, and Mohamed Kissi. Feature selection using an improved chi-square for arabic text classification. *Journal of King Saud University - Computer and Information Sciences*, 32, 05 2018.

2. Hugh Cartwright. Swarm intelligence. by james kennedy and russell c eberhart with yuhui shi. morgan kaufmann publishers: San francisco, 2001. 43.95. xxvii + 512 pp. isbn 1-55860-595-9. *The Chemical Educator*, 7:123–124, 04 2002.

3. Noel Codella, Quoc-Bao Nguyen, S. Pankanti, David Gutman, Brian Helba, Allan Halpern, and John Smith. Deep learning ensembles for melanoma recognition in dermoscopy images. *Ibm Journal of Research and Development*, 61, 06 2017.

4. Asit Das, Sunanda Das, and Arka Ghosh. Ensemble feature selection using bi-objective genetic algorithm. *Knowledge-Based Systems*, 123, 02 2017.

5. Nuhu Ibrahim, H.A. Hamid, Shuzlina Rahman, and Simon Fong. Feature selection methods: Case of filter and wrapper approaches for maximising classification accuracy. *Pertanika Journal of Science and Technology*, 26:329–340, 01 2018.

6. Gregory Krauss. An introduction to neural networks: J.a. anderson (mit press, cambridge, ma, 1995, 672 p., price: Us$ 55.00). *Electroencephalography and Clinical Neurophysiology*, 99:99, 07 1996.

7. Zhihui Lai, Dongmei Mo, Wai Wong, Yong xu, Duoqian Miao, and David Zhang. Robust discriminant regression for feature extraction. *IEEE Transactions on Cybernetics*, PP:1–13, 10 2017.

8. Yuexiang Li and Linlin Shen. Skin lesion analysis towards melanoma detection using deep learning network. *Sensors*, 18, 03 2017.

9. Parham Moradi and Mehrdad Rostami. Integration of graph clustering with ant colony optimization for feature selection. *Knowledge-Based Systems*, 84, 04 2015.

10. Kourosh Neshatian, Mengjie Zhang, and Peter Andreae. A filter approach to multiple feature construction for symbolic learning classifiers using genetic programming. *Evolutionary Computation, IEEE Transactions on*, 16:645–661, 10 2012.

11. Feiping Nie, Heng Huang, Xiao Cai, and Chris Ding. Efficient and robust feature selection via joint ℓ2, 1-norms minimization. pages 1813–1821, 01 2010.

12. B.Gireesha Obaiahnahatti and James Kennedy. A new optimizer using particle swarm theory. pages 39 – 43, 11 1995.

13. Il-Seok Oh, Jin-Seon Lee, and Byung-Ro Moon. Hybrid genetic algorithms for feature selection. *IEEE Transactions on Pattern Analysis and Machine Intelligence*, 26:1424–37, 12 2004.

14. S.M.Jai Sakthi, Mirualini Palaniappan, and Chandrabose Aravindan. Automated skin lesion segmentation of dermoscopic images using grabcut and k-means algorithms. *IET Computer Vision*, 12, 07 2018.

15. Rahul Sivagaminathan and Sreeram Ramakrishnan. A hybrid approach for feature subset selection using neural networks and ant colony optimization. *Expert Systems with Applications*, 33:49–60, 07 2007.

16. Aureli Soria-Frisch. Andries p. engelbrecht (university of pretoria), computational intelligence: An introduction, john wiley & sons ltd., west sussex, england, 2002, isbn 0-470-84870-7. *Applied Soft Computing*, 7:628–629, 03 2007.

17. Yong Xu, Zhong Zuofeng, Jian Yang, Jane You, and David Zhang. A new discriminative sparse representation method for robust face recognition via l_2 regularization. *IEEE Transactions on Neural Networks and Learning Systems*, PP:1–10, 06 2016.

18. Jihoon Yang and Vasant Honavar. Feature subset selection using a genetic algorithm. *Intelligent Systems and Their Applications, IEEE*, 13:44–49, 04 1998.

19. Long Zhang, Linlin Shan, and Jianhua Wang. Optimal feature selection using distance-based discrete firefly algorithm with mutual information criterion. *Neural Computing and Applications*, 28, 09 2017.

11 Hybrid Deep Learning Model to Diagnose Covid-19 on its Early Stages Using Lung CT Images

Kadambari K V and Lavanya Madhuri Bollipo
Dept. of Computer Science and Engineering,
National Institute of Technology, Warangal, India

CONTENTS

11.1 INTRODUCTION

December 2019 is the time in which the entire world got struck with the global threat called Corona Virus, a virus which created a drastic impact on nations around the world [12, 4]. With respect to the COVID cases found on Center-for-Systems Science and Engineering (CSSE) present in Johns-Hopkins University [6], many people in 187 different countries over the globe are affected by COVID-19 and so far the death ratio seems to be huge. The symptoms in COVID-19 patients surfaced after 4 to 7 days from the time of virus infection. During this time frame, the affected patients spread the disease to their neighoring persons unknowingly, which resulted in a drastic increase of affection ratio. However, the ratio of symptoms varies according to the climate scenarios as well, such as in American countries the symptoms are noted on 7-day interval and in Asian countries the symptoms are raised in four days. Primarily, the complexity in COVID-19 disease identification is, the properties of

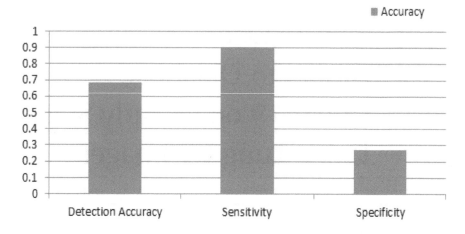

Figure 11.1 RT-PCR testing accuracy ranges.

virus changes over time and making diagnosis challenging. Thus, there is a need for early detection of COVID-19 disease before the symptom arises in order to reduce the spread of transmission.

According to WHO (World Health Organization), the benchmark test for diagnosing COVID-19 is the polymerase chain reaction (PCR). However, the limited availability of equipment and longer time consumption caused the early COVID-19 diagnosis to be a daunting task. Another challenging issue is raise of false negative rates on further testing with respect to PCR device [1, 11]. The clinical survey conducted in Wuhan (China) with more than 1000 COVID-19 people states that the PCR test can predict the disease accuracy range of 90% sensitivity, 27% of specificity and 68% of detection accuracy ratio [1]. Figure 11.1 illustrates the graphical estimations of PCR-based testing accuracy ranges. Thus, an effective alternate solution to detect the Corona Virus in its initial stages is based on Digital Image Processing scheme such as Chest X Rays, Computed Tomography (CT) and so on that are used by the doctors [3, 5]. It is proven that Computed Tomography of Lung images provide best results on test cases for identification of COVID-19 virus [7]. During the current COVID-19 pandemic, the use of deep learning-based approaches is of high utility to produce fast and accurate diagnose. Several researchers analysed the CT image-based COVID-19 detection and attained comparatively good results with respect to traditional classification algorithms such as CNN, Support Vector Machine (SVM), etc [9, 2, 10]. However, most of these models suffer from certain issues like prediction level, time constraints, accuracy manipulations [8].

The present work proposes a novel Hybrid Deep Learning Model (HDLM) that is an integration of CNN, SVM and XGBoost algorithm. The proposed HDLM algorithm is used to classify subjects into COVID and non-COVID using CT lung scan images. The proposed algorithm uses CNN by replacing the last layer with the new dense layer of SVM as output. In addition to this, the trained labels from the SVM is applied to the XGBoost algorithm to attain the better accuracy ratio.

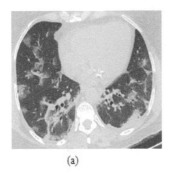
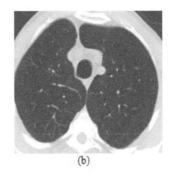

(a)   (b)

Figure 11.2   (a) COVID-19 lung image and (b) non-COVID-19 lung image.

The rest of this paper is organized as follows: Section 11.2 gives the detailed data description and analysis. The proposed methology and algorithm is explained in Section 11.3. Experimental results are shown and discussed in results section and conclusion and future work is summarized in conclusion section.

11.2 DATA PREPARATION

The model is experimented on the COVID-19 dataset (CT scan images obtained from [[13]]). Lung CT image Dataset is taken into account with the variation of more than 4000 lung CT image samples based on COVID as well as normal controls. These images are an accumulation of several real-world people lung CT scan images acquired from clinical environments with the ratio of around 2000 COVID images and the remaining normal CT images without any infection. The data images are pre-processed for model building. Figure 11.2 illustrates the structure of COVID and non-COVID images.

11.2.1 IMAGE PRE-PROCESSING

The images are scaled properly under fixed norms before processing. The input images are optimized accordingly to the defined colour level, and the dimensionality of the input images is reduced by removing unwanted features from the image portions. These metrics will reduce the time consumption and provide proper training process of images. The images are transformed from RGB colour format to Grayscale format, these colour variations are used for different processing models. Pre-processing involves reading input image from the drive to numpy array and then normalizing the image by dividing the matrix into 255 pixel ratio. These processed features are then stored in an array unit for further estimations. The common parameters used for image pre-processing are image width, image height, number of channels, input structure, number of classes, epochs and the batch size. Based on these parameters, the image pre-processing is handled over the proposed approach of deep learning model.

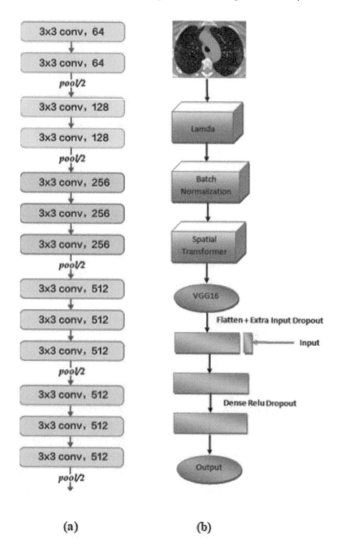

(a)   (b)

Figure 11.3   (a) VGG16 feature extraction process and (b) full architecture.

11.2.2 FEATURE EXTRACTION

The proposed approach of HDLM uses a set of thirteen different layers as shown in Figure 11.3, the VGGNET comprises of the extracted features. However, there are many pre-trained models available but in this proposed approach, the most recent VGG16 model is considered as it is simple and powerful processing model which can improve the training performance ratio. Figure 11.3 shows the architecture of convolutional model with different bit frequencies such as 64, 128, 256 and 512.

Based on these metrics, the features are extracted from the given input images and the resultant metrics are manipulated according to the process of further classification.

11.3 PROPOSED METHOD

Lung CT scans are considered as a better modality to identify the Corona Virus infection at early stages. This work proposes a new Classification algorithm called Hybrid Deep Learning Model (HDLM) that uses lung CT images as input and produces an outcome of classification of subject as COVID or non-COVID. The HDLM is integrated with CNN, Support Vector Classifier and the XGBoost classification models. These techniques are integrated together by means of replacing the last layer of the CNN and define the dense layer by means of Support Vector Classifier where the training samples are moderated with respect to XGBoost algorithm. The last three layers of proposed HDLM in contrast to basic CNN model are modified accordingly. The first layer is the flattened layer which is derived from the output of the VGG16 layers. Following this layer is the dropout layer, and the next two layers are dense layers after each Dropout layer with a gradual decrease in depth. All these features are correlated which produce a powerful classification algorithm called Hybrid Deep Learning Model (HDLM), in which it predicts the COVID-19 disease from the Lung CT images.

The following steps are involved in model building:

1. Load dataset, processed into memory. The digital COVID lung CT images are in RGB color format.

2. Implement network architecture as designed by the proposed scheme.

3. Cross-validate the proposed model with binary accuracy, precision score with threshold, Recall score with threshold, Fbeta score with beta and threshold.

4. Check and verify data generator, model checkpoint, model loss function, training model with training parameters, logging training/validation loss and training/validation accuracy.

5. Save the best training and testing model.

CNN consists of collection of convolution kernels, in which every neuron in that part acts as a kernel. The operation of CNN is similar to co-operative operation while the kernel layers are symmetric. These kernels are used to divide the input image into separate blocks or else small pieces, in which these portions are indicated as Receptive Portions. These small portions of the images are used for feature extraction. The kernel optimizes the images using certain weighting principles. Equation (11.1) shows the operational view of the convolutional neural network.

$$f_l^k(\mathrm{p,q}) \leftarrow \sum_c^{0-n} \cdot \sum_x^y \mathrm{i}(x,y)^{\mathrm{n}} \cdot e_l^k(\mathrm{u,v}) \tag{11.1}$$

where $i(x,y)^n$ indicates the input image element and the term i indicates the elements that are multiplied by the element descriptor e with respect to the indexing of kth convolutional layer kernel of the lth layer. The feature mapping details of the convolutional layer is expressed as: $f_l^k \leftarrow (f_l^1, f_l^2, f_l^3, \ldots, f_l^n) \cdot (p, q)$, where the p and q indicates the lower and upper most element ranges of the convolutional function. Once the features are extracted from the input image, the initial location specifications of the input image are considered to be of less priority. The pooling layer is added after the local operations in the receptive portion's nearest neighbour and it emits the outcome of the local image locations. The Equation (11.2) illustrates the view of pooling.

$$z_l^k \leftarrow G_p \left(F_l^k \right) \tag{11.2}$$

where z_l^k represents the pooled layer feature map functionality with respect to the kth convolutional layer kernel of the lth layer, F_l^k indicates the feature map and G_p defines the category of pooling operation. The CNN is activated by many activation functions such as Sigmoid, Tanh, Maxout, Swish, Rectified Linear Unit (ReLU), etc. In the proposed approach, the integration of ReLU and Sigmoid are used and this kind of activation functions are used to accelerate the learning process and it is activated by following equation.

$$T_l^k \leftarrow F_a \left(F_l^k \right) \tag{11.3}$$

where F_l^k indicates the outcome of a convolution operation, in which it is assigned to the activation function Ga to attain the transformed non-linear form of result T_l^k for the lth layer. The batch normalization process is utilized to eliminate the co-variance shifted over generated feature maps in the convolutional layer. An internal co-variance shifting operation is a transformation over distributed layer hidden unit's value which reduces the time consumption ratio of the convolutional training model. It also provides some awareness regarding the feature maps and the associated parameters. Equation (11.4) gives the transformed feature map co-variance of the proposed model.

$$N_l^k \leftarrow x = \frac{F_l^k - \mu_B}{\sqrt{\beta_b^2 n_{-\Omega}}} \tag{11.4}$$

where N_l^k represents the normalized feature map variant, μ_B and β_b^2 indicate the mean and dense function variables of the feature map associations for a batch of convolution layer metrics, respectively. The dropout function is used to eliminate the layers of the CNN. In CNN, many correlated and non-correlated schemes are merged together resulting in model over-fitting. To avoid this, the DROPOUT functionality is incorporated with proper approximations. The SVM is introduced to identify the optimal hyper plane regions $f(x, y) \leftarrow W.X + b^2$ of the input, separating the dataset into two different classes with respect to COVID-19 and non COVID-19 with the associated metrics of $X \in R^m$. Equation (11.5) is used to learn the parameters specified in optimal hyper-plane object W based on the feature map acquired.

$$\min \frac{1}{p} \mid W \|_2^2 + C \sum_{i=1}^{P} \text{Max} \left(0, 1 - Y \left(\begin{matrix} n \\ i = 1 \end{matrix} \right) W^T X(i) + b^2 \tag{11.5}$$

Where $\|W\|_2^2$ indicates the Euclidean Normalization operation as well as it is usually indicated as L_2 Normal Form, $W^T X(i) + b^2$ indicates the prediction function of the proposed HDLM, Y indicates the actual label, C is the penalty parameter, in which it is acquired during hyper-plane tuning. The logical trained labels will be applied to the XGBoost classifier model to predict the COvid-19 disease with high accuracy estimation. Equation (11.6) defines the determination of class-labels for the outcome $Y(i)$.

$$Y(i) \leftarrow \lambda(Y) \leftarrow \sum_{k=1}^{n} f(i) \subseteq F \qquad (11.6)$$

Where F indicates the frequency term with respect to $F \leftarrow f_1, f_2, f_3, \cdots, f_i$. The frequency factor $\lambda(Y)$ is defined from a running variable k from 1 to n. The proposed HDLM defines the Convolutional state of XGBoost with respect to the time complexity, reduces the error ratio and improves the accuracy range as well. Figure 11.4 shows the sequential model of the CNN, where the last layer still remains with the same Dense as dense (1) with the activation of Sigmoid and ReLU model. But the proposed HDLM approach in Figure 11.5 is deviated with respect to the CNN last layer. HDLM dense layer changes to dense (2) from dense (1) to create an impact over the resulting accuracy.

11.3.1 ALGORITHM STEPS

Algorithm: Hybrid Deep Learning Model (HDLM)
  Input: Dataset with COVID and Normal Lung Images
  Output: Classify the COVID and Normal Images with Proper Accuracy Ratio

1. Import the required libraries: dataset optimization (keras-optimizers), pre-processing with image data generation (keras-ImageDataGenerator), convolution principles (keras-layers-Conv2D, MaxPooling2D) and associated density metrics.

2. Acquire the dataset from the drive to process.

3. Specify the size of the image with fixed metrics. Img.Size(64X64);

4. Perform feature extraction for the images 0 to n-1. This process requires the parameters such as the dataset directory/drive, sample count of images and the batch size.

   Pseudocode

   Function Feature-xtract(drive, cnt, batch)
   {
   Initialize an object to acquire the rescaled images with respect to the size ratio of 1/255.
   Acquire the images from the directory based on the specified batch size and set the target scaling size as 32X32.
   Initialize the image count as 0.

```
Model: "sequential"
```

Layer (type)	Output Shape	Param #
conv2d (Conv2D)	(None, 254, 254, 32)	896
activation (Activation)	(None, 254, 254, 32)	0
max_pooling2d (MaxPooling2D)	(None, 127, 127, 32)	0
conv2d_1 (Conv2D)	(None, 125, 125, 32)	9248
activation_1 (Activation)	(None, 125, 125, 32)	0
max_pooling2d_1 (MaxPooling2	(None, 62, 62, 32)	0
conv2d_2 (Conv2D)	(None, 60, 60, 64)	18496
activation_2 (Activation)	(None, 60, 60, 64)	0
conv2d_3 (Conv2D)	(None, 58, 58, 250)	144250
activation_3 (Activation)	(None, 58, 58, 250)	0
conv2d_4 (Conv2D)	(None, 56, 56, 128)	288128
activation_4 (Activation)	(None, 56, 56, 128)	0
average_pooling2d (AveragePo	(None, 28, 28, 128)	0
conv2d_5 (Conv2D)	(None, 26, 26, 64)	73792
activation_5 (Activation)	(None, 26, 26, 64)	0
average_pooling2d_1 (Average	(None, 13, 13, 64)	0
conv2d_6 (Conv2D)	(None, 11, 11, 256)	147712
activation_6 (Activation)	(None, 11, 11, 256)	0
max_pooling2d_2 (MaxPooling2	(None, 5, 5, 256)	0
flatten (Flatten)	(None, 6400)	0
dense (Dense)	(None, 32)	204832
dropout (Dropout)	(None, 32)	0
dense_1 (Dense)	(None, 1)	33
activation_7 (Activation)	(None, 1)	0

```
Total params: 887,387
Trainable params: 887,387
Non-trainable params: 0
```

Figure 11.4 CNN sequential model.

Raise the "For Loop" to extract the content of the image by using batch size, shape and image normalization principles.
Concatenate the image batches and return the final variants for further process.
}

5. Train the extracted features with respect to proper labels and its associated classes.

```
Model: "sequential"
_____
Layer (type)                 Output Shape              Param #
=================================================================
conv2d (Conv2D)              (None, 254, 254, 32)      896
_____
activation (Activation)      (None, 254, 254, 32)      0
_____
max_pooling2d (MaxPooling2D) (None, 127, 127, 32)      0
_____
flatten (Flatten)            (None, 516128)            0
_____
dense (Dense)                (None, 32)                16516128
_____
dropout (Dropout)            (None, 32)                0
_____
dense_1 (Dense)              (None, 1)                 33
_____
activation_1 (Activation)    (None, 1)                 0
=================================================================
Total params: 16,517,057
Trainable params: 16,517,057
Non-trainable params: 0
_____
```

Figure 11.5   HDLM sequential model.

6. Associate the labels with the testing path to extract the test image features and the size.
   Lbls← list-dir(test path);
   length[test-features];

7. Import the tensorflow packages and define the model with respect to the layers of the proposed approach.

   Pseudocode

   tensorflow import system-models.
   tensorflow import img-layers;
   tensorflow import img-optimizers.
   Define the model based on image shape and specified lables.
   define Model(shape, lbls)
   {
   model.append ← Sequential-Model();
   Associate the convoltional model with respect to the shape and Conv2D layer.
   Include the convolutional models such as Relu, Max-Pooling, Flatten and Dense.
   Activate these models based on sigmoid.
   return the created model for further processing.
   }

8. Train the model with extracted features, shape and associated label length.

9. Create a hybrid form of CNN by changing the last layer density as 2 instead of classical model last layer density 1.
   Model-add[Dense(2)];

10. Create a proposed Model with sequential features

11. Cross-Validate the HDLM with testing featureswhich are selected at random.

12. Calculate the training level accuracy and loss metrics.

13. Extract the features from the last layer of CNN.
    model.feat← Model [inputs←model-input,outputs←model.getLayer[dense-1]-output];
    feat(train) ← model.feat-prediction(train(features));
    feat(val)←model(feat-prediction(test features));
    feat(test) ← model(feat-prediction(test features));

14. Apply the Support Vector Model and remove the last layer of CNN.
    Import the required libraries for SVC.
    SVM←SVC(kernel←RBF);
    SVM.FIT[feat(train),Arg=Max(train-labels,axis-1)];

15. The trained CNN variable called feat(train) is associated with the SVC classifier.

16. Analyse the prediction score of the hybrid model of CNN and SVC.

17. Associate the trained labels to the XGBoost Classification logic.
    Import the required libraries for XGBoost.
    xb←xgb-XGBClassifier();
    xb-fit[feat(train),Arg-Max(train-labels,axis-1)];

18. Estimate the accuracy levels of these combined approach and return to the user end.
    xb-Score[feat(train),Arg-Max(train-labels,axis-1)];

11.4   RESULTS AND DISCUSSION

This work proposes a novel deep learning scheme called Hybrid Deep Learning Model. It encompasses classification algorithms: CNN, SVM and XGBoost.

Table 11.1 shows the comparative study. Figure 11.6 illustrates the graphical representation of proposed Hybrid Deep Learning Model Training and Testing Loss ratio with respect to number of epochs and Figure 11.7 illustrates the graphical representation of proposed Hybrid Deep Learning Model Training and Testing Accuracy ratio with respect to number of epochs. Apart from achieving good training accuracy, the model also shows the declining training loss (from Figure 11.6). The model

Table 11.1
Validation of Proposed HDLM

Algorithm	Accuracy (%)
CNN	95%
CapsNet	89%
Proposed HDLM	98%

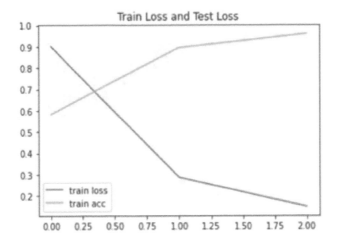

Figure 11.6   HDLM training and testing loss ratio.

converges to better classification accuracy in less number of epochs. Figure 11.8 to Figure 11.11 illustrate the classification accuracy and the training, testing loss of CNN. The accuracy achieved by vanilla CNN is 95% in more number of epochs. Figure 11.12 gives the confusion matrix values of capsule network of COVID-19 dataset. In order to perform experimental evaluation of the models under consideration, various performance metrics like Accuracy, AUC value of the ROC curve, F1-Score, Sensitivity, Specificity, Precision, and Recall are used in this work. The definitions for each of these performance metrics are given below.

True-Positives (TP) – These are the correctly predicted positive values, i.e., the value of actual class is yes and the value of predicted class is also yes.

True-Negatives (TN) – These are the correctly predicted negative values, i.e., that the value of actual class is no and value of predicted class is also no.

False-positives and false-negatives, these values occur when your actual class contradicts with the predicted class.

False-Positives (FP) – When actual class is no and predicted class is yes.

False-Negatives (FN) – When actual class is yes but predicted class in no.

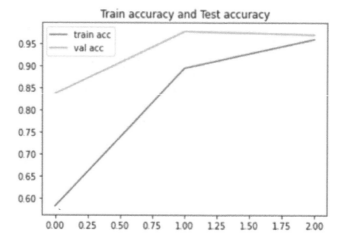

Figure 11.7 Shows the HDLM training and testing accuracy.

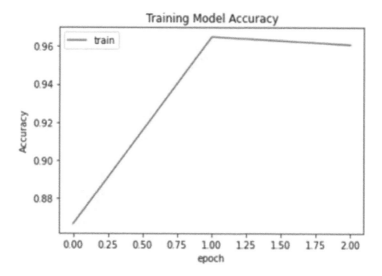

Figure 11.8 Shows the CNN training accuracy.

Accuracy is perhaps the most intuitive performance measure. It is simply the ratio of correctly predicted observations.

$$\text{Accuracy} = \frac{TP+TN}{TP+FP+FN+TN};$$

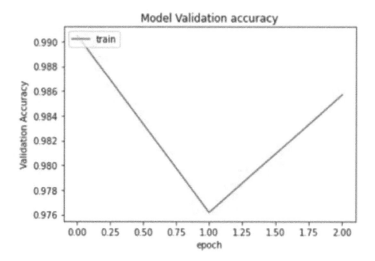

Figure 11.9 Shows the CNN model validation accuracy.

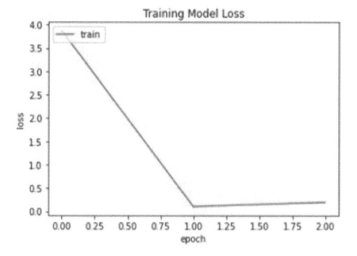

Figure 11.10 Shows the CNN training model loss.

Precision looks at the ratio of correct positive observations.

$$\text{Precision} = \frac{TP}{TP+FP};$$

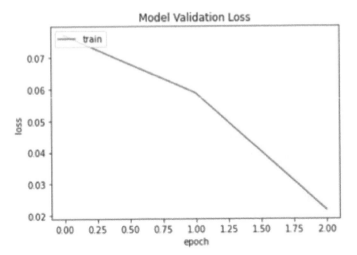

Figure 11.11 Shows the CNN training validation loss.

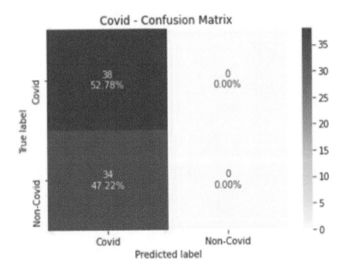

Figure 11.12 Shows the confusion matrix of the proposed model.

Recall is also known as sensitivity or TP rate. It's the ratio of correctly predicted positive events.

$$Recall = \frac{TP}{TP + FN};$$

The F1 Score is the weighted average of Precision and Recall. Therefore, this score takes both FP and FN into account.

Intuitively, it is not as easy to understand as accuracy, but F1 is usually more useful than accuracy, especially if you have an uneven class distribution. It works best if FP and FN have similar cost. If the cost of FP and FN are very different, it's better to look at both Precision and Recall.

$$\text{F1-Score} = \frac{2*(Recall*Precision)}{(Recall+Precision)};$$

Sensitivity refers to the test's ability to correctly detect ill patients who do have the condition.

$$\text{Sensitivity} = \frac{TP}{TP+FN};$$

Specificity relates to the test's ability to correctly reject healthy patients without a condition.

$$\text{Specificity} = \frac{TN}{TN+FP}$$

Using these metrics, the accuracy of capsule network is 89%.

From the above figures, it can be noted that the classification accuracy of HDLM is 98% (viz., is higher than the other two models) and is achieved in less number of epochs. Thereby making it a new promising technique in automatic diagnosis of COVID-19 disease prediction in early stages.

11.5 CONCLUSION AND FUTURE WORK

This work proposes a novel ensemble classification algorithm called HDLM to diagnose the COVID-19 disease efficiently. The proposed technique integrates CNN, Support Vector Classifier and the XGBoost algorithm. The last layer of the CNN is eliminated and it is replaced with SVC along with the density level of 2 as well as the training labels acquired from the SVC model is applied to the training model of XGBoost algorithm in order to attain the high classification accuracy of 98%. The proposed HDLM is cross-validated with the traditional CNN and CapsNet algorithms. The results show that the proposed model proves its efficiency by means of model validation and accuracy estimations with low loss ratio. In future, the work can further be extended to apply on some real-world datasets with large number of samples and adding some time constraint functions.

REFERENCES

1. Tao Ai, Zhenlu Yang, Hongyan Hou, Chenao Zhan, Chong Chen, Wenzhi Lv, Qian Tao, Ziyong Sun, and Liming Xia. Correlation of chest ct and rt-pcr testing for coronavirus disease 2019 (covid-19) in china: a report of 1014 cases. *Radiology*, 296(2):E32–E40, 2020.

2. Mohamed Abd Elaziz, Khalid M Hosny, Ahmad Salah, Mohamed M Darwish, Songfeng Lu, and Ahmed T Sahlol. New machine learning method for image-based diagnosis of covid-19. *Plos one*, 15(6):e0235187, 2020.

3. Hao Feng, Yujian Liu, Minli Lv, and Jianquan Zhong. A case report of covid-19 with false negative rt-pcr test: necessity of chest ct. *Japanese Journal of Radiology*, 38(5):409–410, 2020.

4. Chaolin Huang, Yeming Wang, Xingwang Li, Lili Ren, Jianping Zhao, Yi Hu, Li Zhang, Guohui Fan, Jiuyang Xu, Xiaoying Gu, et al. Clinical features of patients infected with 2019 novel coronavirus in wuhan, china. *The Lancet*, 395(10223):497–506, 2020.

5. Chunqin Long, Huaxiang Xu, Qinglin Shen, Xianghai Zhang, Bing Fan, Chuanhong Wang, Bingliang Zeng, Zicong Li, Xiaofen Li, and Honglu Li. Diagnosis of the coronavirus disease (covid-19): rrt-pcr or ct? *European Journal of Radiology*, 126:108961, 2020.

6. Meg Miller. 2019 novel coronavirus covid-19 (2019-ncov) data repository: Johns hopkins university center for systems science and engineering. *Bulletin-Association of Canadian Map Libraries and Archives (ACMLA)*, (164):47–51, 2020.

7. Geoffrey D Rubin, Christopher J Ryerson, Linda B Haramati, Nicola Sverzellati, Jeffrey P Kanne, Suhail Raoof, Neil W Schluger, Annalisa Volpi, Jae-Joon Yim, Ian BK Martin, et al. The role of chest imaging in patient management during the covid-19 pandemic: a multinational consensus statement from the fleischner society. *Radiology*, 296(1):172–180, 2020.

8. Sachin Sharma. Drawing insights from covid-19-infected patients using ct scan images and machine learning techniques: a study on 200 patients. *Environmental Science and Pollution Research*, 27(29):37155–37163, 2020.

9. Feng Shi, Jun Wang, Jun Shi, Ziyan Wu, Qian Wang, Zhenyu Tang, Kelei He, Yinghuan Shi, and Dinggang Shen. Review of artificial intelligence techniques in imaging data acquisition, segmentation, and diagnosis for covid-19. *IEEE Reviews in Biomedical Engineering*, 14:4–15, 2020.

10. Siham Tabik, Anabel Gomez-Rios, Jose Luis Martin-Rodr guez, Ivan Sevillano-Garcia, Manuel Rey-Area, David Charte, Emilio Guirado, Juan-Luis Suarez, Julian Luengo, MA Valero-Gonzalez, et al. Covidgr dataset and covid-sdnet methodology for predicting covid-19 based on chest x-ray images. *IEEE Journal of Biomedical and Health Informatics*, 24(12):3595–3605, 2020.

11. Alireza Tahamtan and Abdollah Ardebili. Real-time rt-pcr in covid-19 detection: issues affecting the results. *Expert Review of Molecular Diagnostics*, 20(5):453–454, 2020.

12. Chen Wang, Peter W Horby, Frederick G Hayden, and George F Gao. A novel coronavirus outbreak of global health concern. *The Lancet*, 395(10223):470–473, 2020.

13. Jinyu Zhao, Yichen Zhang, Xuehai He, and Pengtao Xie. Covid-ct-dataset: a ct scan dataset about covid-19. *arXiv preprint arXiv:2003.13865*, 2020.

12 Impact of Machine Learning Practices on Biomedical Informatics, Its Challenges and Future Benefits

Hina Bansal, Hiya Luthra and Ankur Chaurasia
Amity Institute of Biotechnology, Amity University
Uttar Pradesh, Sector 125, Noida-201303, Uttar Pradesh, India

CONTENTS

DOI: 10.1201/9781003246688-12

12.1 INTRODUCTION

Bioinformatics domain can be bifurcated into numerous branches, depending upon the materials used for experimentations for learning purposes. When it comes to the research in biomedical field and healthcare domain, then the only word that comes into our minds is "Bioinformatics." Since it forms the base of knowledge in numerous domains. Day by day there is an increase in the salient data throughout the world. Due to these 2 major problems arises, firstly the management and the storage of data that is accumulating bit by bit and secondly drawing out some advantageous particulars through such humongous data. Moreover, the major stumbling block in biomedical informatics is the buildout of specific methods, tools and software regarding the transfiguration for such heterogenous data into gaining some solid knowledge-based evidence [1]. Biomedical informatics aids to analyse the health conditions of patients and the efficiency of healthcare processes to help clinicians, researchers and scientists to improve human health. It focusses on identifying trends in the data discovered through computational tools and techniques. The expansion of big data in biomedical and health field has driven the need of new effective analysis technology. In this context, Machine learning (ML) is the fastest growing field in Health Informatics (HI) having the greatest application. In this chapter, we have discussed the application, challenges and future benefits of ML to improve medical diagnoses, disease analyses and pharmaceutical development. Whether its "Biomedical Informatics" or "Bioinformatics" both the domains include the following components: biological sciences, IT, information processing and computer science. But still there is a very thin line which makes the objectives for both the domains different in public healthcare sector and research in the clinical field. When it comes upon bioinformatics, the researchers in this domain predominantly pursue the implementation provided by computer technology in response to supervise, administer, handle and comprehend or understand the humongous volumes of data received from biological sectors. Whereas, on the other hand biomedical informatics has a tangent towards giving the solutions to the issues like high healthcare costs, reduction of medical errors, and improved decision making with the help of medical data and history of an individual on the basis of details provided through using bioinformatics. In a nutshell we can say that biomedical

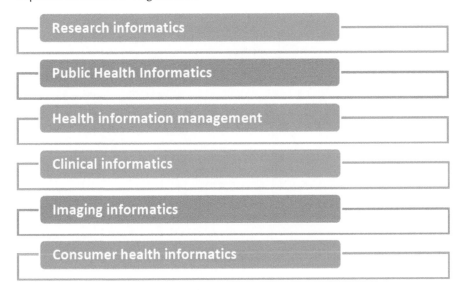

Figure 12.1 A branches of biomedical informatics.

informatics results are more or less dependent upon bioinformatics results. One must also include while talking about bioinformatics and biomedical informatics that the major goal of bioinformatics is to analyse the genomic data and then in biomedical informatics, we use the datasets provided by using bioinformatics concepts in order to find legitimate solutions to the obstacles faced in the healthcare sector [2].

12.2 BIOMEDICAL INFORMATICS

In order to take steps forward in the domain of human health, one must focus on biomedical informatics which is a division of health informatics. This field is taking it headways to shoot up through numerous advancements in the field of biomedicine. Biomedicine is a discipline which have a bearing on the fundamentals of science of the nature, particularly biochemistry & biology to medical management & medicine. Perhaps a question arises that why is there a need for biomedical informatics? The answer to this question is that there is a use of big data and conventional research to obtain clinical insights, detecting the disease, medical care and foretell newer heights of medical and scientific queries. Remarkable advancements have been seen in the field of sequencing of DNA and Genomics which is made possible through the cloud-based supercomputing [3]. The various branches of biomedical informatics are represented in Figure 12.1.

12.2.1 RESEARCH INFORMATICS

Research informatics Public Health Informatics Health information management Clinical informatics Imaging informatonsumer health informatics

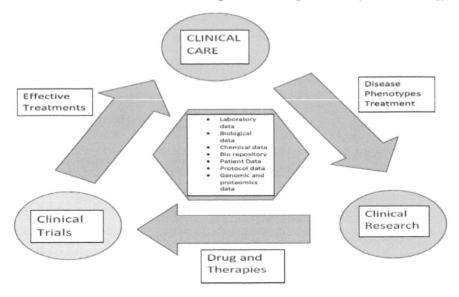

Figure 12.2 A branches of biomedical informatics.

Research Informatics is mainly used for keeping the data in reserve, organizing it, and to report the data which helps in the research processes (Figure 12.2). The systems used in research informatics can be utilised for doing basic research in science and research in the clinical field. The required data can be collected from various healthcare labs, hospitals and from different countries to maximize the outcomes of translational research. This process put a spurt on the translation of fundamentals of science into healthcare practices [4]. According to NIH clinical research informatics has 3 major divisions:

a) Research solely based on humans

b) Epidemiology and the science of behaviour

c) Research related to health services

12.2.2 PUBLIC HEATH INFORMATICS

Public health informatics or better known as PHI is a methodological implementation of computer science, and IT in the fields of healthcare to provide avoidance and readiness to fight against a certain disease. There are two major implementations of PHI:

- To boost the health of an individual, the whole populations health is taken care of
- To put a stop towards injuries and diseases alterations are done in the sectors which has a high risk of affecting the mass.

Moreover, PHI is extremely advantageous for collecting, managing and operating the data collected under public health. The distinct characteristics of PHI sets an example amongst different fields of biomedical informatics. These properties include prevention of certain ailments amongst the population, recognizing the main aim by interfering immensely and working in harmony within the governmental frame. The realm of PHI consists of conceptuality, plan, buildout, classification, conservation and assessment of information, inspecting and systems of information which are in accordance with public healthcare [5].

12.2.3   HEALTH INFORMATION MANAGEMENT

The main objective of health information management is to analyse, store and protect the standard condition of details regarding the patient's health. These details could be stored in either format. It can be fully paper based or completely digitalized or can be a combination of the two. The position of a health information management employee is to make sure that the experience of a patient is smooth and hurdle free whether it's a visit to the doctor or claiming the insurance. Regardless of the issue HIM manages to secure an obstacle free path for a patient to cure himself under the surveillance of doctors and nurses. HIM has also a key role to play when it comes to the security and privateness of the patient. Moreover, HIM is also responsible for the retention of medical records of a patient and its transformation to digital formats with providing the statistics of public healthcare and administrating the advancements [6].

12.2.4   CLINICAL INFORMATICS

Clinical informatics or better known as CI is a subspecialized area of medicine. CI has a wide- ranging reach towards numerous practices done in the field of medicine. Moreover, there are speedy advancements made in this domain, so it becomes quite hard to define CI into a single definition. Although according to the American Medical Informatics services (AMIA), CI is an implementation of the amalgamation between information technology and medical management. CI is a coalescence of the following sectors:

1. The science of information

2. Information system

3. Direction and organisation

The main goal of CI is to attain functional and valuable facts through the transformation of the data. CI and information technology aren't identical to one another. In order of priority to receive data regarding healthcare, IT is one of the numerous ways to obtain the required information orderly.

12.2.5 IMAGING INFORMATICS

Imaging Informatics or Medical Imaging Informatics (MII) is an expansion, implementation and analysis of the information provided through IT for the purpose of medical imaging for healthcare. MII already prevails at an elementary level during the application of radiology. It starts when a physician asks for an imaging study to the point where studies are been conducted to plan the treatment for a certain patient. There are innumerable benefits that can be achieved through MII processes and its products which could be implemented at certain visionary levels. It's been over 100 years since radiology has become an integral part of the healthcare domain. However, there are certain advancements that had been made over the decade in the field of radiology including imaging informatics which has changed the face of future for this sector. Information management is a crucial aspect of imaging informatics when aligned with radiology. The need of imaging informatics had arisen when there was some up-gradation needed for the improvement in dependability of the healthcare with the utilization of data. To attain the precision in medical imaging there must be a solid workflow which runs behind the scenes. In this case two essential factors are responsible for the smooth and precise imaging results which are the mining of data form various radiological database and diagnostics which are completely computer assisted.

There are several areas where imaging informatics is applied with the help of information provided by the computer systems:

1. Surgery

2. Cardiological disease treatment

3. Study of pregnant females

4. Female medicine and study

12.2.6 CONSUMER HEALTH INFORMATICS

Consumer informatics is an offshoot of medical informatics. It primarily keeps up with the demand of information required by the consumer; it also executes the course of action for making the statistics available for the consumer; and incorporates the preferences of the consumer for medical knowledge into the systems. Consumer health informatics has its stand at the junction of other sections like health education, nursing informatics, healthcare and communication science. Consumer health informatics is growing expeditiously in the domain of biomedical informatics. In today's era where the information is the key for the quest, consumer health informatics is helping grow the healthcare sector with the speed of light.

12.3 MACHINE LEARNING

ML techniques have its relevance to numerous domains in computational biology which can further result in getting useful facts from the data. ML is a mathematical methodology that enables one to find trends and recognize the historical record through the study of data. ML is one of the most common forms of AI and data science. It is a significant technique that is fundamental to many approaches to AI, and there are various versions of it. In healthcare, the most important application of classical ml is precision medicine – deciding what medical protocols are likely to function on a patient based on patient conditions and the treatment context [7]. The overwhelming majority of ml and precision medicine applications require a research sample about which the result variable (e.g., initiation of a disease) is known; this is called supervised learning.

More sophisticated technologies including neural networks have been used in research on health care for several decades and have been used for categorization reasons, such as determining whether a patient will experience a specific disease [8]. The machine's capabilities are likened to the way neurons process messages, but the analogy between the computer and the brain is not as much of a deal as other people believe.

Deep learning is used in neuroscience, also known as artificial intelligence, for the purpose of enhancing diagnostic imagery or for machine vision and picture detection. Deep learning is now being extended to radiomics, or any scientifically significant imaging characteristics that we cannot discern through our own, such as tumors and other lesions. Both radiomics and deep learning are used to interpret the imaging results. Using only a single procedure has a better accuracy in identifying cancerous tumors than with the usage of older models of screening technology [9].

12.3.1   UNSUPERVISED MACHINE LEARNING

Frequently, one is fascinated by recognizing clusters and/or anomalies within a given information set. This can be an ordinary field of application for unsupervised ml. The calculation employs a few strategies to distinguish information that's comparative by a few degrees (often the expanse between data points) to gather information into clusters, or to detect anomalies or exceptions. This can be done without a human telling the algorithm what to search for. Therefore, automation is one of the many big advantages [10].

12.3.2   SUPERVISED MACHINE LEARNING

In other such instances, the optimal outcome is known for a fraction of the data, and the algorithm can benefit from these data clusters to generate results for data for which the results were previously unclear. Typical functions are the operations of sorting and regression. In operations such as classification, examples of already computed data can used to teach the algorithm to classify the newer data and assign similar types of input data under the same class. Regression is a basic method to teach the algorithm that the output data is generated from the input data. In any case, the algorithm learns by adapting the internal parameters to figure out, based on the example presented, what output the individual would like to apply to a given input [11].

12.3.3 REINFORCEMENT LEARNING

Any days you realize what you are searching for, but don't know where to search. For reinforcement learning activities, well-suited implies that those tasks are appropriate for the same. Although attempts to treat wellbeing, the dream is that a machine program would be able to help us have more "youthful" (young more like the terms intended), and better-looking skin without surgical procedures [12].

12.4 CHALLENGES FACED BY MACHINE LEARNING

ML can be utilized in many ways to overcome the hurdles faced in biological domain. Figure 12.2 depicts the use of computational techniques that are being used to solve the biological complications. These difficulties have been classified majorly in six domains: Genomics, Proteomics, Microarrays, System Biology, Evolution and Text mining.

There are many deleterious health effects on the lives of a human being. Generally, the healthcare specialists collect the required clinical data from the patient and then anchorage the suitable facts from the population to gain the facts to cure a certain disease a patient is suffering from. Therefore, one can say that the clinical data here has a crucial role to play to treat a patient. The propensity of ml to acquire facts from the humongous amount of data when paired with health care can create wonders for the benefit of the society. The root cause of the challenges faced in the domain of healthcare and ml is the main aim of the health care which is to provide the support and care and not to acquire any kind of analysis from the data generated via the medical systems. Throughout the years, AI has been continuously focusing on hospitals and medical approaches to eventually achieve a stage where patients may be tested correctly and handled with as minimal pain as possible. There are many early rule-based drug industry systems (such as SCRA and MEDS) that have shown several interesting insights into illness but were not accepted in practical operation. Although human doctors which use these to diagnose, they were not up to speed with the AI produced diagnoses and had a low level of integration with the real patient records. There are some major challenges that one must look out for when designing the ml projects for healthcare (Figure 12.3). It can be the relation between the cause and the effect, the absence of certain aspects in the clinical data, and resulting in reliable outcomes. These aspects are key for the framework modelling (example: supervised and unsupervised).

Coming on to the recent situations all around the world, ml was proved to be quite helpful in order to help the human race to come out from such a dreadful phase. A real-time example for this would be the various mathematical models that were developed at the beginning of the pandemic based on the principals of regression and outbreak analysis to try and predict the likely trajectory followed by number of COVID-19 cases.

However, these models only serve as methods to estimate the severity of the problem and are far from accurate in predicting the overall curve of disease as it may not always showcase an exponential rise. It is unlikely that accurate mathematical

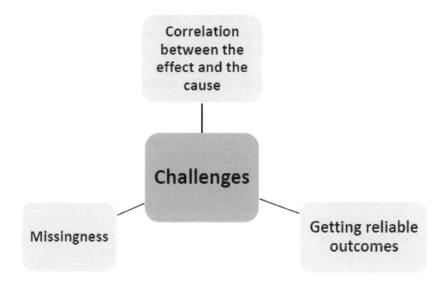

Figure 12.3 Difficulties faced while building ML algorithms for medical management and health maintenance.

models can be developed on very rudimentary form of unprocessed data which ultimately depends on variables out of our control. However, Artificial Intelligence (AI) has showcased promise in the field of developing predictive models. As ml, big data and data science improve the seemingly unlimited capabilities of AI, and as more and more data are generated, superior predictive models can be developed with the assistance of AI so that a better understanding of similar problems can be achieved [13]. Few are listed below:

12.4.1 FINDING THE LINK BETWEEN CAUSE AND EFFECT

In health care, the learning is solely dependent upon the observational data that is received by the agent which could be slightly influenced since the agent's way for choosing the policy of actions could be a different approach. Due to such hinderances, there is a challenge faced for building models that are trained to answer questions which are casual. Let's discuss about a situation where the researchers had found that the patients who were suffering from pneumonia also had asthma were regressively treated for the infection which in turn lowered the mortality rate of subgroup of population [14]. So, to conclude we can say that if a certain model predicts that asthma can be fatal then it will acquire a knowledge that asthma is protective. Although if one more attribute is added to the model to perceive the amount of care needed, then the model will find the amount of risk of death caused by asthma.

The above given example is apt for proving the very fact that for the evaluation of treatments the casual model works at its best. Moreover, it could also lead in the making of predictive models which is far better since it doesn't give any harmful predictions [15]. The research had already been done in detail regarding the disadvantages of statistical learnings for the answering of casual questions [16].

12.4.2   FINDING THE MISSINGNESS IN THE MODELS FOR HEALTHCARE

Despite the inclusion of all the essential variables in the datasets regarding healthcare, it is inevitable that many annotations would be missing. The absolute complete data is near to a myth due to the very factors of cost and volume. In the domain of ML, the incomplete data has caught the lam light and it has become a very popular topic to study about in statistics [17]. It is extremely common to find the missingness in the data which is collected on a daily basis in the healthcare field. Whereas, we have EHR'S also known as electronic health records which has detailed key information which in turn helps in the development of prediction models which are clinically equipped. Now the missingness in such vital data could be the arousal of many challenges faced in the making of disease prediction models.

12.4.3   APPROPRIATE CHOICES OF OUTCOMES THAT COULD BE RELIABLE

The defining of certain tasks is a crucial role to obtain outcomes that are reliable. There are majorly three points to remember for the consideration of outcomes.

1. The creation of outcomes which are reliable

One must consider numerous resources to collect the data for the formation of labels since EHR's are very poor when it comes to labels which are structured.

1. To understand the results or the outcomes in a clinical way

A crucial part for understanding a certain disease in a present situation is the fundamental medical definitions. Therefore, as the time evolves, both the understanding and the definitions also evolve.

1. The importance of label leakage

Let's say we need to calculate the mortality rate of a certain hospital. For that we need to have all the records up to date till the point of death. Now, what if the ventilator of a certain patient is turned off (predict the fatality). This happens in cases where the family allows to withdraw from the treatment where they feel there are no possibilities of recovery. And an ml model which is being carelessly trained to respond on such situation could have high prediction rate with absolutely no utility.

12.5   INTEGRATING MACHINE LEARNING INTO BIOMEDICAL INFORMATICS

Since we all are aware of the very fact that there is a tremendous increase in healthcare data which has circumscribed several sequences fluctuating from molecular level, individual level to a mass population level which in turn had provided the analogy amongst countless establishments in public health domain with escalating high frequency and deepness. Such data is fetching golden opportunities for speeding up the fundamental scientific findings and enabling patient-centred and harm-reduction based solutions for health care. However, it is exceedingly strenuous to convert humongous data to some knowledgeable information through the conventional method of analysis. This successively results in the making of contemporary analysis techniques, which are tremendously well regulated in order to retrieve meaningful information by the use of ml and optimisation. Broadly speaking there are major two applications of ml integrated with biomedical informatics. Firstly, the main focus is on the discovery of knowledge which is acquired by the analyzation of archival data which subsequently provides the insights regarding the unanswered questions of the past. To attain this, the following methods are used:

1. Data statistical modelling

2. Visualization

3. Trend reporting

4. Correlation analysis

Secondly, one more implementation is that we take a dataset which is known, or we can say a dataset which is majorly used for training purposes and has the responsive values and input data feature in sequence to form a model which can make predictions using data which is unseen or also known as the dataset used for testing.

Since the amount of clinical data generated day by day is escalating at a very high rate, it becomes easy to get complex datasets with high volumes these days. But the problem arises when this data has to be converted into some constructive clinical information. Therefore, there is a substantial stipulation for algorithms which includes ml with data mining. This way the need to transform clinical data or healthcare related data to some factual information is resolved. Moreover, it provided a holistic approach to the clinicians to take better decisions for the sake of health care. The concoction of ml and optimization has seen impeccable outcomes for the humankind.

There are numerous difficulties that is faced when it comes to biomedical datasets. Firstly, there is an immense amount of data which certainly is voluminous, dimensionally big, disproportionate, noisy and incomplete. Due to such difficulties, there is an immense need for forming of numerically optimised algorithms concurrently with ml based algorithm. For instance, a major dispute in biomedical informatics is the obstacle that arises with the data. This data can be poorly structured or completely unstructured or it can be noisy and not understandable at the same time. Revisiting the fundamental ml methods is so much intriguing as it consists of topics like classification, clustering, regression, and reduction of the dimension. These methods could be used to reach newer heights in order to pass all the hurdles that is coming across in the arising domain of biomedical informatics.

In order to make a highly efficient predictive model which mainly tackles with recognition and carrying away of surplus data, which is available in a certain dataset, an exercise is needed which is known as feature selection. It is certainly required to maintain the volume and the dimension of the data which ultimately springs up with the number of attributes or features included. We can say that feature selection is more target oriented since it minimizes the magnitude of the dataset through its selection procedure of choosing a single subset in order to build a logical model. Therefore, one can say that minimal features are required to build a model which has uncomplicated interpretations to maximize the use of the result [18].

The impact of ML is enormous on biomedical informatics. Following are the major sectors where this integration is useful in the public healthcare domain (Figure 12.4, Figure 12.5):

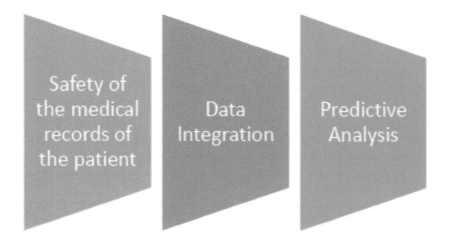

Figure 12.4 An edge achieved through ML and biomedical informatics in the public health-care domain.

12.5.1 DATA STORAGE

Electronic health records or better known as EHR'S are extremely proficient in managing and optimizing operations, cost reduction of healthcare and other related facilities and care of the patient with the help of AI. For instance, natural language processing is used by the doctors for facilitating the clinical records and notes to avoid manual labour. The algorithms used in ML can also be used to upgrade the systems for EHR'S to help the clinicians through clinical decision support, analysis of the image and integration of technologies in telehealth.

12.5.2 INTEGRATION OF DATA

We can assume that if there are some hinderances in the healthcare data, then the predictions made by ML algorithms won't be as accurate as it should be. This will

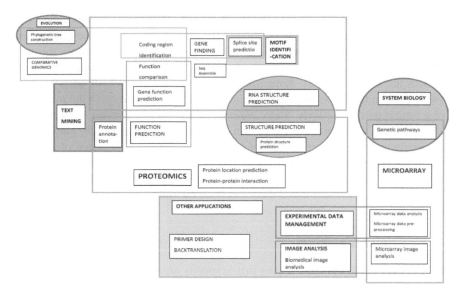

Figure 12.5 Applications of ML in biomedical informatics.

successively result in making poor decision in the healthcare domain. The maintenance of the data is done by the professionals in the informatics field. In order to attain the data integrity an informatics professional must analysation, classification, accumulation and cleaning of the healthcare data.

12.5.3 PREDICTIVE ANALYSIS

The amalgamation of ML, biomedical informatics and prediction analysis has a wide scope to ameliorate the functions in healthcare, transformation of healthcare tools to make better decisions and assist in the improvement of patient results. Predictive analytics is the future of healthcare but only if biomedical informatics and ML goes hand in hand.

ML is the answer to all the unanswered questions in the field of public healthcare. On some instances ML can be helpful in predicting whether a drug is capable to treat a disease or not majorly referring to cancer. Whereas, in other situations, ML can also be used for systems in enhancing the knowledge by showcasing certain variables which has a co-relation between some elements of the system. Few important applications on ML.

12.5.4 MEDICAL DIAGNOSES

A typical utilization of ml could be the predictions or diagnoses of any kind of disease or measurements. For instance, research had been done in the field of psychopathology to predict the mood swings of patients by recording the data through their smartphone usage. A crucial problem that exists in the neurological sciences sector is untangling the neural functions and activities to come to the conclusions of all the unanswered questions of brain. This implementation is certainly the most advantageous since it assists in the advancements for building prosthetics. Prosthetic devices are mainly used by those whose one (or more) of the body parts is paralysed which in turn takes the signals from the brain in order to assist the movement of a certain part of the body. There are innumerable ongoing research taking place all around the world to crack the perfect algorithm and ml apparently plays a vital role in this step towards the future of prosthetics. There is a profusion of issues in the following healthcare domains: cancer, precision medicine and medical testing. In the above-mentioned domain's, the main aim is to get the appropriate predictions. Therefore, it is often said that if the focus is on the quality of prediction, then ml is the key to it.

12.5.5 DISEASE ANALYSES

Latterly, many studies have shown that scientists are designing numerous models that are automated through several models of supervise learning. It is a fact that if one could get diagnosed as early as possible, then the chances of fatality could be controlled due to a certain disease. In recent times, many models were being developed for the automatic diagnosis of numerous ailments like diabetes, COVID-19 and cancer [19]. Moreover, there are many scientists who have developed mobile apps with the use of ml to diagnose various diseases [20]. There are also some advancements made in terms of the risk prediction from a certain condition. Meaningly, some apps developed by the researchers which in turn helps in the prediction of risk from a particular disease. Although there are some speculations made by the researchers that some deep learning models have slightly better performance in comparison with some ml models regarding the automated diagnosis of disease. With the help of data, ML has also assisted in resolving few problems occurring due to corona virus.

For example, in China the number of infected patients were calculated, and the duration of corona virus period was also estimated through numerous ml and mathematical models. Research was held by Bullock, in which several datasets, software, tools and applications were examined so that artificial intelligence could show its strength against corona virus [21]. Moreover, it is impeccably important to put ML and AI against corona virus since it will help us in the timely prediction of the disease. Additionally, researchers are using various ML algorithms to prove the linkage between diabetes and corona virus [22]. In fact, there is certain methodology one can use to make an ML model for diagnosis of a certain disease (Figure 12.6).

12.5.6 PHARMACEUTICAL DEVELOPMENT

There are significant advancements done in area of drug discovery and development. Many pharma organisations had taken advantage of ML algorithms to reach newer heights in the area of drug discovery. Here the use of ML algorithms could be the prediction of following factors in certain compounds used in drug discovery process:

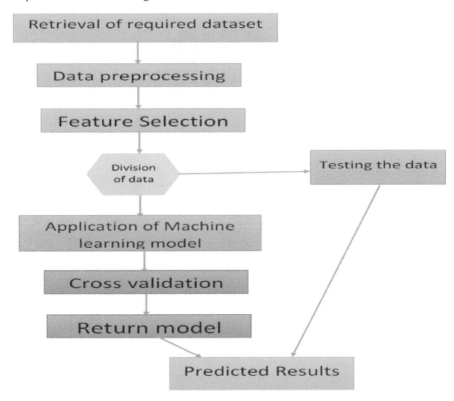

Figure 12.6 Basic methodology for making a disease prediction app/model pharmaceutical development.

1. Physical

2. Biological

3. Chemical

There are many ways in which ml models could use in. Likewise, finding new ways in which a drug could be used, prediction of interaction amongst protein and drug, finding the effectiveness of a certain drug and optimization of the compounds and molecules bioactivity. Majorly three algorithms are used when it comes to drug discovery and development. Firstly, support vector machine (SVM), secondly Random forest (RF) and thirdly Naïve Bayesian (NB) [23]. A fundamental process followed in drug discovery has been mentioned below (Figure 12.7). ML algorithms are applicable in all the four steps to make the process easier, faster and efficient.

Extensively used methodology in the field of Drug Discovery (every stage of this cycle could incorporate the algorithms of ML).

All the ML algorithms and deep learning algorithms walk hand in hand when it comes to the methods applied in drug discovery, as mentioned in Figure 12.7. Lets say for instance in target discoveries through mining proteomics, finding of

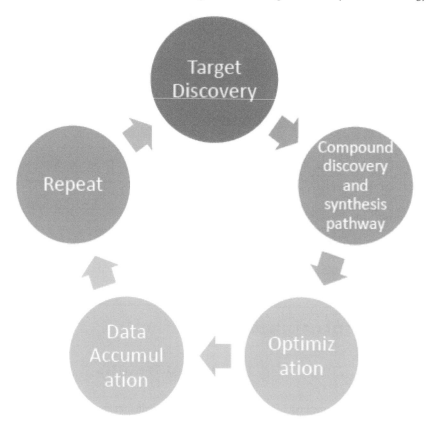

Figure 12.7 Extensively used methodology in the field of drug discovery (every stage of this cycle could incorporate the algorithms of ML).

minute molecules in a form of candidate in the step of lead discoveries in synthesis pathway, and the interpretation of humongous results in assay in data accumulation step. ML algorithms basically bind the process in a synchronous manner in order to get appropriate results [24].

12.5.7 STANDARDIZATION OR BENCH MARKING

Generally, the main goal is not only for the prediction and description but also to generate ml models which could easily be taught and understood. Benchmarking is something that can be used in ML which could be extremely practical. During the time of evaluation of a ML model, there is a problem which arises due to the confusions in detecting the errors. A judgement needs to be made while evaluating the model which is whether the error caused are because of the noise or the incompetency of the ML model. Therefore, let's say a man-made model brings out something unlike to that of ML benchmark, then it can be assumed that its due to the misguided modelling or the missingness of the key fundamentals. Whereas, if a ML model build upon human instincts is extremely close to that of ML benchmark, then the results turn out to be more meaningful.

12.5.8 COMPREHENSION

ML could also play a vital role when it comes to understanding. For example, let's say there is a system which has certain variables to it but it isn't stated that those variables share a non-linear or a linear relation. Therefore, here comes the part where ML has its utmost importance. With the help of ML, one can define that if the details are stored in a signal, which doesn't require the information regarding the relation between variables [25].

12.5.9 CLINICAL GENOMICS

A chief domain in bioinformatics is genomics. Day by day there is an escalation seen in the sequences. Therefore, there is a colossal requirement of analysation of this data in order to get useful facts to act upon. A primary course of action could be the extraction of the location and composition of the gene [26]. With the advancements happening around this field, the recognition of regulatory components and non-coding RNA genes is also made possible [27]. Moreover, the secondary structure of RNA is also predicted with the details of the sequence given.

12.5.10 PROTEOMICS

Let's say the gene is the mother of all the information but the protein has a vital role to play to bring this information to life. The 3-D structure of the protein is a crucial and indispensable attribute in its functional capacities. In this sector, the major application of ML is the prediction of protein structure. Proteins are extremely complex macromolecules. Therefore, we can make out that the construction or we can say the structure received is humongous. Due to this, the process becomes intricate and cumbersome. Here is the need for effective techniques of optimization.

12.5.11 MANAGING EXPERIMENTAL DATA

Computational techniques can also be useful to organize the complex data received for the experimentations done in healthcare domain. An example for such data could be microarray. There are two major problems when it comes to such complex data. Firstly, the data pre-processing and secondly the analysation of the data. Moreover, the prototypical implementation could be identification of a pattern, network induction and classification of the same.

12.5.12 SYSTEM BIOLOGY

This field is another example which portrays the collaboration of BI and ML. All the physiological processes that occur on a cellular level are extremely complex. Here, the computational techniques come into picture. Since it helps in building models for genetic networks, metabolic pathway network and networks for organic processes.

12.6 TOOLS & SOFTWARE

Many advancements are happening all around the globe with the amalgamation of ml and biomedical informatics. In the table given below, the progress around different parts of the globe regarding ml is mentioned (Table 12.1) [28]:

Table 12.1

Evolution around the world in the field of ml and biomedical informatics

Tool/software	Application	Description
IBM Watson Genomics	Diagnosis of ailments	With the help of integration of cognitive computing including tumor sequencing which are genome based
P1vital's PReDicT	Diagnosis of ailments	Providing medication under general clinical situations
Project Hanover by Microsoft	Discovery of Drugs	Use of Ml for technological advancements in cancer treatment and personalised medicine
InnerEye by Microsoft	Imaging Diagnosis	Growth of imaging tools for diagnosis to provide better analysis of image
IBM Watson Oncology	Personalised Medicine	Treatment for every patient will be different according to the needs and the medical history to provide more options for the treatment.
Somatix	Behavioural Modifications	An app which helps a person to make an account to all the unconscious choices we make in our day to day lives.
Google's Cloud Vision API	Maintenance of healthcare records	Provides an ease in maintain the healthcare records
ProMED-mail	Prediction of any kind of outbreaks in future	It is a platform which provides with the information regarding the evolution of diseases and monitors the formation of new diseases. This way they provide the actual statistics about any upcoming outbreak

12.7 FUTURE BENEFITS

Public health care is a crucial sector since it has so much to offer for the wellbeing of the population and also it is one of the highest revenues earning fields for most nations around the world. Day by day the requirement of ml for the growth in the healthcare domain is increasing due to the need. This leads us to the escalation of the graph when we talk about ML applications in public health care. The time is not far when we witness the alliance between analysis, data, and high-tech technological advancements. Which will certainly benefit the thousands of patients all around the world. We are very close to witness the dawn of an era where applications based on ML will become a daily part of our day to day lives. These advancements will ultimately result in increasing chances of curing certain diseases and will also help in the preventive measures. Moreover, newer treatments will also be discovered with the help of ML. What a world it would be if one can access the patient's data from any system around the world. The possibilities are endless in this sector.

In the last few months of the COVID-19 pandemic, a shift has been seen towards the virtual healthcare. This is a big move in the sector of health care in order to provide authentic, dependable, error-free technologies to upgrade the services provided under virtual medicine. These are the times where we expect ml to play an indispensable role to accommodate ourselves with the "New Normal". With the assistance of such mechanism, we are able to amplify the experience which is being shared by a patient and the one who is providing the care. Due to such change, there is a shift which can be seen in terms of patient care from real-world settings to a whole virtual space. In recent times, the world has witnessed the power of pattern matching. One can say that pattern matching is a by-product of ml algorithms. It basically understands the patterns of behaviour of the patient and subsequently provides better solutions to take care of the sick. This in turn escalates the need for virtual care since there is also an increase the magnitude of care provided to a certain population. The major aim for the amalgamation of ML and biomedical informatics is to provide an impeccable, smooth and hassle-free experience to both the patient and the one who is providing the care.

12.8   CONCLUSION

ML is the fast-moving domain in computer science and biomedical informatics is the future for public–healthcare since it promises to provide diagnoses, analyses, and development in the field of pharmacy. Although a lot of development is still needed for ML in biomedical informatics to gain the benefits for humankind. This field also requires research in all the domains like visualization to AI and data science. Moreover, the complexities of the challenges faced in this sector can only be resolved with research which should have a greater reach in all the co-related fields. The world is very close to catch the sight of the new sunrise in biomedical informatics and ML.

REFERENCES

1. Larranaga, P., Calvo, B., Santana, R., Bielza, C., Galdiano, J., Inza, I., Lozano, J.A., Armananzas, R., Santafé, G., Pérez, A. and Robles, V., 2006. Machine learning in bioinformatics. *Briefings in bioinformatics*, 7(1), pp. 86–112.

2. Maojo, V. and Kulikowski, C.A., 2003. Bioinformatics and medical informatics: collaborations on the road to genomic medicine?. Journal of the American Medical Informatics Association, 10(6), pp. 515–522.

3. Shaun Sutner, https://searchhealthit.techtarget.com/definition/biomedical-informatics.

4. Y. Quintana, C. Safran, in Global Health Informatics, 2017 Embi, P. J., & Payne, P. R. (2009). Clinical research informatics: challenges, opportunities and definition for an emerging domain. *Journal of the American Medical Informatics Association : JAMIA*, 16(3), 316–327.

5. Hassan A. Aziz, A review of the role of public health informatics in healthcare, Journal of Taibah University Medical Sciences, Volume 12, Issue 1, 2017.

6. What is Health Information Management & Why Is It Important? https://nearterm.com/what-is-health-information-management-why-is-it-important.

7. Aronis, J.M., and Provost, F.J. (1997) Increasing the efficiency of data mining algorithms with breadth-first marker propagation. Proceedings of the Third International Conference on Knowledge Discovery and Data Mining, Heckerman, D., Mannila, H., and Pregibon, D. (eds.). Menlo Park, CA: AAAI Press, pp. 119–122.

8. Krefl, D. and Seong, R.K., 2017. Machine learning of Calabi-Yau volumes. Physical Review D, 96(6), p. 066014.

9. Ruehle, F., 2017. Evolving neural networks with genetic algorithms to study the String Landscape. Journal of High Energy Physics, 2017(8), pp. 1–20.

10. Carifio, J., Halverson, J., Krioukov, D. and Nelson, B.D., 2017. Machine learning in the string landscape. Journal of High Energy Physics, 2017(9), pp. 1–36.

11. Deloitte Insights *State of AI in the enterprise.* Deloitte, 2018. www2.deloitte.com/content/dam/insights/us/articles/4780_State-of-AI-in-the-enterprise/AICognitive Survey2018_Infographic.pdf. [Google Scholar]

12. Lee, S.I, Celik S, Logsdon BA, et al. (2018). A machine learning approach to integrate big data for precision medicine in acute myeloid leukemia. *Nat Commun* 2018;9:42. [PMC free article] [PubMed] [Google Scholar]

13. Saxena, N., Gupta, P., Raman, R. and Rathore, A.S., 2020. Role of data science in managing COVID-19 pandemic. Indian Chemical Engineer, 62(4), pp. 385–395.

14. Cooper, G.F., Abraham, V., Aliferis, C.F., Aronis, J.M., Buchanan, B.G., Caruana, R., Fine, M.J., Janosky, J.E., Livingston, G., Mitchell, T. and Monti, S., 2005. Predicting dire outcomes of patients with community acquired pneumonia. Journal of biomedical informatics, 38(5), pp. 347–366.

15. Schulam, Peter, and Suchi Saria. Reliable decision support using counterfactual models. Advances in neural information processing systems 30 (2017).

16. Pearl, Judea. Theoretical impediments to machine learning with seven sparks from the causal revolution. arXiv preprint arXiv:1801.04016 (2018).

17. Ding, Peng, and Fan Li. Causal Inference. Statistical Science 33, no. 2 (2018): 214–237.

18. Huang, Shuai, Jiayu Zhou, Zhangyang Wang, Qing Ling, and Yang Shen. Biomedical informatics with optimization and machine learning. EURASIP Journal on Bioinformatics and Systems Biology 2017, no. 1 (2016): 1–3.

19. Sidey-Gibbons, A. M. Jenni, and C. J. Sidey-Gibbons, Machine learning in medicine: a practical introduction,' BMC Medical Research Methodology, vol. 19, 2019.

20. T. Davenport and R. Kalakota, The potential for artificial intelligence in healthcare, Future Healthcare Journal, vol. 6, no. 2, pp. 94–98, 2019.

21. J. Bullock, Mapping the landscape of artificial intelligence applications against COVID-19, 2020, https://arxiv.org/abs/2003.11336.

22. Hussain, B. Bhowmik, and N. C. Moreira, COVID-19 and diabetes: knowledge in progress, Diabetes Research and Clinical Practice, vol. 162, Article ID 108142, 2020.

23. A.M. Jenni Sidey-Gibbons and C.J. Sidey-Gibbons. Machine learning in medicine: a practical introduction. BMC Medical Research Methodology, 19.

24. Davenport, Thomas, and Ravi Kalakota. The potential for artificial intelligence in healthcare. Future healthcare journal 6, no. 2 (2019): 94.

25. Kording, K.P., Benjamin, A., Farhoodi, R. and Glaser, J.I., 2018, January. The roles of machine learning in biomedical science. In Frontiers of Engineering: Reports on Leading-Edge Engineering from the 2017 Symposium. National Academies Press.

26. Mathé, C., Sagot, M.F., Schiex, T. and Rouzé, P., 2002. Current methods of gene prediction, their strengths and weaknesses. Nucleic acids research, 30(19), pp.4103-4117.

27. Aerts, S., Van Loo, P., Moreau, Y. and De Moor, B., 2004. A genetic algorithm for the detection of new cis-regulatory modules in sets of coregulated genes. Bioinformatics, 20(12), pp. 1974–1976.

28. Patel, L., Shukla, T., Huang, X., Ussery, D.W. and Wang, S., 2020. Machine learning methods in drug discovery. Molecules, 25(22), p. 5277.

13 Recognition of Types of Arrhythmia: An Implementation of Ensembling Techniques Using ECG Beat

Arshpreet Kaur, Kumar Shashvat
DIT University, Dehradun, Uttrakhand, India

Hemant Kr. Soni
Department of Computer Science and Engineering,
National Institute of Technology, Delhi, India

CONTENTS

13.1 INTRODUCTION

Arrhythmia is a heart condition which can be detected by the interpretation of an electrocardiogram (ECG) signal [13]. Despite the fact that arrhythmia can be effectively treated and many types are risk free, if ignored, it can even lead to lethal conditions such as stroke or heart failure. Heart conditions can also be the consequence

DOI: 10.1201/9781003246688-13

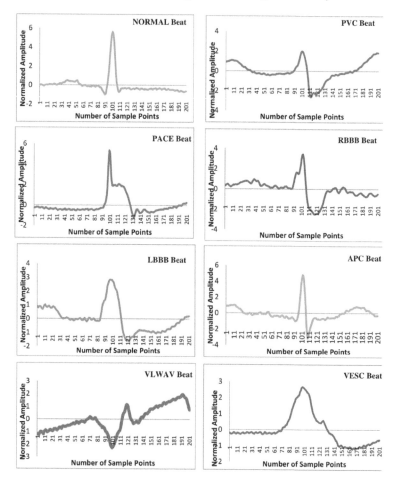

Figure 13.1   Types of heart beat.

of COVID-19 [6] [12]. It is divided into several different types of classes, ranging to cause mild to acute heart issues. Electrocardiogram (ECG) is used to sense abnormalities in the human heart. It works characteristically by sensing and intensifying the tiny electrical changes on the skin that are caused, when the heart muscle "depolarizes" through individual heartbeat. It, on the whole, works by the interpretation of electrical activity of the heart, apprehended over time and superficially recorded by skin electrodes [11], [8]. Figure 13.1 presents the diagrammatic descriptions of all different types of arrhythmic beats.

13.2   LITERATURE REVIEW

Different methods to automate the recognition of arrhythmia have been explored. The goal remains to correctly classify different types of arrhythmia. In [14], 200 points are extracted around the prominent R-peak; these 200 points are used as one ECG beat sample; one of the other feature R-R interval is also appended to the existing feature vector. These features are then further reduced using PCA and LDA feature reduction techniques. Probabilistic neural network is used for classification. The authors have achieved an accuracy of about 99.71%. They have classified the heartbeats into eight arrhythmia classes. In [15], the authors used a technique with auto sparse auto-encoders with softmax regression model. They have used two fifty data points around the prominent R-peak as a feature vector. They have classified the beats into six types and achieved an accuracy of 99.45%.

In [2], convolutional neural network (CNN) has been used for the classification of two major classes of shockable and non-shockable ECG signals at an instance. MITB, CUDB and VFDB datasets were used. Firstly, they were standardized in context to frequency, and synthetic data were also created to overcome class imbalance problem. Ground truth annotations provided by their respective database were used to segment signals. Z-score normalization was carried out before feeding these signals to CNN network where 93.18% accuracy was achieved using an only 2s length of the signal. In [4], the authors have extracted the features with the help of discrete wavelet transform (DWT). DWT is a popular technique used as a preprocessing method in various works such as in [5]. The classification method used was random forest. They were able to achieve accuracy of about 99.8% on the MIT-BIH arrhythmia dataset. In [13], the authors implemented dual tree complex wavelet transform for feature extraction. They worked with 256 samples around the R-peak. Using multi-layer back propagation neural network, the authors achieved an accuracy of 94.64%. In our problem formulation, we mainly focused on the development of effective classification algorithm to differentiate all eight types of arrhythmia using ensemble models.

13.3 METHODS

The complete methodology followed in this work is explained in Figure 13.2. In this paper, ECG arrhythmia recordings are acquired from publically available dataset available at [10]. In feature extraction and normalization step, 201 data points at 360Hz sampling rate around the prominent R-peak (100 points on either side) are extracted as a one ECG beat sample. To further normalize, Z-score normalization technique is used to standardize the beat sample. Since random forest and XG Boost both take features randomly from all set of features, we need not apply any feature reduction technique.

13.3.1 DATA ACQUISITION

The data used in this study is acquired from publically available database in [10]. The MIT-BIH database contains forty-eight half hour records of forty-seven subjects considered by the BIH Arrhythmia Laboratory starting from the year 1975 till 1979. The aim is to classify eight types of ECG beats Table 13.1.

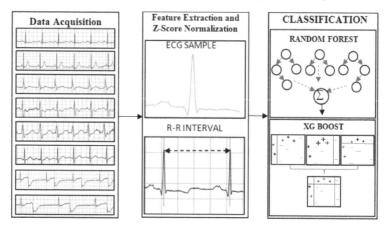

Figure 13.2 Working flow diagram of the proposed methodology.

Table 13.1
Distribution of the eight types of ECG beats

Type	File Number	Total Number of Each Type
NORMAL	100, 101, 103, 105, 108, 112, 113, 114, 115, 117, 121, 122, 123, 202, 205, 219, 230, 234	3600
PVC	106, 119, 200, 203, 208, 213, 221, 228, 233	2460
	201	
	210	
	215	
	116	
PACE	102, 104, 107, 217	800
RBBB	118, 124, 212, 231	800
LBBB	109, 111, 207, 214	800
APC	209, 222, 232	
	220	764
	223	
VLWAV	207	472
VESC	207	104

13.3.2 PREPROCESSING AND FEATURE EXTRACTION

In this step, the 201 points (100 on either side) around the R-peak are extracted from the signal. ECG beat is sampled at 360 Hz sampling rate. In this work, we have utilized Z-score to reduce amplitude variance and offset effect from each ECG beat sample.

13.3.3 FEATURE REDUCTION

From the total of "n" features $\sqrt{n}$ features randomly in both the algorithms. Random forest and XGBoost both acquire the features randomly from the available set of features. This ability allows both algorithms to work on high dimensional databases exceptionally.

13.3.4 RANDOM FOREST CLASSIFIER

Random forest is an ensemble method which constructs a strong learner with the assistance of weak learners. Divide and conquer approach is used to improve its performance. A group of weak learners approaches together to create a strong learner which results in low variance and improved performance. It has the ability to handle large datasets with higher dimensionalities. It is one of the algorithm that has been deployed for classification of arrhythmia in different ECG classes [7],[9], others include artificial neural network such as in [16] and convolution neural network such as in [1].

Random forest can efficiently handle the missing values while maintaining the accuracy of missing data. It is a supervised learning technique, which targets both regression and classification. In this algorithm, more the number of trees, more robust our algorithm becomes, leading to better results. It builds up multiple decision trees using algorithms such as Gini Index, Information Gain, etc. Information gain method has been used here. In case of classification, we predict the class label by the voting of all the working decision trees. Algorithm for random forest:

1. Pick X random annals from the database.

2. Construct a decision tree based on these X records.

3. Choose the quantity of trees in the specific algorithm and repeat steps 1 and 2.

4. The final value can be calculated by taking the mode of all the values predicted by all the trees in forest for a classification problem.

13.3.5 XGBOOST CLASSIFIER

XGBoost is a supervised learning model where training data are used to make predictions. Prediction value can have a different interpretation based on the type of problem, i.e. classification and regression. It also is known as Extreme Gradient Boosting [3]. It is discernible from random forest classifier as there is dependency of trees in this classifier; wherein they are independent in random forest algorithm.

To improve the misclassifications in this model, trees are built in an iterative manner one after the other. It fits the consecutive trees where each solves the net error of the prior trees. Though it can consume more training time, it remains good model training performance because of the cache optimization. Additionally, XGBOOST is a fast parallel tree, because of its design which stands fault tolerant of the distributed setting. It handles millions of the samples on a single node using top-down approach. It holds many other advantages, such as parallel computing, which enable it to use all the cores of our machine. It also handles missing values internally by capturing the trend.

Eta is taken as 0.3 and 0.1; the default value of eta is 0.3. Gamma is taken as 0, which is its default value. Gamma controls the regularization and prevents overfitting. Max depth is taken as 6, which is the depth of the tree. In our work, we take objective as multi:softprob which is for multiclass classification using softmax objective, and it returns predicted class probabilities. And we take eval_metrc as Logloss which Negative Likelihood and evaluate model's accuracy on validation data.

13.4 SIMULATION SETUP

To complete the entire computation, Windows 8 environment is used, with the help of analysis tools like R studio and MATLAB. We have used the quad-core i-3 processor for the implementation. Implementation of XGBoost is done in R with the help of R package; XGBoost library is being used. Xgb.train is a training function, and it receives xgb.Dmatrix object as input. In general parameters, we use gbtree as booster type and let n threads to take its default value which is the maximum number of cores available in the machine for parallel processing. In booster parameters we have parameters such as n rounds, eta, gamma, max depth, etc. N rounds are the number of trees to grow. We try n rounds with different values like 50,100, 200, etc.

13.5 RESULTS AND DISCUSSION

The results of the work are presented and explained below in the Table 13.2, Table 13.3 and Table 13.4. To evaluate the performance, we have used confusion matrix. The confusion matrix is the table that describes the performance of a classification model on the test data for which the true values are known.

> Sensitivity: It is also known as true positive rate/ recall/ probability of detection. It measures the section of the samples that are correctly classified:

$$\text{Sensitivity} = \frac{TP}{TP + FN}$$

> Specificity: It is also known as true negative rate. It is defined as the section of the negatives that are appropriately categorized as negative:

$$\text{Specificity} = \frac{TN}{TN + FN}$$

> Accuracy: Classification accuracy is the ratio of correct predictions to a total number of predictions made:

$$\text{Accuracy} = \frac{TP + TN}{TP + TN + FP + FN}$$

Table 13.2
Results with both classifiers

	RANDOM FOREST		XGBOOST	
	SPECIFICITY (%)	SENSITIVITY (%)	SPECIFICITY (%)	SENSITIVITY (%)
NORMAL	99.89	99.94	99.94	100
PVC	99.16	99.70	99.43	99.7
PACE	98.13	100	100	100
RBBB	99.8	100	99.74	100
LBBB	100	100	100	100
APC	99.73	99.9	100	99.95
VLWAV	96.42	99.84	96.17	99.9
VESC	95	100	98.18	100
ACCURACY	99.50		99.60	

Table 13.3
Results with random forest classifier

Iteration	50	100	200	500
Accuracy(%)	99.40	99.48	99.50	99.48

Table 13.4
Results with XGBOOST classifier

Iteration	50	100	200	500	500
Max depth	6	6	6	6	6
Eta	0.3	0.3	0.3	0.1	0.3
Subsample	0.8	0.8	0.8	0.8	0.8
Accuracy(%)	99.50	99.52	99.60	99.60	99.54

Data division: Nine thousand eight hundred samples of eight ECG beat types from the dataset have been selected, from which half of them are selected as training set and other half is selected as a testing set. We have divided the data into two sets, training set and testing set equally (i.e., 50% training examples and 50% testing examples). Evaluation of the model on the basis of sensitivity as well as sensitivity has also been considered. With both the classifiers, random forest classifier performs best with 200 iterations, though at 100 iterations it performs quite nicely. From Table 13.5 we can observe that our algorithm has performed better than the most with XGboost performing the best.

Table 13.5
Comparison table

Author (Year)	ECG Signals Obtained From	Approaches	Length of ECG Points	Classes	Accuracy (in %)
U. Rajendra Acharya 2017 [2]	MITDB, VFDB, CUDB	Convolution neural network + 10-fold cross-validation	500 (2 seconds)	2	93.18
Jeen-Shing Wang 2013 [14]	MITDB	Z-score+ LDA + PNN	200 points	8	93
Jeen-Shing Wang 2013 [14]	MITDB	Z-score + PCA + PNN	200 points	8	98.42
Jianli Yang 2017 [8]	MITDB	Stack sparse autoencoder for feature extraction + softmax classifier	250 points (train and test on the same patient)	6	99.45
Zerina Masetic 2016 [9]	MITDB	AR Burg + C.45		2	99.93
Zerina Masetic 2016 [9]	MITDB	AR Burg + KNN		2	99.7
Zerina Masetic 2016 [9]	MITDB	AR Burg + SVM		2	96.13
Zerina Masetic 2016 [9]	MITDB	AR Burg + ANN		2	95.92
Zerina Masetic 2016 [9]	MITDB	AR Burg + RF		2	100
Nahit Emanet 2009 [4]	MITDB	DWT + random forest	256 points	5	99.8
Manu Thomas 2015 [13]	MITDB	DTCWT + morphological features+ multi-layer back propagation neural network	256 points	5	94.64
Jose Antonio Gutiérrez-Gnecch 2017 [5]	MITDB	Wavelet transform + probabilistic neural network	2160 points	8	92.75
Proposed Work	MITDB	Z-score + random forest	201 points	8	99.50
Proposed Work	MITDB	Z-score + XGBoost	201 points	8	99.60

13.6 CONCLUSION AND FUTURE WORK

In this work, we have proposed an effective ECG arrhythmia classification arrangement that uses time domain features around the R-peak and R-R interval as a characteristic feature. Our algorithm has performed efficiently and provided better accuracy as compared to other techniques. Its outstanding presentation designates that the algorithm possesses the potential to be installed in the handheld devices to detect, indicate and save the life from various abnormalities of the heart.

REFERENCES

1. U Rajendra Acharya, Hamido Fujita, Oh Shu Lih, Yuki Hagiwara, Jen Hong Tan, and Muhammad Adam. Automated detection of arrhythmias using different intervals of tachycardia ecg segments with convolutional neural network. *Information Sciences*, 405:81–90, 2017.

2. U Rajendra Acharya, Hamido Fujita, Shu Lih Oh, U Raghavendra, Jen Hong Tan, Muhammad Adam, Arkadiusz Gertych, and Yuki Hagiwara. Automated identification of shockable and non-shockable life-threatening ventricular arrhythmias using convolutional neural network. *Future Generation Computer Systems*, 79:952–959, 2018.

3. Tianqi Chen and Carlos Guestrin. Xgboost: A scalable tree boosting system. In *Proceedings of the 22nd ACM SIGKDD International Conference on Knowledge Discovery and Data Mining*, pages 785–794, 2016.

4. Nahit Emanet. Ecg beat classification by using discrete wavelet transform and random forest algorithm. In *2009 Fifth International Conference on Soft Computing, Computing with Words and Perceptions in System Analysis, Decision and Control*, pages 1–4. IEEE, 2009.

5. Jose Antonio Gutiérrez-Gnecchi, Rodrigo Morfin-Magana, Daniel Lorias-Espinoza, Adriana del Carmen Tellez-Anguiano, Enrique Reyes-Archundia, Arturo Méndez-Patiño, and Rodrigo Castañeda-Miranda. Dsp-based arrhythmia classification using wavelet transform and probabilistic neural network. *Biomedical Signal Processing and Control*, 32:44–56, 2017.

6. Sk Sarif Hassan, Ranjeet Kumar Rout, Kshira Sagar Sahoo, Nz Jhanjhi, Saiyed Umer, Thamer A Tabbakh, and Zahrah A Almusaylim. A vicenary analysis of sars-cov-2 genomes. *Cmc-Computers Materials & Continua*, pages 3477–3493, 2021.

7. R Ganesh Kumar, YS Kumaraswamy, et al. Investigating cardiac arrhythmia in ecg using random forest classification. *International Journal of Computers and Applications*, 37(4):31–34, 2012.

8. Eduardo José da S Luz, William Robson Schwartz, Guillermo Cámara-Chávez, and David Menotti. Ecg-based heartbeat classification for arrhythmia detection: A survey. *Computer Methods and Programs in Biomedicine*, 127:144–164, 2016.

9. Zerina Masetic and Abdulhamit Subasi. Congestive heart failure detection using random forest classifier. *Computer Methods and Programs in Biomedicine*, 130:54–64, 2016.

10. George B Moody and Roger G Mark. The impact of the mit-bih arrhythmia database. *IEEE Engineering in Medicine and Biology Magazine*, 20(3):45–50, 2001.

11. Keerthi G Reddy, PA Vijaya, and S Suhasini. Ecg signal characterization and correlation to heart abnormalities. *International Research Journal of Engineering and Technology (IRJET)*, 4:1212–1216, 2017.

12. Ranjeet Kumar Rout, Saiyed Umer, Sabha Sheikh, Sanchit Sindhwani, and Smitarani Pati. Eightydvec: a method for protein sequence similarity analysis using physicochem-ical properties of amino acids. *Computer Methods in Biomechanics and Biomedical Engineering: Imaging & Visualization*, pages 1–11, 2021.

13. Manu Thomas, Manab Kr Das, and Samit Ari. Automatic ecg arrhythmia classification using dual tree complex wavelet based features. *AEU-International Journal of Electron-ics and Communications*, 69(4):715–721, 2015.

14. Jeen-Shing Wang, Wei-Chun Chiang, Yu-Liang Hsu, and Ya-Ting C Yang. Ecg arrhyth-mia classification using a probabilistic neural network with a feature reduction method. *Neurocomputing*, 116:38–45, 2013.

15. Jianli Yang, Yang Bai, Feng Lin, Ming Liu, Zengguang Hou, and Xiuling Liu. A novel electrocardiogram arrhythmia classification method based on stacked sparse auto-encoders and softmax regression. *International Journal of Machine Learning and Cy-bernetics*, 9(10):1733–1740, 2018.

16. Sung-Nien Yu and Kuan-To Chou. Integration of independent component analysis and neural networks for ecg beat classification. *Expert Systems with Applications*, 34(4):2841–2846, 2008.

14 Feature Selection, Machine Learning and Deep Learning Algorithms on Multi-modal Omics Data

Supantha Das
Department of Information Technology, Academy of Technology,
Hooghly, West Bengal, India

Soumadip Ghosh
Department of Computer Science and Engineering,
Institute of Engineering & Management, Kolkata, West Bengal, India

Saurav Mallik
Center for Precision Health, School of Biomedical Informatics,
The University of Texas Health Science Center at
Houston, Houston, TX, USA

Guimin Qin
Department of Computer Science and Engineering,
Xidian University, Xian, Shaanxi, China

CONTENTS

DOI: 10.1201/9781003246688-14

14.1 INTRODUCTION

Artificial intelligence (AI) in biomedical space is regarded as one of crucial aspects for the enhancement of medical service. Feature selection denotes the process for choosing most relevant data among high dimensional biological dataset and implementation of such process by dint of various approaches of machine learning as well as deep learning explores one of imperative techniques employing AI technologies in several fields of bioinformatics [46]. Different strategies dealing with feature selection are prominently used for the analysis of gene (DNA sequence possessing biological instructions) expression and incorporation of such strategies in medical environment has expanded the region of several biomedical researches [47] including researches on lung cancer, breast cancer, COVID-19 and many other diseases. Furthermore, the employment of feature selection principles for researches on omics data (genome, epigenome, etc.) acquired from multi-modal datasets is essential to comprehend correlation among biological entities and determine biomarkers [6] for complex diseases. The key purpose of feature selection focuses on the scheme of finding relevant features from original dataset discarding useless, redundant features and accomplishment of such scheme by means of machine learning which is indispensible element in order to develop constructive predictive models concerned with the studied context for achieving enhanced performance [33][24][53]. Machine learning (ML) belonged to AI provides computing systems the ability of gaining learning power rather than equipped by explicit programming. Though traditional ML for computing machines is responsible for thinking like human beings, it is unable to make any future decision based on the acquisition of data extracted from any image or video like human brain. To eradicate such problem of ML, Deep Learning (DL) which is a subset of ML has emerged in the field of AI. DL employs the concept of artificial neural network consisting of multiple layers with a view to simulate human brain by acquiring relevant values from large dataset including any image and video for taking future decision [16]. This research article not only reveals fundamental concepts regarding feature selection, ML, DL and many more, but also represents succinct overview of recent research works on the development of feature selection-based model by means of ML as well as DL depicting their generalized views. Moreover, such brief overviews also enlarge the research knowledge about the utilization of feature selection for several recent works like cancer research, workings with multi-omics data and even COVID detection.

14.2 SUPERVISED LEARNING

Supervised learning denotes a machine learning approach by training machine using labelled data and in accordance with such labelled data machines are enabled to predict correct output or intended variable. Supervised machine learning implies the presence of a supervisor which can train machine using labelled data, i.e. data already tagged with truthful and acceptable answers in order to predict correct outcome for new other data inputted to machine [23][26]. Rather than taking all features formed from biological training data into account only subset of such features will

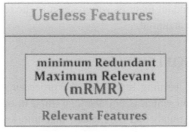

Figure 14.1   Feature selection using mRMR.

Table 14.1
Different supervised feature selection techniques

Filtering method [23]	Acquisition on of top ranked features according to the characteristics of data by applying methods like Chi-Square, Mutual information ANOVA F-value and many more separately on each feature.
Wrapper method [54]	Searching and collecting subset of pertinent features in the context of intended or target variable like forward feature selection process, etc.
Embedded method [33]	Assemblage of filtering and wrapper methods for embedding features into learning model (Example: LASSO, elastic net, etc.)

be deemed to lead to the learning model and also discard non-informative irrelevant features [29]. With a view to acquire relevant features supervised machine learning policy employs the concept of information gain or mRMR (depicted in Figure 14.1) or other key strategies with respect to inputted trained data. Finally those biologically informative selected features provide viable guidance in order to create the classifier for prediction. Table 14.1 provides concise overview of different feature selection approaches. Filter methods [23] ignores feature dependencies hidden inside the dataset due to incorporation of specific machine learning strategy separately on each feature and thereby contains less computational performance compared with other supervised feature selection methodologies. Though wrapper method encounters the limitation of filtering approach for the perspectives of computation performance, it possesses high computation time and suffers from over-fitting problems but embedded policy contains less over-fitting and low computational time compared to wrapper method [34].

Minimum redundancy maximum relevance (mRMR) [38] denotes the selection mode of relevant features widely applied in the sphere of cancer diagnosis. According to operational policy of mRMR (depicted in Figure 14.1) generally less-informative redundant features showing limited relevance are discarded and most valuable non-redundant features with respect to the required and intended outcome of application are acquired.

14.3 UNSUPERVISED LEARNING

Unsupervised machine learning represents the procedure of training machine by means of unlabelled data which is not also classified in order to obtain the pattern concealed behind such unlabelled data. Such machine learning scheme trains the machine with a view to create groups or clusters [30] from the unlabelled data set in accordance with likeness, any hidden pattern and also the differences among data. Unsupervised learning [22][16][48] approach indicates the absence of any supervisor and facilitates the machine to learn without any prior training of applied dataset. Feature selection strategy implemented by such machine learning policy implies the formation of clusters where each cluster is responsible for assembling similar features together and focuses on selection process of pertinent representative feature for every cluster [41][27].

14.4 SEMI-SUPERVISED LEARNING

The acquisition of data considered for supervised learning must be labelled and such labelling procedure especially for huge data leads to an expensive technique. Though consideration of unlabelled data for unsupervised learning scheme indicates comparatively inexpensive process, limitation of application spectrum is regarded as most widespread drawback for unsupervised learning scheme. The notion of semi-supervised learning has emerged in machine learning environment with a view to wipe out the shortcomings of supervised and unsupervised approaches. Semi-supervised learning incorporates the characteristics of supervised learning as well as unsupervised learning principles and considers less amount of labelled data along with high amount unlabelled data. Such machine learning policy generates clusters based on unsupervised algorithms for homogeneous unlabelled data and after that accomplishes the job of labelling remaining unlabelled data by dint of existing limited amount of labelled data [12]

Semi-supervised GAN (SGAN) is the modification of network architecture of generative adversarial network (GAN) [25] comprised of a generator and a discriminator (or a classifier) [29]. According to traditional GAN, the generator is used to generate an image relying on convolution neural network from random input vector with noise in problem domain and discriminator is responsible for determining whether input image is original from dataset or artificially produced image from generator. Simply the generator tries to create an image imitating original one and discriminator evaluates the performance of generator. Such battle held between two adversaries provides viable ways to system to learn from trained data and produce image possessing similar characteristics of trained data. SGAN [17] trains discriminator not only in unsupervised mode like traditional GAN but also in supervised mode for achieving enhanced performance with respect to accuracy and machine training time. Hence, the concept of SGAN is widely employed in many applications

Table 14.2

Layers of convolution neural network (CNN)

Convo layer	Comprised of learnable filters(kernels) and employs convolution operations to mine features
Pooling layer	Dwindle the dimension of convolved feature map
Fully connected layer	Concerned with classification process

including computer vision and detection of the presence of tumor cells by means of comparing with the images of normal and healthy human cells [11].

14.5   DEEP LEARNING

The operational principle of ML has to rely on relevant features acquired manually from input image in order to generate predictive model. On the other hand, deep learning posses the ability of obtaining relevant features from an input image in automatic fashion with a view to select appropriate features from image, text and other functions including the recognition of objects and also distinguishing the image from other identified objects, etc. with higher accuracy. DL employs the artificial neural network (ANN) with a view to imitate the working principle of human brain. ANN stimulated by workings of various neurons in biological system incorporates multiple layers to maintain enhanced accuracy level in performance of the model formed by DL [21][45]. Hence, such machine learning scheme is endowed with the term deep. DL is able to classify relevant features from text, image, voice and video data due to the presence of such layers which comprise ANN, and every layer learns by means of transforming its input to output and feeds that transformed or processed output to its next layer in order to achieve final output with supreme accuracy. In case of recognizing any image by DL, initial layers are responsible for detecting edges and tail ending layers learns about the identification of complex shape of the object to finally recognize the object. Owing to having such characteristics, DL possesses different applications including disease prediction [18][10] with other medical researches, automated driving and speech translation. Convolution neural network (CNN)[also known as ConvNets] regarded widely applied deep learning technique contains the ability of extracting pertinent features from inputted image owing to being endowed with multiple layers perceptrons which enable machine to perceive like human brain [60]. On account of possessing this characteristic, approaches of CNN are drastically used in several cognitive activities as well as biomedical research works including cancer diagnosis and even COVID detection also. Table 14.2 reveals most essential layers along with their functions briefly.

Table 14.3

Omics Types and their functions

Genomics	Analyzes genome (DNA sequence of a cell) and also considers all genes along with their interaction within living organism.
Transcriptomics	Deals with transcriptome(RNA transcripts from DNA) which is concerned with rRNA, tRNA, mRNA and miRNA.
Proteomics	Concerns proteome (protein sequence in cell)
Epigenomics	Considers epigenomic modification of DNA sequence
Metabolomics	Accounts for metabolome (concerned with cellular metabolic activities)

14.6   MULTI-OMICS DATA

Omics data in the sphere of molecular biology are deemed for the recognition and delineation of biological molecules which are responsible for the formation along with activities of cells or tissues. Study of omics data is also performed with a view to analyze entire genes and their correlation as well as biochemical actions of any living organism. Hence, such data are expansively preferred for biomedical researches, Table 14.3 shows the functions of major five omics data in brief. Multi-omics (also acknowledged as integrative omics or pan-omics) denotes the approach of analysing integrated data acquired from various single-omic types and such integration is profoundly requisite for achieving superior accuracy for medical diagnosis rather than focusing on a single-omics data [31]. The method of multi-omics data integration [42] is also necessary to accomplish of feature selection in order to come across new biomarkers [6].

14.7   LITERATURE SURVEY

Bioinformatics field in recent years contains various research articles exploring the performance of feature selection and such articles reveal new approaches for medical study in order to diagnose diseases. Feature selection carried out through several supervised machine learning approaches including SVM, mRMR, etc. exposes new frameworks of acquiring most pertinent features for improving biomedical researches. This section provides concise overview of some of such research articles. Authors in [35] proposed a technique concerned with the enhancement of working principle of SVM-RFE with a view to classify non-redundant relevant genes for feature selection with improved accuracy for biomedical research. A scheme is proposed for feature selection based on combining rankings of features in [43]. The integrating scheme of elastic net and probabilistic SVM is applied in order to decompose multi-class classification problem along with high-dimensional data into binary classification problems in [59]. Authors in [29] demonstrate feature selection mechanism using support vector machine taking recursive feature elimination concept into account. LinSun et al. [50] focused on entropy to implement feature selection strategy,

Table 14.4

Some of research works on feature selection using machine learning

Yuan et al. [59]	2020	Detection of cancer from data of platelets
LinSun et al. [50]	2019	Entropy based feature selection for gene expression
Amazona Adorada et al. [2]	2018	Breast cancer
Zifa Li et al. [29]	2018	Selection of key genes causing complex diseases
D. Pavithra et al. [37]	2017	Genes responsible for cancer

whereas D. Pavithra et al. [37] deployed the concept of mutual information as well as genetic algorithm on cancer microarray dataset for the accomplishment of feature selection with superior accuracy for cancer research. Amazona Adorada et al. [2] presented a framework capable of acquiring relevant features using the approach of supervised learning on feature selection advantageous for breast cancer research. In order to minimize wrapper evaluation time for acquiring top ranked features, Markov blanket technique was incorporated by Aiguo Wang et al. in [54]. Authors in [38] employed the technique for the analysis of gene expression data by dint of mRMR. Table 14.4 exposes some of recent works on feature selection by dint of supervised learning algorithm for bio-medical space. A generalized approach related with feature selection using supervised machine learning scheme is depicted in Figure 14.2, and it reveals the insight of a learning model by dint of mRMR, RFE, SVM to lead to a classifier for achieving fruitful results. Some research articles focused on unsupervised feature selection techniques for medical researches. In [1], a clustering technique based on PAM is employed in order to extract and recognize top ranked genes regarding gene ontology (GO). This technique does not provide a viable approach for automatic determination of dimensionality of proper gene space, it is able to shrink the dimension of considered gene dataset and improving accuracy level. Xiucal Ye et al. [58] proposed adaptive unsupervised feature selection (AUFS) capable of recognizing gene signatures maintaining enhanced accuracy for non-small-cell lung cancer (NSCLC).

Authors in [36] represents the usage of the unsupervised learning processes like t-distributed stochastic neighbor embedding along with the concept of auto encoders and self-organizing maps, whereas an unsupervised learning procedure formed by two phases consisting of filtering approach and clustering strategy based on autoencoder for classifying the sample data of cancer is shown in [4]. Jocelyn Gal et al. in [15] deployed eveal metabolomic-based clusters in order to select features from the patients of breast cancers and move them towards the road to recovery. V. Sesha Srinivas et al. [49] showed a technique of employing differential evolution for the purpose of selecting clusters comprised of subset of appropriate features of genes. An iterative unsupervised machine learning approach leading to clusters based on several biological records of patients including heritability, etc. for medical

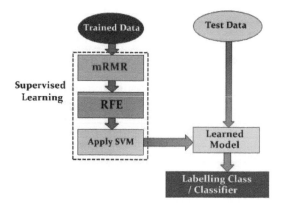

Figure 14.2 Working steps of supervised machine learning for feature selection.

Table 14.5
Feature selection (by unsupervised learning) for medical research

Uzmal et al. [4]	2020	Classification of cancerous cells
Xiucai Ye et al. [58]	2020	Non-small-cell lung cancer (NSCLC)
Jocelyn Gal et al. [15]	2019	Recovery from breast cancer
Stefan Nitica et al. [36]	2018	Breast cancer detection
Sudipta Acharya et al. [1]	2017	Genetic information acquired from Gene Ontology
Sunanda Das et al. [13]	2016	Adoption of Genetic Algorithm for gene expression

treatment was represented by Christian Lopez et al. in [30]. Saul Solorio-FernÃndez et al. [48] developed feature evaluating spectrum technique (USFSM) considering numerical as well as non-numerical dataset related with heart, liver and dermatology. Genetic Algorithm (GA) generating automatic creation of optimal clusters for extracting most valuable features was mentioned by Sunanda Das et al. in [13]. Table 14.5 represents a brief outline of some research works on feature selection implemented by the principles of unsupervised learning strategies on biological area in recent years. Figure 14.3 shows the common operational view regarding the implementation of feature selection through the selection of most relevant genes from the clusters of top ranked genes in unsupervised learning.

Jisha Augustine et al. [8] developed semi-supervised techniques operated in two stages through clustering and iterative fashioned for acquiring the comprehensive delineation regarding regulatory connections among genes with high prediction accuracy. Hamid Reza Hassanzadeh et al. in [20] considered multi-modal data obtained from RNA sequence and created a graph-based pipeline endowed with the principles of semi-supervised learning in order to forecast the survival probability of cancer patients.

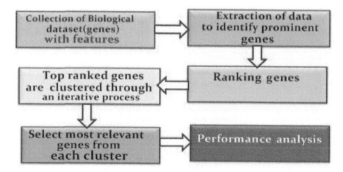

Figure 14.3 Working steps of unsupervised machine learning for feature selection.

Table 14.6
Feature selection (by semi-supervised learning) for biomedical zone

Yi Zhou et al. [63]	2019	Segmentation of medical images
Xin Yang et al. [57]	2019	Medical image synthesis
Kamran Ghasedi Dizaji et al. [17]	2018	Gene expression inference
Hamid Reza Hassanzadeh et al. [20]	2016	Prediction of cancer survival

Xin Yang et al. proposed the approach of synthesising medical image through SGAN for speeding up the progress flow of medical researches in [57]. Kamran Ghasedi Dizaji et al. developped SGAN-based framework for leading to a gene interference network considering cheap unlabelled dataset of genes for flourishing biomedical research dealing with genes [17]. Jun Chin Ang et al. [7] expressed existing research deeds regarding the relevance of semi-supervised learning schemes for the procedure of finding proper genes essential for the advancement of biological research activities. Some research works on semi-supervised learning based feature selection for biological region are briefly represented in Table 14.6. Notion of deep learning approaches for the accomplishment of feature selection has enlarged the application space of feature selection in the ambience of medical researches. Deep learning approaches are regarded as viable means in order to accomplish the deed of segmenting biological images. Employment of principles of this key learning approach of artificial intelligence for choosing key features has accelerated the deed of disease detection as well as medical treatment. Several research works in this context illustrate the frameworks used for medical diagnosis relying on deep learning schemes and authors of such research articles expressed improved performance of proposed works for biomedical aspects. Nabendu Bhui et al. [9] represented one framework formed by deep learning principles and utilized the idea of autoencoder to achieve superior result for the perspectives of feature selection. Russul Alanni et al. [5] adopted deep learning strategies to acquire most relevant genes for

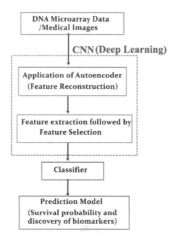

Figure 14.4   Generalized procedure of employing deep learning for feature selection.

increasing the flow of cancer research whereas Jan Zrimec et al. [64] considered deep learning algorithms for gene regulatory formation accountable for gene expression to realize the existence of cancerous cells. Min Zeng et al. in [61] represented the technique of learning protein–protein interaction (PPI) in an automated fashion in order to discover proper biological features for disease detection. With a view to attain supreme accuracy level for performance perspectives, Omar Ahmed et al. [3] proposed the consideration of deep neural network accompanied by recurrent neural network(RNN) along with the conception of CNN for gene expression analysis needed to commence medical treatment of cancer patients. Yawen Xiao et al. in [56] mentioned an algorithm concerned with deep learning along with the idea of stacked auto encoder and SVM in order to form to a learning model advantageous for breast cancer diagnosis. Figure 14.4 depicts the common operational scheme for such research articles dealing with the appliance of feature selection in working space of bioinformatics.

Few research works in recent years for the application of feature selection using deep learning schemes in the sphere of biomedical research are represented briefly in Table 14.7.

A.F. Syafiandini et al. [51] represented the approach towards the identification of biomarkers using Deep Boltzmann Machines (DBM) which is multimodal and drastically applied for observing the growth of cancerous cell, therby increasing survival probability of patients. Xiaoyu Zhang et al. [62] developed a model called OmiVAE in accordance with deep learning approaches for integrating high dimensional multiomics data to acquire proper features equipped with low dimensions for research activities on cancer. Dibyendu Bikash Seal et. al in [42] proposed a predictive model relying on Deep Denoising Auto-encoder (DDAE) to analyze gene expression and mingle multi-omics data for assimilating the growth and functioning behaviour of cancerous cells. By means of incorporating the principles of mRMR on biomedical

Table 14.7

Feature selection (by deep learning) for biomedical area

Nabendu Bhui et al. [9]	2020	Usage of auto encoder for gene expression
Jan Zrimec et al. [64]	2020	Determination of gene expression level
Omar Ahmed et al. [3]	2019	Classification of gene expression for medical diagnosis
Russul Alanni et al. [5]	2019	Recognition of genes responsible for cancer
Yawen Xiao et al. [56]	2018	Diagnosis of breast cancer

data the task of amplifying positive rate of recognizing epigenetic biomarkers was presented by Saurav Mallik et al. in [32].

Several researches on COVID-19 taking feature selection scheme into consideration in the recent world are also accountable for exploring the immense impact of artificial intelligence as well as machine learning approaches on modern civilization [39]. RajuVaishya et al. [52] focused on the employment of artificial intelligence as well as feature selection technique for COVID-19 researches. Ryan Yixiang Wang et al. in [55] illustrated an approach of predicting COVID-19 cases by dint of co-morbidity correlated Single Nucleotide Polymorphisms and machine learning policies for analysis of genetic data. S.Sen et al. in [44] mentioned about a procedure of feature selection approaches applied in two stages concerning CT images of chest with a view to produce a classifier dealing with COVID-19 prediction. Jawad Rasheed et al. in [40] developed a framework comprised of various methods of machine learning in order to diagnose COVID-19 taking X-Ray images of chest into account. For the purpose of classifying COVID-19 concerned image et al. considered voting classifier algorithms in [14]. Figure 14.5 shows general steps employed for COVID-19 detection using feature selection approaches.

14.8 DISCUSSION

Newly developed approaches in research works briefly outlined in literature survey of this article are indicative of the advancement of the relevance of artificial intelligence as well as feature selection implemented in various fashioned by means of several machine learning policies in human health. Selection of most relevant genes must be an essential mean to recognize existence of disease in human bodies. In this context we have found 500 features showing maximum relevance and minimum redundancy from 23,630 entries regarding scRNA-seq data by dint of mRMR technique. It is also indispensable information to attain the knowledge of suitable feature selection practices on the road to detect correct diseases and also recuperate patients from diseases. Table 14.8 represents some of recent research activities in order to accelerate the biomedical researches on some complex diseases.

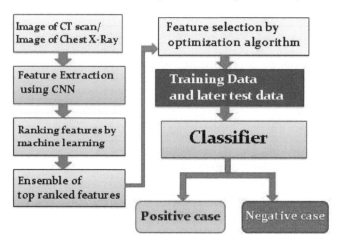

Figure 14.5 Employment of feature selection for COVID-19 detection.

Table 14.8
Succinct overview of some research articles (2020) regarding feature selection

Proposed Technique	Dataset	Detection of Disease	Performance(in accuracy)
Determination of feature subset through correlation along with recursive feature elimination [19]	Medical data (Patient's Age, Gender, Glucose status, smoking status, etc.) from kaggle dataset	Cardio Vascular	73.125%
Classification using self normalizing neural (SNN) network along with feature selection [28]	Medical data of LUAD patients, OV patients, LIHC patients, BRCA patients from TCGA database	Pan Cancer	68.90% using 1693 features
Feature selection using unsupervised learning employing auto encoder [56]	Breast Cancer Wisconsin Diagnostics(BCWD) and Surface Enhanced Raman Spectroscopy (SERS) data set	Breast Cancer	93.5%

14.9　CONCLUSION

This chapter demonstrates the utility of feature selection technique and also provides the widespread views of the accomplishment of this technique through existing biomedical research articles briefly outlined in the literature survey which will create ample opportunities for being acquainted with the advancement of medical research studies. Adoption of future AI technologies as well as future approaches regarding feature selection for biomedical researches will be our future directives.

REFERENCES

1. Sudipta Acharya, Sriparna Saha, and N Nikhil. Unsupervised gene selection using biological knowledge: application in sample clustering. *BMC Bioinformatics*, 18(1):1–13, 2017.

2. Amazona Adorada, Ratih Permatasari, Panji Wisnu Wirawan, Adi Wibowo, and Adi Sujiwo. Support vector machine-recursive feature elimination (svm-rfe) for selection of microrna expression features of breast cancer. In *2018 2nd International Conference on Informatics and Computational Sciences (ICICoS)*, pages 1–4. IEEE, 2018.

3. Omar Ahmed and Adnan Brifcani. Gene expression classification based on deep learning. In *2019 4th Scientific International Conference Najaf (SICN)*, pages 145–149. IEEE, 2019.

4. Feras Al-Obeidat, Abdallah Tubaishat, Babar Shah, Zahid Halim, et al. Gene encoder: a feature selection technique through unsupervised deep learning-based clustering for large gene expression data. *Neural Computing and Applications*, pages 1–23, 2020.

5. Russul Alanni, Jingyu Hou, Hasseeb Azzawi, and Yong Xiang. Deep gene selection method to select genes from microarray datasets for cancer classification. *BMC Bioinformatics*, 20(1):1–15, 2019.

6. Nada Almugren and Hala M Alshamlan. New bio-marker gene discovery algorithms for cancer gene expression profile. *IEEE Access*, 7:136907–136913, 2019.

7. Jun Chin Ang, Andri Mirzal, Habibollah Haron, and Haza Nuzly Abdull Hamed. Supervised, unsupervised, and semi-supervised feature selection: a review on gene selection. *IEEE/ACM Transactions on Computational Biology and Bioinformatics*, 13(5):971–989, 2015.

8. Jisha Augustine and AS Jereesh. Gene regulatory network inference: A semi-supervised approach. In *2017 International conference of Electronics, Communication and Aerospace Technology (ICECA)*, volume 1, pages 68–72. IEEE, 2017.

9. Nabendu Bhui, Pintu Kumar Ram, and Pratyay Kuila. Feature selection from microarray data based on deep learning approach. In *2020 11th International Conference on Computing, Communication and Networking Technologies (ICCCNT)*, pages 1–5. IEEE, 2020.

10. V Chandrasekar, V Sureshkumar, T Satish Kumar, and S Shanmugapriya. Disease prediction based on micro array classification using deep learning techniques. *Microprocessors and Microsystems*, 77:103189, 2020.

11. Tatjana Chavdarova and François Fleuret. Sgan: An alternative training of generative adversarial networks. In *Proceedings of the IEEE Conference on Computer Vision and Pattern Recognition*, pages 9407–9415, 2018.

12. Asma Chebli, Akila Djebbar, and Hayet Farida Marouani. Semi-supervised learning for medical application: A survey. In *2018 International Conference on Applied Smart Systems (ICASS)*, pages 1–9. IEEE, 2018.

13. Sunanda Das, Shreya Chaudhuri, and Asit K Das. Optimal set of overlapping clusters using multi-objective genetic algorithm. In *Proceedings of the 9th International Conference on Machine Learning and Computing*, pages 232–237, 2017.

14. El-Sayed M El-Kenawy, Abdelhameed Ibrahim, Seyedali Mirjalili, Marwa Metwally Eid, and Sherif E Hussein. Novel feature selection and voting classifier algorithms for covid-19 classification in ct images. *IEEE Access*, 8:179317–179335, 2020.

15. Jocelyn Gal, Caroline Bailleux, David Chardin, Thierry Pourcher, Lun Jing, Jean-Marie Guignonis, Jean-Marc Ferrero, Renaud Schiappa, Emmanuel Chamorey, and Olivier Humbert. Unsupervised machine learning methods reveal metabolomic based clusters in breast cancer patients, 2019.

16. Pilar Garca-Diaz, Isabel Sanchez-Berriel, Juan A Martinez-Rojas, and Ana M Diez-Pascual. Unsupervised feature selection algorithm for multiclass cancer classification of gene expression rna-seq data. *Genomics*, 112(2):1916–1925, 2020.

17. Kamran Ghasedi Dizaji, Xiaoqian Wang, and Heng Huang. Semi-supervised generative adversarial network for gene expression inference. In *Proceedings of the 24th ACM SIGKDD International Conference on Knowledge Discovery & Data Mining*, pages 1435–1444, 2018.

18. Intisar Rizwan I Haque and Jeremiah Neubert. Deep learning approaches to biomedical image segmentation. *Informatics in Medicine Unlocked*, 18:100297, 2020.

19. Najmul Hasan and Yukun Bao. Comparing different feature selection algorithms for cardiovascular disease prediction. *Health and Technology*, 11(1):49–62, 2021.

20. Hamid Reza Hassanzadeh, John H Phan, and May D Wang. A multi-modal graph-based semi-supervised pipeline for predicting cancer survival. In *2016 IEEE International Conference on Bioinformatics and Biomedicine (BIBM)*, pages 184–189. IEEE, 2016.

21. Nour Eldeen M Khalifa, Mohamed Hamed N Taha, Dalia Ezzat Ali, Adam Slowik, and Aboul Ella Hassanien. Artificial intelligence technique for gene expression by tumor rna-seq data: a novel optimized deep learning approach. *IEEE Access*, 8:22874–22883, 2020.

22. Nimrita Koul and Sunilkumar S Manvi. A scheme for feature selection from gene expression data using recursive feature elimination with cross validation and unsupervised deep belief network classifier. In *2019 3rd International Conference on Computing and Communications Technologies (ICCCT)*, pages 31–36. IEEE, 2019.

23. C Arun Kumar, MP Sooraj, and S Ramakrishnan. A comparative performance evaluation of supervised feature selection algorithms on microarray datasets. *Procedia Computer Science*, 115:209–217, 2017.

24. Pramod Kumar, Vandana Mishra, and Subarna Roy. Machine learning framework: Predicting protein structural features. In *Soft Computing for Biological Systems*, pages 121–141. Springer, 2018.

25. Lan Lan, Lei You, Zeyang Zhang, Zhiwei Fan, Weiling Zhao, Nianyin Zeng, Yidong Chen, and Xiaobo Zhou. Generative adversarial networks and its applications in biomedical informatics. *Frontiers in Public Health*, 8:164, 2020.

26. Moh Abdul Latief, Titin Siswantining, Alhadi Bustamam, and Devvi Sarwinda. A comparative performance evaluation of random forest feature selection on classification of hepatocellular carcinoma gene expression data. In *2019 3rd International Conference on Informatics and Computational Sciences (ICICoS)*, pages 1–6. IEEE, 2019.

27. Junghye Lee, In Young Choi, and Chi-Hyuck Jun. An efficient multivariate feature ranking method for gene selection in high-dimensional microarray data. *Expert Systems with Applications*, 166:113971, 2021.

28. Junyi Li, Qingzhe Xu, Mingxiao Wu, Tao Huang, and Yadong Wang. Pan-cancer classification based on self-normalizing neural networks and feature selection. *Frontiers in Bioengineering and Biotechnology*, 8, 2020.

29. Zifa Li, Weibo Xie, and Tao Liu. Efficient feature selection and classification for microarray data. *PloS one*, 13(8):e0202167, 2018.

30. Christian Lopez, Scott Tucker, Tarik Salameh, and Conrad Tucker. An unsupervised machine learning method for discovering patient clusters based on genetic signatures. *Journal of Biomedical Informatics*, 85:30–39, 2018.

31. Koushik Mallick, Saurav Mallik, Sanghamitra Bandyopadhyay, and Sikim Chakraborty. A novel graph topology based go-similarity measure for signature detection from multi-omics data and its application to other problems. *IEEE/ACM Transactions on Computational Biology and Bioinformatics*, 2020.

32. Saurav Mallik, Tapas Bhadra, and Ujjwal Maulik. Identifying epigenetic biomarkers using maximal relevance and minimal redundancy based feature selection for multi-omics data. *IEEE Transactions on Nanobioscience*, 16(1):3–10, 2017.

33. Jianyu Miao and Lingfeng Niu. A survey on feature selection. *Procedia Computer Science*, 91:919–926, 2016.

34. MM Mohamed Mufassirin and Roshan G Ragel. A novel filter-wrapper based feature selection approach for cancer data classification. In *2018 IEEE International Conference on Information and Automation for Sustainability (ICIAfS)*, pages 1–6. IEEE, 2018.

35. Anu J Nair, Rizwana Rasheed, KM Maheeshma, LS Aiswarya, and K R Kavitha. An ensemble-based feature selection and classification of gene expression using support vector machine, k-nearest neighbor, decision tree. In *2019 International Conference on Communication and Electronics Systems (ICCES)*, pages 1618–1623, 2019.

36. Ştefan Niţicǎ, Gabriela Czibula, and Vlad-Ioan Tomescu. A comparative study on using unsupervised learning based data analysis techniques for breast cancer detection. In *2020 IEEE 14th International Symposium on Applied Computational Intelligence and Informatics (SACI)*, pages 000099–000104. IEEE, 2020.

37. D Pavithra and B Lakshmanan. Feature selection and classification in gene expression cancer data. In *2017 International Conference on Computational Intelligence in Data Science (ICCIDS)*, pages 1–6. IEEE, 2017.

38. Milos Radovic, Mohamed Ghalwash, Nenad Filipovic, and Zoran Obradovic. Minimum redundancy maximum relevance feature selection approach for temporal gene expression data. *BMC Bioinformatics*, 18(1):1–14, 2017.

39. Jawad Rasheed, Alaa Ali Hameed, Chawki Djeddi, Akhtar Jamil, and Fadi Al-Turjman. A machine learning-based framework for diagnosis of covid-19 from chest x-ray images. *Interdisciplinary Sciences: Computational Life Sciences*, 13(1):103–117, 2021.

40. Jawad Rasheed, Akhtar Jamil, Alaa Ali Hameed, Fadi Al-Turjman, and Ahmad Rasheed. Covid-19 in the age of artificial intelligence: A comprehensive review. *Interdisciplinary Sciences: Computational Life Sciences*, pages 1–23, 2021.

41. Barnali Sahu, Satchidananda Dehuri, and Alok Kumar Jagadev. Feature selection model based on clustering and ranking in pipeline for microarray data. *Informatics in Medicine Unlocked*, 9:107–122, 2017.

42. Dibyendu Bikash Seal, Vivek Das, Saptarsi Goswami, and Rajat K De. Estimating gene expression from dna methylation and copy number variation: A deep learning regression model for multi-omics integration. *Genomics*, 112(4):2833–2841, 2020.

43. Borja Seijo-Pardo, Vernica Boln-Canedo, Iago Porto-Diaz, and Amparo Alonso-Betanzos. Ensemble feature selection for rankings of features. In *International Work-Conference on Artificial Neural Networks*, pages 29–42. Springer, 2015.

44. Shibaprasad Sen, Soumyajit Saha, Somnath Chatterjee, Seyedali Mirjalili, and Ram Sarkar. A bi-stage feature selection approach for covid-19 prediction using chest ct images. *Applied Intelligence*, pages 1–16, 2021.

45. Rahul K Sevakula, Vikas Singh, Nishchal K Verma, Chandan Kumar, and Yan Cui. Transfer learning for molecular cancer classification using deep neural networks. *IEEE/ACM Transactions on Computational Biology and Bioinformatics*, 16(6):2089–2100, 2018.

46. K Aditya Shastry and HA Sanjay. Machine learning for bioinformatics. In *Statistical Modelling and Machine Learning Principles for Bioinformatics Techniques, Tools, and Applications*, pages 25–39. Springer, 2020.

47. Shailendra Singh et al. A novel algorithm to preprocess cancerous gene expression dataset for efficient gene selection. In *2017 2nd International Conference for Convergence in Technology (I2CT)*, pages 632–635. IEEE, 2017.

48. Saul Solorio-Fernandez, Jose Fco Martinez-Trinidad, and J Ariel Carrasco-Ochoa. A new unsupervised spectral feature selection method for mixed data: a filter approach. *Pattern Recognition*, 72:314–326, 2017.

49. V Sesha Srinivas, A Srikrishna, and B Eswara Reddy. Automatic clustering simultaneous feature subset selection using differential evolution. In *2018 5th International Conference on Signal Processing and Integrated Networks (SPIN)*, pages 468–473. IEEE, 2018.

50. Lin Sun, Lanying Wang, Weiping Ding, Yuhua Qian, and Jiucheng Xu. Feature selection using fuzzy neighborhood entropy-based uncertainty measures for fuzzy neighborhood multigranulation rough sets. *IEEE Transactions on Fuzzy Systems*, 29(1):19–33, 2020.

51. Arida Ferti Syafiandini, Ito Wasito, Setiadi Yazid, Aries Fitriawan, and Mukhlis Amien. Cancer subtype identification using deep learning approach. In *2016 International Conference on Computer, Control, Informatics and Its Applications (IC3INA)*, pages 108–112, 2016.

52. Raju Vaishya, Mohd Javaid, Ibrahim Haleem Khan, and Abid Haleem. Artificial intelligence (ai) applications for covid-19 pandemic. *Diabetes & Metabolic Syndrome: Clinical Research & Reviews*, 14(4):337–339, 2020.

53. S Vanjimalar, D Ramyachitra, and P Manikandan. A review on feature selection techniques for gene expression data. In *2018 IEEE International Conference on Computational Intelligence and Computing Research (ICCIC)*, pages 1–4. IEEE, 2018.

54. Aiguo Wang, Ning An, Jing Yang, Guilin Chen, Lian Li, and Gil Alterovitz. Wrapper-based gene selection with markov blanket. *Computers in Biology and Medicine*, 81:11–23, 2017.

55. Ryan Yixiang Wang, Tim Qinsong Guo, Leo Guanhua Li, Julia Yutian Jiao, and Lena Yiqi Wang. Predictions of covid-19 infection severity based on co-associations between the snps of co-morbid diseases and covid-19 through machine learning of genetic data. In *2020 IEEE 8th International Conference on Computer Science and Network Technology (ICCSNT)*, pages 92–96. IEEE, 2020.

56. Yawen Xiao, Jun Wu, Zongli Lin, and Xiaodong Zhao. Breast cancer diagnosis using an unsupervised feature extraction algorithm based on deep learning. In *2018 37th Chinese Control Conference (CCC)*, pages 9428–9433. IEEE, 2018.

57. Xin Yang, Yi Lin, Zhiwei Wang, Xin Li, and Kwang-Ting Cheng. Bi-modality medical image synthesis using semi-supervised sequential generative adversarial networks. *IEEE Journal of Biomedical and Health Informatics*, 24(3):855–865, 2019.

58. Xiucai Ye, Weihang Zhang, and Tetsuya Sakurai. Adaptive unsupervised feature learning for gene signature identification in non-small-cell lung cancer. *IEEE Access*, 8:154354–154362, 2020.

59. Lei-ming Yuan, Yiye Sun, and Guangzao Huang. Using class-specific feature selection for cancer detection with gene expression profile data of platelets. *Sensors*, 20(5):1528, 2020.

60. Diyar Qader Zeebaree, Habibollah Haron, and Adnan Mohsin Abdulazeez. Gene selection and classification of microarray data using convolutional neural network. In *2018 International Conference on Advanced Science and Engineering (ICOASE)*, pages 145–150. IEEE, 2018.

61. Min Zeng, Fuhao Zhang, Fang-Xiang Wu, Yaohang Li, Jianxin Wang, and Min Li. Protein–protein interaction site prediction through combining local and global features with deep neural networks. *Bioinformatics*, 36(4):1114–1120, 2020.

62. Xiaoyu Zhang, Jingqing Zhang, Kai Sun, Xian Yang, Chengliang Dai, and Yike Guo. Integrated multi-omics analysis using variational autoencoders: Application to pan-cancer classification. In *2019 IEEE International Conference on Bioinformatics and Biomedicine (BIBM)*, pages 765–769. IEEE, 2019.

63. Y. Zhou, X. He, L. Huang, L. Liu, F. Zhu, S. Cui, and L. Shao. Collaborative learning of semi-supervised segmentation and classification for medical images. *2019 IEEE/CVF Conference on Computer Vision and Pattern Recognition (CVPR)*, pages 2074–2083, 2019.

64. Jan Zrimec, Filip Buric, Mariia Kokina, Victor Garcia, and Aleksej Zelezniak. Learning the regulatory code of gene expression. *Frontiers in Molecular Biosciences*, 8, 2021.

Index